Allergenic Pollen

Mikhail Sofiev • Karl-Christian Bergmann
Editors

Allergenic Pollen

A Review of the Production, Release, Distribution and Health Impacts

Editors
Mikhail Sofiev
Finnish Meteorological Institute
Erick Palmenin Aukio 1, 00561
Helsinki, Finland

Karl-Christian Bergmann
Foundation German Pollen
Information Service
Luisenstrasse 2, 10117
Berlin, Germany

ISBN 978-94-007-4880-4   ISBN 978-94-007-4881-1 (eBook)
DOI 10.1007/978-94-007-4881-1
Springer Dordrecht Heidelberg New York London

Library of Congress Control Number: 2012948980

© Springer Science+Business Media Dordrecht 2013
This work is subject to copyright. All rights are reserved by the Publisher, whether the whole or part of the material is concerned, specifically the rights of translation, reprinting, reuse of illustrations, recitation, broadcasting, reproduction on microfilms or in any other physical way, and transmission or information storage and retrieval, electronic adaptation, computer software, or by similar or dissimilar methodology now known or hereafter developed. Exempted from this legal reservation are brief excerpts in connection with reviews or scholarly analysis or material supplied specifically for the purpose of being entered and executed on a computer system, for exclusive use by the purchaser of the work. Duplication of this publication or parts thereof is permitted only under the provisions of the Copyright Law of the Publisher's location, in its current version, and permission for use must always be obtained from Springer. Permissions for use may be obtained through RightsLink at the Copyright Clearance Center. Violations are liable to prosecution under the respective Copyright Law.
The use of general descriptive names, registered names, trademarks, service marks, etc. in this publication does not imply, even in the absence of a specific statement, that such names are exempt from the relevant protective laws and regulations and therefore free for general use.
While the advice and information in this book are believed to be true and accurate at the date of publication, neither the authors nor the editors nor the publisher can accept any legal responsibility for any errors or omissions that may be made. The publisher makes no warranty, express or implied, with respect to the material contained herein.

Printed on acid-free paper

Springer is part of Springer Science+Business Media (www.springer.com)

This book is dedicated to Yoav Waisel and Alicia Stach, both dear colleagues whose untiring efforts in pollen and pollen related research inspired many of the authors of this book. They were always very active members of our COST action as well as highly respected and beloved colleagues in our field.

We the authors dearly miss them.

Foreword

Allergic rhinitis was not well described before John Bostock's paper, whereas asthma was identified before Christ. Although rhinitis may have been recognized in the Antiquity, the "rose catarrh" was probably the first description of allergic rhinitis in the sixteenth century by Vallerianus and later by Botallo (1565). John Bostock described precisely what he called "catarrhus aestivus" and later became "hay fever" and then "allergic rhinitis." However, John Bostock was unable to find that pollen was the cause of the disease.

First described by Nehemia Grew (1641–1712) in 1682 (*The Anatomy of Plants*), pollen gained attraction as trigger of pollinosis in the early nineteenth century in Great Britain; a handful of people suffered with hay fever.

Today, allergic pollen are the main trigger of allergic rhinoconjunctivitis, asthma and the oral allergy-syndrome in about 15 million people in Europe. The ever-increasing prevalence of pollen allergy in Europe and climate change further stimulate the interest in allergic pollen, both in the public and scientific community.

There are still many questions on the production, monitoring, distribution, forecast and health impact of pollen, and this research needs an interdisciplinary working process – fulfilled at present as best as possible by a group of scientists from across Europe who are members of the COST action group.

This group has summarized their knowledge – the result being an excellent concentrated update on allergic pollen, which is the finest of its kind in the field of pollen research at present. This book is an example of what can be achieved by the support of interdisciplinary research in Europe by the European Community.

Jean Bousquet

Acknowledgements

We would also like to acknowledge Ingrid van Hofman and Matthias Werchan for their efforts in organising, formatting and helping with the groundwork of this book.

Contents

1   **Introduction**... 1
     Lorenzo Cecchi

2   **Pollen Sources**... 9
     Carsten Ambelas Skjøth, Branko Šikoparija, Siegfried Jäger,
     and EAN-Network

3   **The Onset, Course and Intensity of the Pollen Season**............... 29
     Åslög Dahl, Carmen Galán, Lenka Hajkova, Andreas Pauling,
     Branko Sikoparija, Matt Smith, and Despoina Vokou

4   **Monitoring, Modelling and Forecasting of the Pollen Season**..... 71
     Helfried Scheifinger, Jordina Belmonte, Jeroen Buters,
     Sevcan Celenk, Athanasios Damialis, Chantal Dechamp,
     Herminia García-Mozo, Regula Gehrig, Lukasz Grewling,
     John M. Halley, Kjell-Arild Hogda, Siegfried Jäger,
     Kostas Karatzas, Stein-Rune Karlsen, Elisabeth Koch,
     Andreas Pauling, Roz Peel, Branko Sikoparija, Matt Smith,
     Carmen Galán-Soldevilla, Michel Thibaudon, Despina Vokou,
     and Letty A. de Weger

5   **Airborne Pollen Transport**.. 127
     Mikhail Sofiev, Jordina Belmonte, Regula Gehrig, Rebeca Izquierdo,
     Matt Smith, Åslög Dahl, and Pilvi Siljamo

6   **Impact of Pollen**... 161
     Letty A. de Weger, Karl Christian Bergmann, Auli Rantio-Lehtimäki,
     Åslög Dahl, Jeroen Buters, Chantal Déchamp, Jordina Belmonte,
     Michel Thibaudon, Lorenzo Cecchi, Jean-Pierre Besancenot,
     Carmen Galán, and Yoav Waisel

7   **Presentation and Dissemination of Pollen Information**.............. 217
     Kostas D. Karatzas, Marina Riga, and Matt Smith

Chapter 1
Introduction

Lorenzo Cecchi

Abstract Since the first description of hay fever by John Bostock in 1819, the role of pollen in the pathogenesis of allergic diseases is well established now. Most important allergic diseases are asthma and rhinitis, which affect from 5 to 30% of the population in industrialized countries. The standardization of pollen count protocols and the creation of wide and coordinated networks have provided an invaluable tool for epidemiological and clinical studies and, more recently, for a better understanding of the effects of climate change on plants producing allergenic pollen and on allergic diseases. Significant advances occurred in allergy diagnostic with the use of recombinant molecules. Thanks to new methods for airborne allergen measurement, the same development occurred in aerobiology, thus shedding a new light on the relationship between pollen and allergic diseases.

Keywords Allergic diseases • Asthma • Rhinitis • Pollen • Aerobiology

1.1 Allergenic Pollen

"About the beginning or middle of June in every year … A sensation of heat and fullness is experienced in the eyes …. To this succeeds irritation of the nose producing sneezing. To the sneezing are added a further sensation of tightness of the chest, and a difficulty of breathing" (Bostock 1819). This is the first description of hay fever, published in 1819 by John Bostock. More than 50 years later, in 1873, pollen was recognized as the cause of this clinical picture by Charles Harrison Blackley (1873), who also performed the first skin prick test on his own arm. Since then, thousands

---

L. Cecchi (✉)
Interdepartmental Centre of Bioclimatology, University of Florence,
Piazzale delle Cascine 18, 50144 Florence, Italy
e-mail: lorenzo.cecchi@unifi.it

of papers have been published on this issue and nowadays the role of pollen in the pathogenesis of allergic rhinitis, asthma and conjunctivitis is well established. Epidemiological studies and clinical trials designed with the aim to evaluate the efficacy of antiallergic drugs and specific immunotherapy confirmed early results about the importance of pollen in the seasonal allergic disorders, although inconsistently in some cases. Indeed, several questions are still open.

Aerobiology (from Greek ήρ, aēr, "air"; βίος, bios, "life"; and -λογία, -logia) is a branch of biology that studies organic particles, such as bacteria, fungal spores, pollen grains and viruses, which are passively transported by the air. Aerobiologists played a key role in the understanding of the relationship between allergic diseases and pollen, especially through the standardization of the procedure for the assessment of pollen concentration in the atmosphere. Pollen count has being used for over 50 years for the assessment of allergen exposure both in clinical practice and clinical and experimental studies. The method, proposed by Hirst (1952) is based on the identification and count with a microscope of pollen and spores collected with a volumetric trap and provides the standard for the national networks which are currently covering most of the European continent. This type of airborne particle assessment presents several advantages: it allows a comprehensive evaluation of airborne particles with a wide spectrum of applications; long time-series are now available, which can be used for pollen calendars and for research purposes. To this regard 20–25 years long datasets provide an extraordinary tool for climate change studies, showing both changes in the past decades and providing the basis for modelization of future scenarios (Cecchi et al. 2010). However, proof is insufficient that pollen count is representative for allergen exposure, thus providing the explanation of some controversial results of epidemiological studies aimed at showing the effects of pollen on respiratory allergic diseases, especially on asthma.

1.1.1 Pollen-Related Allergic Diseases

Rhinitis, conjunctivitis and asthma are the typical clinical pictures of allergy to pollen and they often occur in the same patient simultaneously during the pollen season.

Asthma is a chronic inflammatory disease of the airways characterized by recurrent episodes of wheezing, breathlessness, chest tightness and coughing (GINA 2009). Exposure to allergens represents a key factor among environmental determinants of asthma, which also include air pollution (Eder et al. 2006). Allergic rhinitis is clinically defined as a symptomatic disorder of the nose induced by an IgE-mediated inflammation after allergen exposure of the membranes lining the nose. Symptoms of rhinitis include rhinorrhea, nasal obstruction, nasal itching and sneezing which are reversible spontaneously or under treatment (Bousquet et al. 2001). Pathophysiological and clinical studies have strongly suggested a relationship between rhinitis and asthma. However, epidemiology provides the most convincing data, showing that the prevalence of asthma in patients with rhinitis varies from 10 to 40 % depending on the study. Moreover, allergic rhinitis is correlated to, and constitutes a risk factor for, the occurrence of asthma (Bousquet et al. 2008). Taken together, these data have lead to the concept that upper and lower airways may be considered as a unique entity influenced by a common, evolving inflammatory process. Conjunctivitis is also commonly associated to pollen-induced rhinitis.

1.1.2 Mechanisms

Type I allergies are mediated by the production of IgE specific for otherwise harmless environmental substances, most of which are proteins. Most allergenic molecules that elicit IgE-mediated immune responses are derived from plants, animals and fungi.

Sensitization occurs at the site of allergen exposure, such as the airways and skin, but can also occur through the gastrointestinal tract. However, not everybody who is exposed will become sensitized and have allergies. Aside from the individual exposure conditions, there is a high variability in the individual responsiveness to a given allergen dose.

The most important allergen carriers in the outdoor air are pollen – with a diameter between 15 and 60 µm – from anemophilic plants such as trees, grasses and weeds. However, whole pollen grains are too large to penetrate the small airways. Since pollen is able to evoke IgE-mediated allergic reactions within seconds after contact with the mucosa, pollen allergens must be extremely water soluble and readily available. In fact allergen liberation from pollen grains can occur on the mucosal surface of the upper respiratory tract after exposure to pollen (Behrendt and Becker 2001). Symptoms can be explained by the interaction between the antigen and its corresponding IgE antibody and this phase is situated at the end of a cascade of events leading to allergy.

The experimental data of Bacsi et al. (2005, 2006), Boldogh et al. (2005) and Traidl-Hoffman et al. (2002) provide additional evidence that pollen fragments, containing NAD(P)H oxidases and lipid particles, can amplify the immune response by producing reactive oxygen species (ROS) as well as chemo-attractants. To this regard, this pollen-mediated mechanism seems to be able to induce a non-specific inflammatory response at the mucosal level (Traidl-Hoffmann et al. 2009), providing a possible explanation of some data showing an association between pollen count and hospital admissions of nonallergic diseases (Besancenot 2011).

Patients affected by respiratory allergy are subjected to inhalation of aerosols of pollen, pollen fragments, air pollutants and other caustic chemicals. The relative importance of this mixture of pro-inflammatory agents in the airways and their interactions still needs to be clarified.

Numerous studies have shown that air pollution is consistently associated with adverse health effects and it has a quantifiable impact on respiratory diseases, on cardiovascular diseases and stroke (Dockery and Stone 2007). Positive associations have been observed between urban air pollution and respiratory symptoms in both adults and children (D'Amato et al. 2010), and the literature reports of a relation between motor vehicle exhausts and acute or chronic respiratory symptoms in children living near traffic (Ryan et al. 2005). Air pollution can also negatively influence lung development in children and adolescents (Gauderman et al. 2007).

In the so called thunderstorm, asthma fragments of pollen might play a role. A thunderstorm is an extreme weather event with dramatic consequences on respiratory asthma, as showed in the last 15 years with asthma outbreaks during thunderstorms worldwide (D'Amato et al. 2007a). Despite some uncertainties, a mechanism underlining asthma epidemics might be pollen grains that rupture by osmotic shock and release part of their content, including respirable, allergen-carrying starch granules (0.5–2.5 mμ) into the atmosphere (Taylor and Jonsson 2004).

Thunderstorms have often been linked to epidemics of asthma, especially during the grass and molds season (Marks and Bush 2007), even if other pollens might be involved (D'Amato et al. 2008).

1.1.3 Epidemiology

Type I hypersensitivity reactions, such as allergic rhinitis and asthma, are the most common allergic diseases, with current prevalence rates ranging from 5 to 30 % in industrialized countries (Asher et al. 2006).

Pollen allergy has a remarkable clinical impact all over Europe and there is a body of evidence suggesting that the prevalence of respiratory allergic reactions induced by pollens in Europe is on the increase (D'Amato et al. 2007b; ECRHS 1996; ISAAC, 1998; Burney et al. 1997). In fact, for the past 40 years, the prevalence of asthma has increased, and it is still increasing worldwide in parallel with that of allergy (Law et al. 2005) even if recent findings of the phase three of ISAAC study showed the absence of increases or little changes in prevalence of asthma symptoms, allergic rhinoconjunctivitis and eczema for European centres with existing high prevalence in older children (Asher et al. 2006). This data suggests that it might possible be that the increase is coming to an end especially in some western countries (von Hertzen and Morais-Almeida 2005).

Since airborne-induced respiratory allergy does not recognize national frontiers, the study of pollinosis cannot be limited to national boundaries, as obviously happens with most diseases that can be prevented by avoiding exposure to the causative agent. In Europe, the main pollination period covers about half the year, from spring to autumn.

1.1.4 Towards a Molecular Era of Allergology and Aerobiology

Thanks to the development and progress made in the field of recombinant allergens, allergy diagnostic has changed in the last 10–15 years, moving from the use of extracts for both *in vivo* and *in vitro* diagnostic to the so called "Component Resolved

Diagnosis" (CRD), a tool that characterises each patient's IgE antibody profile to individual allergen components, thereby discriminating between genuine sensitisation to certain specific allergen sources and cross-reactivity (Sastre 2010). This new approach is particularly useful for a better identification of patients to be treated with specific immunotherapy. To this regard, companies are now producing extracts for immunotherapy standardized according to the content of major allergens expressed in mcg/ml, as suggested by the last guidelines for sublingual immunotherapy (Canonica et al. 2009).

As expected, Aerobiology is developing in the same direction. Allergen count is based on the measurement of the concentration of single allergenic components in the air, collected with special devices which suck a higher volume of air than the classic Hirst-type pollen trap. After the early observations by Schäppi et al. (1999) a number of papers have been published in the last 10 years. Nowadays we use to call "molecular aerobiology" this new branch of Aerobiology. So far, research focused on grass pollen (Phl p 5 and Lol p 1), birch (Bet v 1), olive (Ole e 1), wall pellittory (Par j 1 and Par j 2) allergens mainly in Australia (Schäppi et al. 1999), Germany (Buters et al. 2008, 2010) and Spain (De Linares et al. 2007, 2010; Jato et al. 2010; Moreno-Grau et al. 2006). In these studies different methods for allergen measurement and collection were used which explains several inconsistencies between the papers published in this field. However, almost all authors indicate that allergen concentration deviates from pollen count and that this is due to a number of factors.

Despite several questions are being left, it is now clear that monitoring the allergen itself in ambient air might be an improvement in allergen exposure assessment. These new methods are already contributing to the clarification of some aspects of exposure to airborne allergens, such as the debated importance of pauci- and sub-micronic particles or the clinical thresholds.

Even if it is the time to move towards the allergen count, we still need "classic" pollen count and the combination of the two methods seems to be the best choice in the medium term. Allergen count is still limited by cross-reactivity between allergens, lack of homogenization of methods and devices, unavailability of long time-data sets and high costs.

References

Asher, M. I., Montefort, S., Bjorksten, B., Lai, C. K., Strachan, D. P., Weiland, S. K., Williams, H., & ISAAC Phase Three Study Group. (2006). Worldwide time trends in the prevalence of symptoms of asthma, allergic rhinoconjunctivitis, and eczema in childhood: ISAAC phases one and three repeat multicountry cross-sectional surveys. *The Lancet, 368*, 733–743.

Bacsi, A., Dharajiya, N., Choudhury, B. K., & Boldogh, I. (2005). Effect of pollen-mediated oxidative stress on immediate hypersensitivity reactions and late-phase inflammation in allergic conjunctivitis. *The Journal of Allergy and Clinical Immunology, 116*, 836–843.

Bacsi, A., Choudhury, B. K., Dharajiya, N., Sur, S., & Boldogh, I. (2006). Subpollen particles: Carriers of allergenic proteins and oxidases. *The Journal of Allergy and Clinical Immunology, 118*, 844–850.

Behrendt, H., & Becker, W. M. (2001). Localization, release and bioavailability of pollen allergens: The influence of environmental factors. *Current Opinion in Immunology, 13*, 709–715.

Besancenot, J.-P., Thibaudon, M., & Cecchi, L. (2011, June). Has allergenic pollen an impact on non-allergic diseases? *European Annals of Allergy Clinical Immunology, 43*(3), 69–76.

Blackley, C. H. (1873). *Experimental researches on the causes and nature of Catarrhus Aestivus (hay-fever and hay-asthma)*. London: Ballière Tindall and Cox. 1873.

Boldogh, I., Bacsi, A., Choudhury, B. K., Dharajiya, N., Alam, R., Hazra, T. K., Mitra, S., Goldblum, R. M., & Sur, S. (2005). ROS generated by a pollen NADPH oxidase provide a signal that augments antigen-induced allergic airway inflammation. *The Journal of Clinical Investigation, 115*, 2169–2179.

Bostock, J. (1819). Case of periodic affection of the eyes and chest. *Medico-Chirurgical Transactions, 10*, 161.

Bousquet, J., Van Cauwenberge, P., Khaltaev, N., Aria Workshop Group, & World Health Organization. (2001). Allergic rhinitis and its impact on asthma. *The Journal of Allergy and Clinical Immunology, 108*(Suppl. 5), 147–334.

Bousquet, J., Khaltaev, N., Cruz, A. A., Denburg, J., Fokkens, W. J., Togias, A., Zuberbier, T., et al. (2008). Allergic Rhinitis and its Impact on Asthma (ARIA) 2008 update (in collaboration with the World Health Organization, GA(2)LEN and AllerGen. *Allergy, 63*(Suppl. 86), 8–160.

Burney, P. G. J., Malmberg, E., Chinn, S., Jarvis, D., Luczynska, C., & Lai, E. (1997). The distribution of total and specific serum IgE in the European community respiratory health survey. *The Journal of Allergy and Clinical Immunology, 99*, 314–322.

Buters, J. T., Kasche, A., Weichenmeier, I., Schober, W., Klaus, S., Traidl-Hoffmann, C., Menzel, A., Huss-Marp, J., Kramer, U., & Behrendt, H. (2008). Year-to-year variation in release of Bet v 1 allergen from birch pollen: Evidence for geographical differences between West and South Germany. *International Archives of Allergy and Immunology, 145*(2), 122–130.

Buters, J. T., Weichenmeier, I., Ochs, S., Pusch, G., Kreyling, W., Boere, A. J., Schober, W., & Behrendt, H. (2010). The allergen Bet v 1 in fractions of ambient air deviates from birch pollen counts. *Allergy, 65*(7), 850–858.

Canonica, G. W., Bousquet, J., Casale, T., Lockey, R. F., & Baena-Cagnani, C. E. (2009). Sub-lingual immunotherapy: World Allergy Organization Position Paper 2009. *Allergy, 64*(Suppl. 91), 1–59.

Cecchi, L., D'Amato, G., Ayres, J. G., Galan, C., Forastiere, F., Forsberg, B., Gerritsen, J., Nunes, C., Behrendt, H., Akdis, K., Dahl, R., & Annesi-Maesano, I. (2010). Projections of the effects of climate change on allergic asthma: The contribution of aerobiology. *Allergy, 65*(9), 1073–1081.

D'Amato, G., Liccardi, G., & Frenguelli, G. (2007a). Thunderstorm-asthma and pollen allergy. *Allergy, 62*(1), 11–16.

D'Amato, G., Cecchi, L., Bonini, S., Nunes, C., Annesi-Maesano, I., Behrendt, H., Liccardi, G., Popov, T., & van Cauwenberge, P. (2007b). Allergenic pollen and pollen allergy in Europe. *Allergy, 62*, 976–990.

D'Amato, G., Cecchi, L., & Liccardi, G. (2008). Thunderstorm-related asthma: Not only grass pollen and spores. *The Journal of Allergy and Clinical Immunology, 121*(2), 537–538.

D'Amato, G., Cecchi, L., D'Amato, M., & Liccardi, G. (2010). Urban air pollution and climate change as environmental risk factors of respiratory allergy: An update. *Journal of Investigational Allergology and Clinical Immunology, 20*(2), 95–102.

De Linares, C., Nieto-Lugilde, D., Alba, F., Díaz de la Guardia, C., Galán, C., & Trigo, M. M. (2007). Detection of airborne allergen (Ole e 1) in relation to Olea europaea pollen in S Spain. *Clinical and Experimental Allergy, 37*(1), 125–132.

De Linares, C., Díaz de la Guardia, C., Nieto Lugilde, D., & Alba, F. (2010). Airborne study of grass allergen (Lol p 1) in different-sized particles. *International Archives of Allergy and Immunology, 152*(1), 49–57.

Dockery, D. W., & Stone, P. H. (2007). Cardiovascular risks from fine particulate air pollution. *New England Journal of Medicine, 356*(5), 511–513.

Eder, W., Ege, M. J., & von Mutius, E. (2006). The asthma epidemic. *New England Journal of Medicine, 355*, 2226–2235.

European Community Respiratory Health Survey. (1996). Variations in the prevalence of respiratory symptoms, self-reported asthma attacks and the use of asthma medications in the European Community Respiratory Health Survey (ECRHS). *The European Respiratory Journal, 9*, 687–695.

Gauderman, W. J., Vora, H., McConnell, R., Berhane, K., Gilliland, F., Thomas, D., Lurmann, F., Avol, E., Kunzli, N., Jerrett, M., & Peters, J. (2007). Effect of exposure to traffic on lung development from 10 to 18 years of age: A cohort study. *The Lancet, 369*, 571–577.

GINA. (2009). Global strategy for asthma management and prevention. Updated 2009. www.ginasthma.org. Accessed on 25 July 2011.

Hirst, J. M. (1952). An automatic volumetric spore trap. *Annals of Applied Biology, 39*, 257–265.

Jato, V., Rodríguez-Rajo, F. J., González-Parrado, Z., Elvira-Rendueles, B., Moreno-Grau, S., & Vega-Maray, A. (2010). Detection of airborne Par j 1 and Par j 2 allergens in relation to Urticaceae pollen counts in different bioclimatic areas. *Annals of Allergy, Asthma & Immunology, 105*(1), 50–56.

Law, M., Morris, J. K., Wald, N., Luczynska, C., & Burney, P. (2005). Changes in atopy over a quarter of a century, based on cross sectional data at three time periods. *British Medical Journal, 330*, 1187–1188.

Marks, G. B., & Bush, R. K. (2007). It's blowing in the wind: New insights into thunderstorm-related asthma. *The Journal of Allergy and Clinical Immunology, 120*, 530–532.

Moreno-Grau, S., Elvira-Rendueles, B., Moreno, J., García-Sánchez, A., Vergara, N., & Asturias, J. A. (2006). Correlation between *Olea europaea* and *Parietaria judaica* pollen counts and quantification of their major allergens Ole e 1 and Par j 1-Par j 2. *Annals of Allergy, Asthma & Immunology, 96*(6), 858–864.

Ryan, P. H., LeMasters, G., Biagini, J., Bernstein, D., Grinshpun, S. A., Shukla, R., Wilson, K., Villareal, M., Burkle, J., & Lockey, J. (2005). Is it traffic type, volume or distance? Wheezing in infants living near truck and bus traffic. *The Journal of Allergy and Clinical Immunology, 116*, 279–284.

Sastre, J. (2010). Molecular diagnosis in allergy. *Clinical and Experimental Allergy, 40*(10), 1442–1460.

Schäppi, G. F., Taylor, P. E., Pain, M. C., Cameron, P. A., Dent, A. W., Staff, I. A., & Suphioglu, C. (1999). Concentrations of major grass group 5 allergens in pollen grains and atmospheric particles: implications for hay fever and allergic asthma sufferers sensitized to grass pollen allergens. *Clinical and Experimental Allergy, 29*(5), 633–641.

Taylor, P. E., & Jonsson, H. (2004). Thunderstorm asthma. *Current Allergy and Asthma Reports, 4*, 409–413.

The International Study of Asthma and Allergy in Childhood (ISAAC). Steering Committee. (1998). Worldwide variation in prevalence of symptoms of asthma, allergic rhinoconjunctivitis and atopic eczema. *The Lancet, 351*, 1225–1232.

Traidl-Hoffman, C., Kasche, A., Jakob, T., Huger, M., Plotz, S., Feussner, I., Ring, J., & Behrendt, H. (2002). Lipid mediators from pollen act as chemoattractants and activators of polymorphonuclear granulocytes. *The Journal of Allergy and Clinical Immunology, 109*, 831–838.

Traidl-Hoffmann, C., Jakob, T., & Behrendt, H. (2009). Determinants of allergenicity. *The Journal of Allergy and Clinical Immunology, 123*(3), 558–566.

von Hertzen, L., & Morais-Almeida, M. (2005). Signs of reversing trends in prevalence of asthma. *Allergy, 60*, 283–292.

Chapter 2
Pollen Sources

Carsten Ambelas Skjøth, Branko Šikoparija, Siegfried Jäger, and EAN-Network

Abstract This chapter reviews what is known about abundance and distribution of the 12 most important aeroallergenic pollens in Europe: *Ambrosia*, *Alnus*, *Artemisia*, *Betula*, Chenopodiaceae, *Corylus*, Cupressaceae/Taxaceae, *Olea*, *Platanus*, Poaceae, *Quercus* and *Urtica/Parietaria*. Abundance is based on 10 years of pollen records from 521 stations of the European Aeroallergen Network that were interpolated into 12 distribution maps covering most of Europe. The chapter compares the distribution maps with other types of distribution maps that are available for selected tree species and discuss two methods for making harmonized pollen source inventories: "bottom-up" and "top-down". Both methods have advantages and disadvantages, and both need to be explored and further developed. Remote sensing has shown to be a valuable method to improve the inventories, especially the use of satellites. The full potential as well as limitations of remote sensing in relation to pollen sources remains to be explored. The review suggests that the most probable way of obtaining inventories of all 12 pollen species is to use top-down methods that use an ecosystem-based approach that for each particular species connects ecological preference, pollen counts and remote sensing.

C.A. Skjøth (✉)
Department for Environmental Science, Aarhus University,
Frederiksborgvej 399, DK-4000, Roskilde, Denmark
e-mail: cas@dmu.dk

B. Šikoparija
Laboratory for Palynology, Department of Biology and Ecology, Faculty of Sciences,
University of Novi Sad, Trg Dositeja Obradovica 2, 21000 Novi Sad, Serbia

S. Jäger
Department of Oto-Rhino-Laryngology, Research Group Aerobiology and Pollen Information,
Medical University of Vienna, Waehringer Guertel 18-20, A-1090, Wien, Austria

EAN-Network
Department of Oto-Rhino-Laryngology, Research Group Aerobiology and Pollen Information
European Aeroallergen Network, (Community name) – Medical University of Vienna,
Waehringer Guertel 18-20, A-1090, Wien, Austria

M. Sofiev and K-C. Bergmann (eds.), *Allergenic Pollen: A Review of the Production, Release, Distribution and Health Impacts*, DOI 10.1007/978-94-007-4881-1_2,
© Springer Science+Business Media Dordrecht 2013

Keywords Pollen inventories • Methods • Top-down • Bottom-up • EAN • Observations • Remote sensing

2.1 Introduction

2.1.1 Gymnosperms and Angiosperms

Pollen is a biological structure functioning as a container, in which is housed male gametophyte generation of the angiosperms and gymnosperms (Moore and Webb 1983). Such a container is an evolutionary adaptation for life out of water because it protects male gametes from adverse atmospheric influence while transferring from anthers to pistils.

The importance of particular pollen grain from allergological point of view depends both on (1) pollen allergological potency and on (2) pollen abundance in the atmosphere. Keeping in mind both of above-mentioned prerequisites, 12 pollen types originating from anemophilous plants are of particular allergological interest: ragweed (*Ambrosia*), alder (*Alnus*), mugwort (*Artemisia*), birch (*Betula*), goosefoots (Chenopodiaceae), hazel (*Corylus*), cypresses including yews (Cupressaceae/Taxaceae), olive (*Olea*), plane tree (*Platanus*), grass (Poaceae), oak (*Quercus*) and wall pellitory (including stinging nettle) (*Urtica/Parietaria*).

The purpose of this chapter is an overview of what is known about pollen source location: Inventories and how they can be constructed.

2.1.2 Inventories

An inventory in environmental science is in general an aggregation of all available material with respect to abundance and distribution of subject on some sort of geographical area (big or large). Within air pollution, this is often related to chemical air pollutants. Such inventories are typically a gridded estimate of the annual release of the pollutant, and they can be used (1) by law makers and advisory bodies for development exposure limits on local, regional or international scale; (2) by atmospheric transport modellers to study processes and make scenarios and finally (3) by forecasters in daily routines to inform the public about the current air quality.

In chemical air pollution, inventories are usually made for anthropogenic sources such as traffic, industry, agriculture, etc., and include pollutants such as nitrogen monoxide + nitrogen dioxide (NO_x), sulphur dioxide (SO_2), ammonia (NH_3), volatile organic compounds (VOC), etc., but some of the inventories can also include emissions from nature (Simpson et al. 1999). Emissions from nature fluctuate to a much higher degree than their anthropogenic counterparts and are therefore often simulated by using advanced models like MEGAN (Model of Emissions of Gases and Aerosols from Nature) (Guenther et al. 2006). Models like MEGAN in general rely on locations of the sources. Similarly, pollen emission models rely on the location

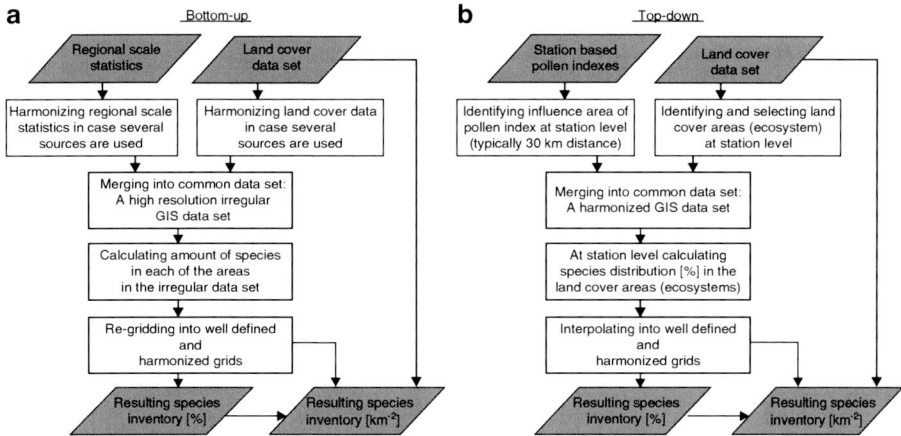

Fig. 2.1 Bottom-up (**a**) and top-down (**b**) approaches for making species inventories for allergy-related plants (see text for description)

of allergenic pollen sources. The location of allergenic pollen sources can be used by aerobiologist in explaining measured levels of pollen concentrations (e.g. Skjøth et al. 2009) as well as in the daily advice of allergenic sufferers on a local scale or recommendations with respect to travelling in between countries (Nillsson and Spieksma 1994).

Predictions of atmospheric concentrations of pollutants are in general carried out using mathematical models. New types of models for allergenic pollen are source-orientated models that have recently been introduced in aerobiology (Helbig et al. 2004; Pasken and Pietrowicz 2005; Schueler and Schlünzen 2006; Skjøth 2009; Sofiev et al. 2006; Vogel et al. 2008). These models use mathematical formulae of atmospheric transport and diffusion to calculate concentrations at various distances from a known source or release site. The character of the source is typically based on an inventory (Fig. 2.1), which can be constructed using bottom-up approaches (Sect. 2.2) or top-down approaches (Sect. 2.3).

The emission inventories are considered among the biggest uncertainties in the application of transport models (Russell and Dennis 2000), and it has been shown that dedicated focus on the inventories and the corresponding release mechanisms (Gyldenkærne et al. 2005) can significantly improve model results and understanding (Skjøth et al. 2004, 2011). In comparison to chemical air pollutants, very limited work has been done with respect to localization and inventorying the sources of allergenic airborne pollen. D'Amato et al. (2007) included information on general pollen source distribution in review concerning allergenic pollen and pollen allergy in Europe. But gridded inventories of allergenic pollen sources are very rare compared to their counterparts in chemical air quality. Additionally, making inventories of pollen sources is a scale-dependent problem (Fig. 2.2) as observed by Skjøth (2009) who used different remote sensing products to identify tree-covered areas. Coarse-resolution data is usually easy to obtain and handle but also introduces a risk of losing valuable information (Fig. 2.2a).

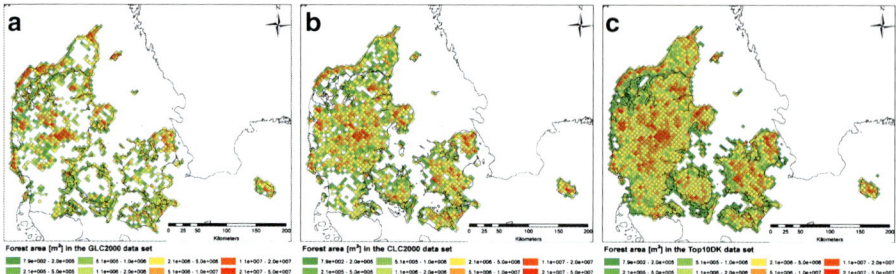

Fig. 2.2 Forest cover gridded to the DEHM-Pollen model domain using (**a**) GLC2000 (global), (**b**) CLC2000 (European) and (**c**) Top10DK(Danish) data sets resulting in 7, 9 and 13% forest cover of Denmark, respectively (Source: Skjøth 2009)

High-resolution data are much more demanding to obtain and analyse (Fig. 2.2c) but may also reveal that in some areas, the majority of sources cannot be identified by using coarse-resolution data. The focus of this chapter is to review what has been done in respect of inventories of airborne pollen sources by:

- Reviewing and discussing available data and methodologies that can be useful for production of pollen source inventories in Europe
- Presenting current knowledge on source locations of the 12 pollen types relevant to allergy
- Suggesting future research and directions for improvement of airborne pollen inventories

2.2 Methodologies for Making Bottom-Up Inventories and Their Application

Bottom-up inventories are typically produced by using statistical analysis of data with the respect to location and amount of the pollutant. For pollen-producing species, this includes location and amount within a given geographical area (Fig. 2.1a). Statistical data with respect to tree distribution and population abundance (in particular for olive, oak, alder and birch) can be obtained from forest inventories and crop databases. The statistical data are then aggregated by using a model – often a simple one – to some sort of gridded dataset. This aggregation also often uses additional information such as land cover information to upscale the distribution information to a larger geographical domain. This information is generally used in forest inventories (Forestry Commission 2001) and has at European scale been applied by Simpson et al. (1999), Köble and Seufert (2001) and Skjøth et al. (2008), where the latter is currently considered the most comprehensive and detailed inventory with respect to allergenic species from forest trees. A related source to land cover data is remote sensing data, either in its original form as digital images of the earth or analysed data such as the Corine Land Cover data set (European Commission 2005). Remote sensing data can typically be used for mapping of relevant ecosystems such as conifer or broadleaved forest but not for distinguishing between allergenic tree species (Table 2.1).

2 Pollen Sources

Table 2.1 Error matrix from a species-specific classification, including forest trees with allergenic pollen

Region/class	Ash	Sycamore	Birch	Beech	Oak	Conifers	Other	Sum	Producer acc
Ash	16	2	11	1	1	0	0	31	51.6
Sycamore	5	50	13	10	1	0	0	79	63.3
Birch	7	26	29	1	0	2	0	65	44.6
Beech	26	87	32	235	75	4	0	459	51.2
Oak	0	9	5	49	461	22	0	546	84.4
Conifers	2	1	1	2	39	644	1	690	93.3
Other areas	0	0	0	0	0	0	200	200	100.0
Sum	56	175	91	298	577	672	201	2,109	
User acc	28.6	28.6	31.9	78.9	79.9	95.8	99.5		

Based on results from Bonde (2009) by using the SPOT5 satellite over a well-defined forest area in Denmark

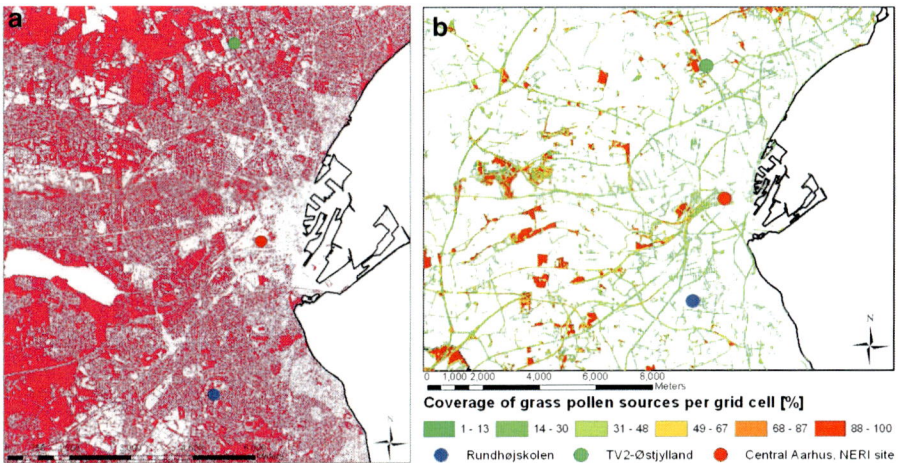

Fig. 2.3 (**a**) Very high-resolution RS image over the city of Aarhus showing grass locations (*pink*) and location of three pollen traps for grass pollen analysis. (**b**) Gridded inventory of grass flowering areas over the city of Aarhus and location of three pollen traps (Source: Skjøth et al. 2010a)

Remote sensing can also be used for mapping ground-based vegetation such as grass areas (Skjøth et al. 2010a), but here, the limitation is that a large fraction of these areas are crops or grass areas that are regularly cut and therefore do not flower. Therefore, management schemes and crop databases can be used in combination with remote sensing in order to identify possible grass flowering areas among the ground-based vegetation (Fig. 2.3).

Finally, *Olea* is a special case as this species is an important crop in Europe with the majority of the trees being located in olive groves, with geographically known locations. The location of these olive groves can with high detail be identified on the

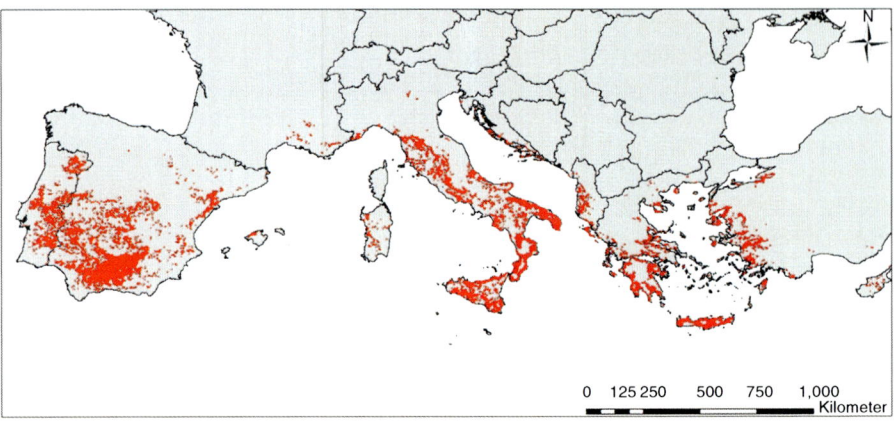

Fig. 2.4 Location of olive groves (*red*) in Europe from the CLC2006 data set

pan-European Corine Land Cover data set CLC2000 (Fig. 2.4) and must therefore be considered the pollen source with the highest accuracy with respect to distribution and amount.

Other sources of airborne pollen such as *Platanus*, Cupressaceae and *Corylus* are to a large degree ornamentals and present in a very limited degree in the main European forests. For these species statistical distribution information, in particular regional scale, are not available. Similarly, statistical information concerning abundancy and distribution for weeds are hardly available. Information can be obtained from sources such as Flora Europaea (Tutin et al. 1964) and the Nobanis network (http://www.nobanis.org/), but these sources deal with presence/absence of plants and not their abundance which makes these sources of limited use for making bottom-up inventories. In addition, Flora Europaea gives plant distribution information on a country-based scale, leading to generalization even in countries with obvious biogeographic diversity such as France, Germany and Switzerland. Usage of local Floras can help overcome this scaling problem, but such publications are unavailable for many areas and are often out of date. Bottom-up inventories for all 12 pollen species do therefore not seem likely to be obtained at a European scale within the near future.

2.3 Methodologies for Making Top-Down Inventories

Conversely to bottom-up approaches, top-down approaches often use a measured quantity as a starting point and then a backwards calculation method for estimating the geographical distribution of the species of interest (Fig. 2.1b). In aerobiology, this information can be obtained from basic results in aerobiology, such as pollen calendars or studies including source-receptor analysis (Peternel et al. 2005). This approach may be aggregated to European scale (Nillsson and Spieksma 1994) on a very coarse resolution using main biogeographical regions such as central Scandinavia or take more advanced methods into account such as land cover information and knowledge of preferred habitats for specific species (Skjøth et al. 2010b). The most widely covered database of information with respect to allergenic species

2 Pollen Sources 15

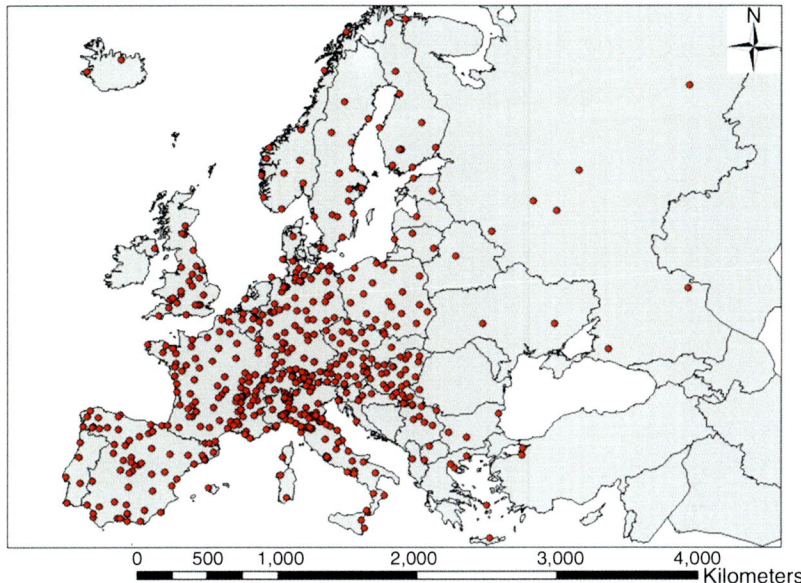

Ean stations used in the calculation of the distribution maps
• Station locations

Fig. 2.5 Location of the 521 EAN stations used in the production of the distribution maps (Figs. 2.6, 2.7, and 2.8)

is the European Aeroallergen Network (EAN), which has measured airborne pollen and calculated annual pollen indexes for all 12 allergenic pollen species at 257–521 different stations in Europe (Fig. 2.5 and Table 2.2).

Here, a simple approach by using the average pollen index of all available annual indexes for the period 2000–2009, simple interpolation between the stations (up to 400 km), buffer zones of 200 km and presence/absence information in Flora Europaea is used to summarize species distribution and abundance according to the European Aeroallergen Network. Additionally, the typical habitat is listed, and the EAN distribution is compared with available large-scale inventories that include geographical coverage and abundance of the species.

2.4 Top-Down and Bottom-Up Information Concerning the 12 Most Allergenic Pollen Types

2.4.1 *Alnus*

Ecological preference: Consists of mainly five species in Europe, where *Alnus incana* and *Alnus glutinosa* are the most common. According to Flora Europaea (Tutin et al. 1964), the species are present in most of Europe, and typical habitats are forest, woodlands and especially for *Alnus glutinosa* wet areas, such as in bogs and streams, where few other species will survive.

Table 2.2 Pollen species, number of data stations with data for each species and coverage of distribution maps (above 20 grains), maximum and mean value in the maps

Taxon	Number of stations	Land area covered [km^2]	Max	Mean
Alnus	468	5,602,964	8,055	1,048
Ambrosia	368	2,974,812	14,590	697
Artemisia	471	5,604,016	2,287	245
Betula	461	6,005,064	32,708	3,782
Chenopodiaceae	430	4,847,776	3,013	211
Corylus	457	5,033,640	3,239	398
Olea	257	1,911,564	51,094	1,253
Platanus	402	3,516,256	23,352	1,084
Poaceae	521	6,171,880	12,353	2,502
Quercus	440	5,272,036	19,587	2,011
Taxaceae + Cupressaceae	430	4,624,496	36,442	3,064
Urtica + Parietaria	471	5,593,080	68,652	4,128

Top-down approach from EAN (Fig. 2.6a): 468 stations have reported pollen indexes with an average of up to 8055. The geographical area is most of Europe from Scandinavia to central Spain and Italy. Data coverage ends in Russia, Ukraine, Romania and Turkey. Highest densities are found in the Boreal region including Poland, Lithuania, Latvia, Estonia, Belarus, Russia and Finland.

Comparisons to bottom-up information: The European scale inventory by Skjøth et al. (2008) suggests *Alnus* coverage over most of Europe from Norway/Finland to central Spain and southern Italy as well as significant coverage in Belarus, Ukraine and Russia in the east. Highest densities are found in the Boreal region from Poland, Lithuania, Latvia and Estonia and medium density in parts of Germany and most of Scandinavia.

2.4.2 Ambrosia

Ecological preference: One native (*A. maritima*) and four naturalized (*A. artemisiifolia*, *A. coronopifolia*, *A. trifida*, *A. elatior*) species could be considered as source of *Ambrosia*-type airborne pollen. *A. maritime* inhabits marine sands of the Mediterranean region, while others prefer riparian and ruderal habitats often colonizing agricultural fields (Hansen 1976).

Top-down approach from EAN (Fig. 2.6b): 368 stations have reported pollen indexes with an average up to 14,590. The geographical area is Central and Eastern Europe ranging from Germany/Poland to Italy and Greece. Highest densities are found in the Carpathian Basin and a few hotspots in the Po (Italy) and Rhone (France) valleys, respectively. There is limited data coverage in Russia and Ukraine, but the measurements suggest peak concentrations in some areas of Ukraine.

Comparisons to bottom-up information: N/A

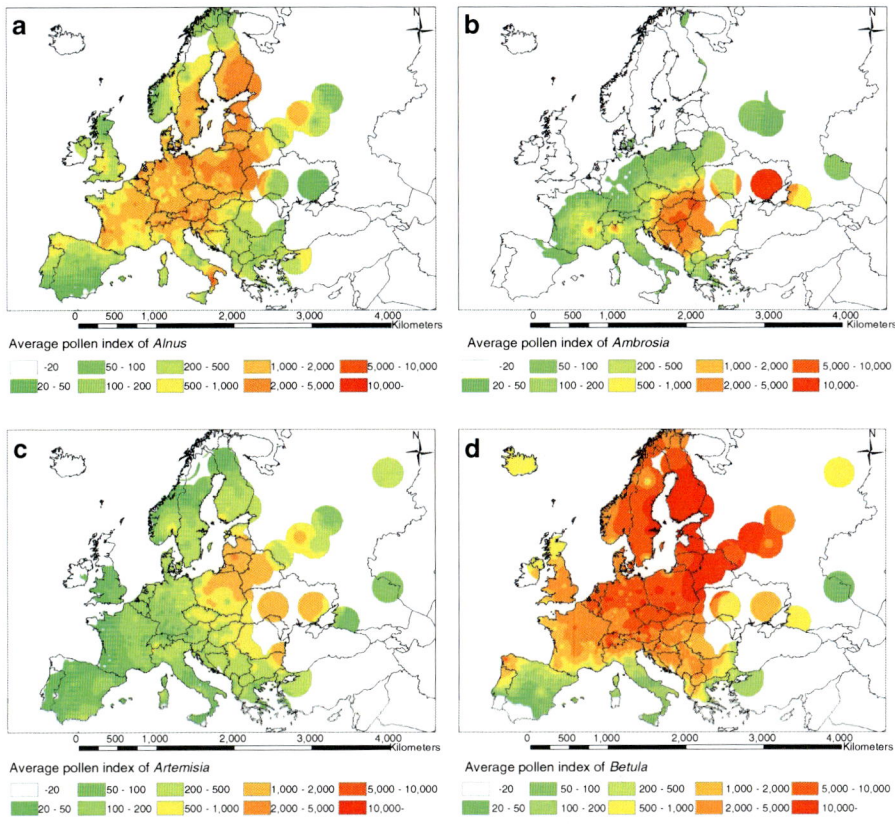

Fig. 2.6 (**a–d**) Average annual pollen index at EAN stations (Table 2.2) based on bilinear interpolation between stations, a maximum distance of 200 km to each station and Flora Europaea to determine presence on national scale

2.4.3 Artemisia

Ecological preference: Numerous species (57) belonging to the genus *Artemisia* are a source of the *Artemisia* airborne pollen type all around Europe (Tutin 1976). The most common species of *Artemisia* in Europe are *A. vulgaris*, *A. annua* and *A. verlotorum* which grow mainly in Southern Europe. All of these are present both in urban and suburban areas (D'Amato et al. 2007).

Top-down approach from EAN (Fig. 2.6c): 471 stations have reported pollen indexes with an average up to 2,287. The geographical area is most of Europe from central parts in Scandinavia to southern Spain and Italy. Highest densities are found in Poland, Lithuania, Latvia and Ukraine and medium densities in Czech Republic, Slovakia, Hungary, Serbia and Romania.

Comparisons to bottom-up information: N/A

2.4.4 Betula

Ecological preference: Betula airborne pollen in Europe originates from four native species (*B. pubescens, B. pendula, B. humilis, B. nana*) and two non-native species (*B. papyrifera, B. utilis*) often planted as ornamentals (Walters 1993).

Top-down approach from EAN (Fig. 2.6d): 461 stations have reported pollen indexes with an average up to 32,708. The geographical area is most of Europe from Scandinavia to central Spain and Italy. Data coverage ends in Russia, Ukraine, Romania and Turkey. Highest densities are found in the Boreal region including Poland, Lithuania, Latvia, Estonia, Belarus, Russia and Finland.

Comparisons to bottom-up information: The European scale inventory by Skjøth et al. (2008) suggest *Betula* coverage over most of Europe from Scandinavia to central Spain and southern Italy as well as significant coverage in Belarus, Ukraine and Russia in the east. Highest densities are found in the Boreal region from Lithuania, Latvia and Estonia and medium density in parts of Germany and Poland.

2.4.5 Chenopodiaceae

Ecological preference: The majority of species are halophytes or ruderals preferring marine habitats, steppe and semi-desert regions (Edmondson 1993). The most widespread species are considered weeds, but there are also agricultural crops such as sugar beet.

Top-down approach from EAN (Fig. 2.7a): 430 stations have reported pollen indexes with an average up to 3,013. The geographical area is Europe excluding most of Scandinavia and the British Isles. Highest densities are found on the Iberian Peninsula and Eastern Europe including the Czech Republic, Slovakia and the countries in the Carpathian Basin.

Comparisons to bottom-up information: N/A

2.4.6 Corylus

Ecological preference: Three species (*C. avellana, C. colurna, C. maxima*) are the source of the *Corylus* pollen type. Although all grow naturally in Europe, many are planted as ornamentals or in nut production fields (Tutin 1993).

Top-down approach from EAN (Fig. 2.7b): 457 stations have reported pollen indexes with an average up to 3,239. The geographical area is most of Europe from Scandinavia to central Spain and Italy. Data coverage ends in Russia, Ukraine, Romania and Turkey. Highest densities are found in Central Europe, especially the Alpine region in France, Switzerland and Austria.

Comparisons to bottom-up information: N/A

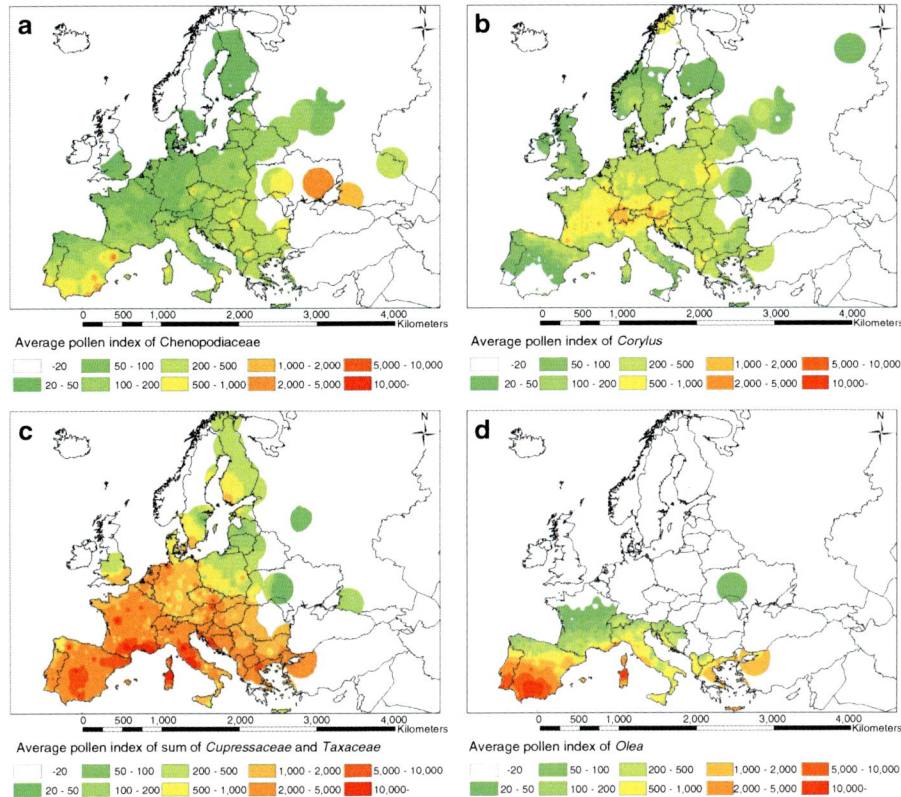

Fig. 2.7 (a–d) Average annual pollen index at EAN stations (Table 2.2) based on bilinear interpolation between stations, a maximum distance of 200 km to each station and Flora Europaea to determine presence on national scale

2.4.7 Cupressaceae/Taxaceae

Ecological preference: One species (*Taxus baccata*) classified in the Taxaceae family and numerous species classified into five genera (*Cupressus*, *Chamaecyparis*, *Juniperus*, *Thuja*, *Tetrachius*) of the Cupressaceae family produce this pollen type. The former grows naturally or is often planted as ornamental all around Europe except east and above 63° N (Moore 1993a). The latter are widely distributed with some being planted as ornamentals, for shelter or for timber (Moore 1993b).

Top-down approach from EAN (Fig. 2.7c): 430 stations have reported pollen indexes with an average up to 36,442. The geographical area is most of Europe from parts of Scandinavia to Spain and Italy. Highest densities are found in the western and southern parts of Europe, while relative low densities are found in the Boreal region.

Comparisons to bottom-up information: N/A

2.4.8 Olea

Ecological preference: Species *Olea europaea* and its cultivar variety are considered as the only source of this pollen type in Europe. The species naturally inhabits dry and rocky places of the Mediterranean region and also at Krim peninsula. *O. europaea* is introduced to southern Switzerland (do Amaral Franco and da Rocha Afonso 1972).

Top-down approach from EAN (Fig. 2.7d): 257 stations have reported pollen indexes with an average up to 51,094. The geographical area is limited to Southern Europe, mainly below the Alpine region. Highest densities are found in southern Spain, and lowest densities are found in central France and areas in the Carpathian Basin with data coverage.

Comparisons to bottom-up information: The CLC2006 data set (Fig. 2.2) with location of olive groves suggests highest densities in southern Spain, Portugal and Italy. Most easterly parts are found in western parts of Turkey, and most northern parts are found in Croatia, France and Italy.

2.4.9 Platanus

Ecological preference: *Platanus* airborne pollen sources in Europe are *P. orientalis* and *P. acerifolia*. Natural habitats are damp woods and streamsides, but both species are commonly planted in much of Europe as roadside trees (Tutin and Edmondson 1993).

Top-down approach from EAN (Fig. 2.8a): 402 stations have reported pollen indexes with an average up to 23,352. The geographical area is most of Europe from parts of Scandinavia to central Spain and Italy. High densities are found in a number of isolated locations near certain large urban areas such as London, Madrid, Milano and Vienna, respectively.

Comparisons to bottom-up information: N/A

2.4.10 Poaceae

Ecological preference: Pollen of this type originates from numerous ubiquitous species (Tutin 1980) that inhabit both natural and artificial grasslands. In addition, many species are cultivated as wheat in agriculture.

Top-down approach from EAN (Fig. 2.8b): 521 stations have reported pollen indexes with an average up to 12,353. The geographical area is Europe and the largest of all pollen species. Data coverage is limited in Belarus, Russia, Ukraine, Romania and Turkey.

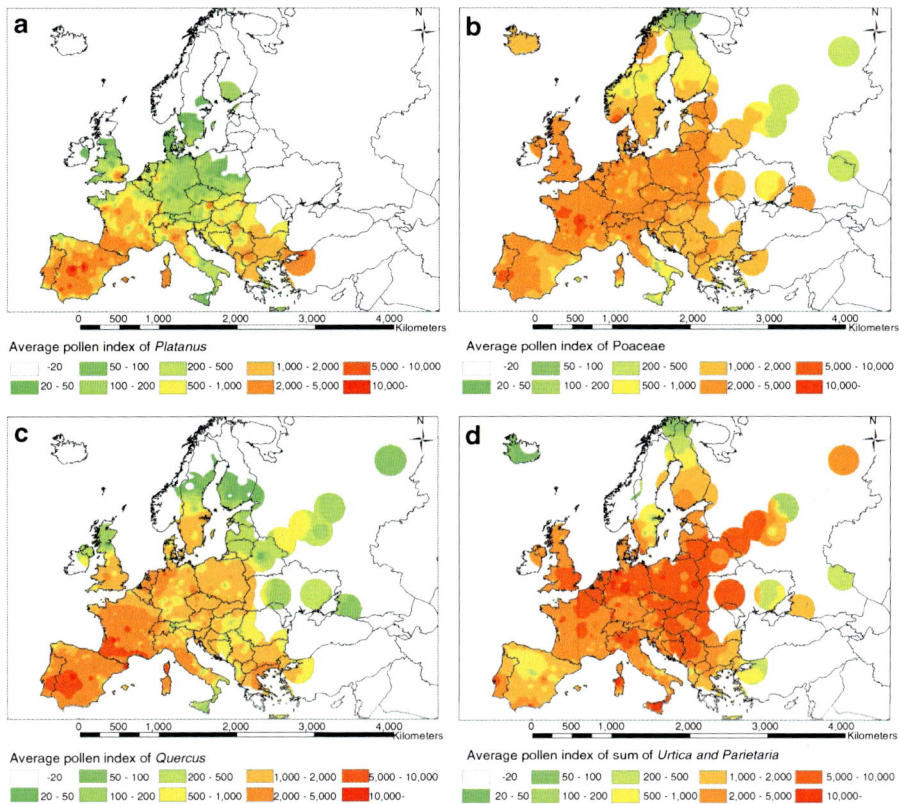

Fig. 2.8 (**a–d**) Average annual pollen index at EAN stations (Table 2.2) based on bilinear interpolation between stations, a maximum distance of 200 km to each station and Flora Europaea to determine presence on national scale

Highest densities are found in a relatively large area from Denmark and the British Isles in the North to the Iberian Peninsula and central Italy.

Comparisons to bottom-up information: N/A

2.4.11 Quercus

Ecological preference: Airborne pollen originates from a number of species (22) distributed all around Europe (Schwarz 1993).

Top-down approach from EAN (Fig. 2.8c): 440 stations have reported pollen indexes with an average up to 19,587. The geographical area is most of Europe from central Scandinavia to southern Spain and Italy. Highest densities are found in southern France and Spain and medium densities in most of Europe from southern Sweden and England in the North to central Italy and Greece in the South.

Comparisons to bottom-up information: The European scale inventory by Skjøth et al. (2008) suggests *Quercus* coverage over most of Europe from southern Sweden in the North to Spain in the South. Additionally, this inventory is divided into species such as *Quercus rubra*, *Q. petraea*, *Q. suber*, etc.

2.4.12 Urtica/Parietaria

Ecological preference: Airborne pollen originates from species classified in two genera *Urtica* and *Parietaria*. The most important are *U. dioica* (Ball 1993) and *P. judaica* that is widespread in ruderal rocky habitats (Ball 1993).

Top-down approach from EAN (Fig. 2.8d): 471 stations have reported pollen indexes with an average up to 68,652. The geographical area is most of Europe from parts of Scandinavia in the North to Spain and Italy in the South. Data coverage ends in Russia, Ukraine, Romania and Turkey. Highest densities are found in a relatively large area in Central Europe including southern England, Belgium, the Netherlands, parts of Germany and Poland. Relatively low densities are found in parts of Scandinavia, Spain and South-eastern Europe.
Comparisons to bottom-up information: N/A

2.5 Overall Conclusions

In general very little is known about location of pollen sources. Most well known are the location of tree species such as *Olea*, *Quercus*, *Alnus* and *Betula*. Here, *Olea* is a special case, as the majority of *Olea* trees are found in olive groves that are mapped with high detail in Europe (Fig. 2.4). *Alnus*, *Betula* and *Quercus* have been mapped using both top-down methods in the EAN (Figs. 2.6a, d, 2.8c) and bottom-up methods (Skjøth et al. 2008). On a European scale, the distribution and density of these three species is very similar, but on a regional to local scale such as over the UK, the differences can be significant.

The remaining species *Ambrosia*, *Artemisia*, Chenopodiaceae, *Corylus*, Cupressaceae, *Platanus*, Poaceae, and *Urtica/Parietaria* have only been mapped with respect to abundancy on a European scale using the top-down approach in the EAN (Figs. 2.6b, c, 2.7a, b, c, 2.8a, b, d). Other inventories such as the NOBANIS (North European and Baltic Network on Invasive Alien Species) network only register presence/absence. At a regional scale, *Ambrosia* has recently been mapped in the Carpathian Basin using an ecosystem-based approach in combination with airborne pollen data and detailed land cover data (Skjøth et al. 2010b). This methodology is likely to be applicable in other regions as well as for other species such as *Artemisia* and *Platanus*, where typical ecosystems can be identified in European scale land cover data sets such as the Corine Land Cover. Other species such as Chenopodiaceae, *Corylus* and *Cupressaceae*, Poaceae and *Urtica/Parietaria* can

Table 2.3 Satellite relevant for mapping of allergenic species on local, regional or European scale, the main properties and availability. "–" without ending means that the satellite is still active

Satellite/instrument	Resolution at nadir	Period	Bands (n)	Swath	Accessibility
Landsat 1–3	79	1972–1983	4	185 km	Free
Landsat 4	30 m	1982	6	185 km	Free
Landsat 5	30 m	1984	6	185 km	Free
Landsat 7/ETM+	15	1999–2003!	6	183 km	Free
NOAA AVHRR	1.1 km	1981–	5	2,500 km	Free
SPOT1	10 m /20 m	1986–1990	1/3	60 km	Free*
SPOT2	10 m /20 m	1990–	1/3	60 km	Free*
SPOT3	10 m /20 m	1993–1997	1/3	60 km	Free*
SPOT4	20 m	1998–	1/3	60 km	Free*
SPOT5	2.5 m/10 m	2002–	1/3	60 km	Free*
Terra (ASTER)	15/30/90	1999–	15	2,330 km	Free
IRS-P6	5.8/23	2003–	1/4	23/141 km	Free*
IKONOS	0.8 m/3.2 m	1999–	1/4	11 km	Commercial
KOMPSAT-2	1 m/4 m	2006	1/4	15 km	Commercial/free*
ENVISAT/MERIS	300 m	2002–2012	15	1,150 km	Free*
GeoEye-1	0.4/1.6	2008	1/4	15 km	Commercial
Formosat-2	2 m/8 m	2004–	1/4	24 km	Commercial/free*

Free*: Available free from the European Space Agency for Category-1 users at www.esa.int or through simple registration
Free: Available free on the internet through Warehouse Inventory Search Tool (WIST), https://wist.echo.nasa.gov/wist-bin/api/ims.cgi?mode=MAINSRCH&JS=1

most likely also be mapped by using an ecosystem-based approach, but it is also likely that the approach then needs an improvement in methods, as the pollen observations in the EAN from several of these species originate from a number of different species with different pollen production and different ecological preference.

Remote sensing has recently been introduced as an additional source of information for mapping of allergenic species. Its use is promising and has by far been explored enough. Nevertheless, existing use of remote sensing has already shown two major limitations. (1) Satellites need at least four channels (3 colour and one near infrared) for a good identification of different types of vegetation but are still not able to distinguish plants and trees at the species level (Table 2.1), which means that satellite products need additional information such as statistics of species distribution in forests or similar ecosystems that can be observed from space. (2) Satellites need medium- to high-resolution spatial coverage in order to correctly identify the majority of the sources over Europe. Pre-calculated global data sets like the GLC2000 (Bartalev et al. 2003; Fritz et al. 2003) do not meet that criteria. The Corine Land Cover or related data sets like Image2000 or JRC forest cover (Schuck et al. 2003) are likely to meet this criteria in most countries. Higher resolution than the CLC2000 data set or JRC forest cover is desired in some areas and required in case the major source is found in urban areas such as the KOMPSAT-2 satellite or commercial satellites like Quickbird satellite (Table 2.3).

In the year 2013, the ESA (European Space Agency) will launch the first of a pair of Sentinel-2 satellites. This pair of satellites is very well designed for the needs in mapping allergological relevant species as they combine a high revisit time, high spatial resolution and multispectral imagery (13 bands) that are well designed for mapping vegetation. This suggests that aerobiology and the mapping of the relevant species can be advanced significantly today by using existing remote sensing products and that these possibilities will be further improved within the next few years.

Overall, the EAN data set is the largest and, for a number of pollen relevant plant species, also the only large-scale data set for mapping abundancy such as *Urtica* and Chenopodiaceae, and Figs. 2.6, 2.7, 2.8 are examples of how distribution maps can be produced (see summary in Table 2.2). However, this methodology also has four obvious limitations. (1) Station coverage is highly variable (Fig. 2.5 and Table 2.2) with some stations having observations during the entire period while others only for 1 year. This means that the distribution maps have a solid database in Central Europe, while other areas such as Eastern Europe have very limited or no data coverage. (2) The applied size of buffer zones can also be questioned. Data from pollen traps is a point measurement, and the introduction of a buffer zone where interpolation and data coverage is valid will introduce an error. It is not known how large this error is. (3) The database will be subject to errors in the data reporting or misclassification of the pollen grains in the microscope. (4) The pollen traps capture pollen within an area, which will be affected by amounts of plants, geographical variation in pollen production and atmospheric transport. One of these four limitations might be the reason to certain high-density hotspots seen in southern Italy for *Alnus* (Fig. 2.6a) or low-density areas seen in Scandinavia for *Betula* (Fig. 2.6d). Another example is the distribution map for *Olea*. In these maps, Hungary is white despite pollen counts of *Olea* being registered in the EAN database. These registrations could potentially be a misdetermination with *Ligustrum* pollen, or the pollen could originate from olives growing in pots because according to Flora Europaea, olive trees are not present in Hungary. Similarly, the Spanish network by definition does not upload *Ambrosia* observation to the EAN data, although measurements indicate small quantities of *Ambrosia*, which is also supported by Flora Europaea, which suggests a presence of *Ambrosia* in Spain. Similarly, the use of buffer zones and interpolations shows *Olea* distribution in Switzerland because Flora Europaea suggests olive trees in Switzerland. As such, these examples show the limitation of this very simple method for making distribution maps. Remote sensing products have been proven as a valuable tool for additional information for mapping pollen species. Many regional scale products are freely available from these satellites, including the Corine Land Cover (Landsat satellite), Globcover (Envisat Satellite) or the GLC2000 (SPOT satellite) or more detailed products such as the JRC forest mapping (Landsat satellite) (Schuck et al. 2003): http://forest.jrc.ec.europa.eu/forest-mapping or the Urban Atlas (SPOT 5): http://www.eea.europa.eu/data-and-maps/data/urban-atlas. However, better information over a specific area can usually be obtained by detailed analysis of remote sensing pictures that are used to produce data sets such as CLC2000 data set (Schuck et al. 2003) or other similar products. However, remote sensing can so far not be used as a stand-alone product for mapping sources on the species level. Remote sensing products (CLC2000, Image2000, etc.) need additional information such as highly detailed ground-based statistics (e.g. Skjøth et al. 2008) in order to apply bottom-up approaches for making inventories. These inventories will then be limited by the geographical coverage of the region with statistics, which means that a large uncertainty will be present in regions such as Ukraine and relatively small in sub-national regions of the UK or Denmark. Most medium to coarse-scale resolution remote sensing images can be obtained free of charge (Table 2.3), while high-resolution satellites such as Quickbird or Ikonos are commercial. These satellites are usually used for dedicated urban scale investigations and are well designed for identifying ornamentals trees including *Platanus*, *Betula* and Cupressaceae. A recent possibility for urban scale mapping – in case allergenic plants are available – is the Kompsat-2 satellite. Kompsat-2 is a high-resolution satellite, and images over a large amount of European cities can be obtained free of charge through Category-1 proposals with the European Space Agency. This possibility however remains to be explored. Finally, remote sensing products can also be used on the large scale in combination with pollen indexes in order to apply top-down approaches for an ecosystem-based method for mapping location of pollen species in Europe (Skjøth et al. 2010b). This methodology will however still be limited by trap coverage and the fact that the local pollen index is influenced by variations in pollen production and atmospheric transport. Nevertheless, the methodology possess a significant potential for detailed mapping of all major pollen species with high detail over Europe including those that be mapped cannot using bottom-up approaches due to insufficient information.

References

Ball, P. W. (1993). *Alnus* Miller. In T. G. Tutin, N. A. Burges, A. O. Chater, J. R. Edmondson, V. H. Heywood, D. M. Moore, D. H. Valentine, S. M. Walters, & D. A. Webb (Eds.), *Flora Europaea* (Psilotaceae to Platanaceae, Vol. 1, pp. 69–70). Cambridge: Cambridge University Press.

Bartalev, S. A., Belward, A. S., Erchov, D. V., & Isaev, A. S. A. (2003). New SPOT4-VEGETATION derived land cover map of Northern Eurasia. *International Journal of Remote Sensing, 24*, 1977–1982.

Bonde, H. (2009). Mapping of trees with high spatial resolution remote sensing data (In Danish: Kortlægning af træer med højt rumligt opløst remote sensing data) Roskilde University, Universitetsvej 1, 02, DK-4000, Roskilde, Denmark.

D'Amato, G., Cecchi, L., Bonini, S., Nunes, C., Annesi-Maesano, I., Behrendt, H., Liccardi, G., Popov, T., & Van Cauwenberge, P. (2007). Allergenic pollen and pollen allergy in Europe. *Allergy, 62*, 976–990.

do Amaral Franco, J., & da Rocha Afonso, M. L. (1972). *Olea* L. In T. G. Tutin, N. A. Burges, A. O. Chater, J. R. Edmondson, V. H. Heywood, D. M. Moore, D. H. Valentine, S. M. Walters, & D. A. Webb (Eds.), *Flora Europaea* (Diapensiaceae to Myoporaceae, Vol. 2, p. 55). Cambridge: Cambridge University Press.

Edmondson, J. R. (1993). Chenopodiaceae. In T. G. Tutin, N. A. Burges, A. O. Chater, J. R. Edmondson, V. H. Heywood, D. M. Moore, D. H. Valentine, S. M. Walters, & D. A. Webb (Eds.), *Flora Europaea* (Psilotaceae to Platanaceae, Vol. 1, pp. 108–130). Cambridge: Cambridge University Press.

European Commission. (2005). Image2000 and CLC2000 products and methods. European commission, Joint Research Center (DG JRC), Institute for Environment and Sustainability, Land Management Unit, I-21020 Ispra (VA), Italy, pp. 152.

Forestry Commission. (2001). National inventory of woodland and trees: Forestry commission, publications, PO Box 25, Wetherby, West Yorkshire, LS23 7EW. pp. 68.

Fritz, S., Bartholome, E., Belward, A., Hartley, A., Stibig, H.-J., Eva, H., Mayaux, P., Bartalev, S., Latifovic, R., Kolmert, S., Roy, P. S., Agrawal, S., Bingfang, W., Wenting, X., Ledwith, M., Pekel, J.-F., Giri, C., Mücher, S., de Badts, E., Tateishi, R., Champeaux, J.-L. & Defourny, P. (2003). The global land cover for the year 2000. European Commission, Joint Research Centre, pp. 41.

Guenther, A., Karl, T., Harley, P., Wiedinmyer, C., Palmer, P. I., & Geron, C. (2006). Estimates of global terrestrial isoprene emissions using MEGAN (Model of Emissions of Gases and Aerosols from Nature). *Atmospheric Chemistry and Physics, 6,* 3181–3210.

Gyldenkærne, S., Ambelas Skjøth, C., Hertel, O., & Ellermann, T. (2005). A dynamical ammonia emission parameterization for use in air pollution models. *Journal of Geophysical Research, 110,* 1–14. doi:10.1029/2004JD005459.

Hansen, A. (1976). Flora Europaea. In T. G. Tutin, N. A. Burges, A. O. Chater, J. R. Edmondson, V. H. Heywood, D. M. Moore, D. H. Valentine, S. M. Walters, & D. A. Webb (Eds.), *Flora Europaea* (Plantaginaceae to Compositae (and Rubiaceae), Vol. 4, pp. 142–143). Cambridge: Cambridge University Press.

Helbig, N., Vogel, B., Vogel, H., & Fiedler, F. (2004). Numerical modelling of pollen dispersion on the regional scale. *Aerobiologia, 20,* 3–19.

Köble, R., & Seufert, G. (2001). Novel maps for forest tree species in Europe. *Proceedings of the 8th European Symposium on the Physico-Chemical Behaviour of Air Pollutants: "A Changing Atmosphere!"*, Torino (Italy), 17–20 September 2001.

Moore, D. M. (1993a). Taxaceae. In T. G. Tutin, N. A. Burges, A. O. Chater, J. R. Edmondson, V. H. Heywood, D. M. Moore, D. H. Valentine, S. M. Walters, & D. A. Webb (Eds.), *Flora Europaea* (Psilotaceae to Platanaceae, Vol. 1, p. 48). Cambridge: Cambridge University Press.

Moore, D. M. (1993b). Cupressaceae. In T. G. Tutin, N. A. Burges, A. O. Chater, J. R. Edmondson, V. H. Heywood, D. M. Moore, D. H. Valentine, S. M. Walters, & D. A. Webb (Eds.), *Flora Europaea* (Psilotaceae to Platanaceae, Vol. 1, pp. 45–48). Cambridge: Cambridge University Press.

Moore, P. D., & Webb, J. A. (1983). *An illustrated guide to pollen analysis* (p. 133). London: Hodder and Stoughton.

Nillsson, S., & Spieksma, F. T. M. (1994). Allergy service guide in Europe. Palynological laboratory Swedish Museum of Natural History, pp. 123. ISBN 91-86510-31-2.

Pasken, R., & Pietrowicz, J. A. (2005). Using dispersion and mesoscale meteorological models to forecast pollen concentrations. *Atmospheric Environment, 39,* 7689–7701.

Peternel, R., Culig, R., Srnec, L., Mitic, B., Vukusic, I., & Hrga, I. (2005). Variation in ragweed (*Ambrosia artemisiifolia* L.) pollen concentrations in central Croatia. *Annals of Agricultural and Environmental Medicine, 12,* 11–16.

Russell, A., & Dennis, R. (2000). NARSTO critical review of photochemical models and modelling. *Atmospheric Environment, 34,* 2283–2324.

Schuck, A., Paivinen, R., Hame, T., Van Brusselen, J., Kennedy, P., & Folving, S. (2003). Compilation of a European forest map from Portugal to the Ural mountains based on earth observation data and forest statistics. *Forest Policy and Economics, 5,* 187–202.

Schueler, S., & Schlünzen, K. (2006). Modeling of oak pollen dispersal on the landscape level with a mesoscale atmospheric model. *Environmental Modelling and Assessment, 11,* 179–194.

Schwarz, O. (1993). *Quercus* L. In T. G. Tutin, N. A. Burges, A. O. Chater, J. R. Edmondson, V. H. Heywood, D. M. Moore, D. H. Valentine, S. M. Walters, & D. A. Webb (Eds.), *Flora Europaea* (Psilotaceae to Platanaceae, Vol. 1, pp. 72–76). Cambridge: Cambridge University Press.

Simpson, D., Winiwarter, W., Borjesson, G., Cinderby, S., Ferreiro, A., Guenther, A., Hewitt, C. N., Janson, R., Khalil, M. A. K., Owen, S., Pierce, T. E., Puxbaum, H., Shearer, M., Skiba, U., Steinbrecher, R., Tarrason, L., & Oquist, M. G. (1999). Inventorying emissions from nature in Europe. *Journal of Geophysical Research, 104,* 8113–8152.

Skjøth, C. A. (2009). Integrating measurements, phenological models and atmospheric models in Aerobiology – creating new concepts within aerobiological integrated monitoring and forecasting Faculty of Science. Copenhagen University, Ph.D. thesis, 124 pp.

Skjøth, C. A., Hertel, O., Gyldenkærne, S., & Ellermann, T. (2004). Implementing a dynamical ammonia emission parameterization in the large-scale air pollution model ACDEP. *Journal of Geophysical Research, 109*, 1–13. doi:10.1029/2003JD003895.

Skjøth, C. A., Geels, C., Hvidberg, M., Hertel, O., Brandt, J., Frohn, L. M., Hansen, K. M., Hedegaard, G. B., Christensen, J., & Moseholm, L. (2008). An inventory of tree species in Europe – an essential data input for air pollution modelling. *Ecological Modelling, 217*, 292–304.

Skjøth, C. A., Smith, M., Brandt, J., & Emberlin, J. (2009). Are the birch trees in Southern England a source of *Betula* pollen for North London? *International Journal of Biometeorology, 53*, 75–86.

Skjøth, C. A., Becker, T., Ørby, P. V., Geels, C., Schlünssen, V., Sigsgaard, T., Bønløkke, J. H., Sommer, J., Søgaard, P., & Hertel, O. (2010a). Urban sources cause elevated grass pollen concentrations. Presented at the 9th International congress on Aerobiology, Buenos Aires, 23–27 August 2010.

Skjøth, C. A., Smith, M., Sikoparija, B., Stach, A., Myszkowska, D., Kasprzyk, I., Radisic, P., Stjepanovic, B., Hrga, I., Apatini, D. R., Magyar, D., Páldy, A., & Ianovici, N. (2010b). A method for producing airborne pollen source inventories: An example of *Ambrosia* (ragweed) on the Pannonian Plain. *Agricultural and Forest Meteorology, 150*, 1203–1210.

Skjøth, C. A., Geels, C., Berge, H., Gyldenkærne, S., Fagerli, H., Ellermann, T., Frohn, L. M., Christensen, J., Hansen, K. M., Hansen, K., & Hertel, O. (2011). Spatial and temporal variations in ammonia emissions – a freely accessible model code for Europe. *Atmospheric Chemistry and Physics, 11*, 5221–5236.

Sofiev, M., Siljamo, P., Ranta, H., & Rantio-Lehtimaki, A. (2006). Towards numerical forecasting of long-range air transport of birch pollen: Theoretical considerations and a feasibility study. *International Journal of Biometeorology, 50*, 392–402.

Tutin, T. G. (1976). *Artemisia* L. In T. G. Tutin, N. A. Burges, A. O. Chater, J. R. Edmondson, V. H. Heywood, D. M. Moore, D. H. Valentine, S. M. Walters, & D. A. Webb (Eds.), *Flora Europaea* (Plantaginaceae to Compositae (and Rubiaceae), Vol. 4, pp. 178–186). Cambridge: Cambridge University Press.

Tutin, T. G. (1980). Gramineae (Poaceae) In T. G. Tutin, V. H. Heywood, N. A. Burges, D. M. Moore, D. H. Valentine, S. M. Walters, & D. A. Webb (Eds.), *FloraEuropaea (Alismataceae to Orchidaceae (Monocotyledones))* (Vol. 5, pp. 118–267). Cambridge: Cambridge University Press.

Tutin, T. G. (1993). *Corylus* L. In T. G. Tutin, N. A. Burges, A. O. Chater, J. R. Edmondson, V. H. Heywood, D. M. Moore, D. H. Valentine, S. M. Walters, & D. A. Webb (Eds.), *Flora Europaea* (Psilotaceae to Platanaceae, Vol. 1, p. 71). Cambridge: Cambridge University Press.

Tutin, T. G., & Edmondson, J. R. (1993). *Platanus* L. In T. G. Tutin, N. A. Burges, A. O. Chater, J. R. Edmondson, V. H. Heywood, D. M. Moore, D. H. Valentine, S. M. Walters, & D. A. Webb (Eds.), *Flora Europaea* (Psilotaceae to Platanaceae, Vol. 1, p. 463). Cambridge: Cambridge University Press.

Tutin, T. G., Heywood, V. H., Burges, N. A., Valentine, D. H., Walters, S. M., & Webb, D. A. (1964). Flora Europea. Cambridge University Press, pp. 2392 (Also available as CD from 2001 and online by the Royal Botanic Garden in Edinburgh).

Vogel, H., Pauling, A., & Vogel, B. (2008). Numerical simulation of birch pollen dispersion with an operational weather forecast system. *International Journal of Biometeorology, 52*(8), 805–814.

Walters, S. M. (1993). *Betula* L. In T. G. Tutin, N. A. Burges, A. O. Chater, J. R. Edmondson, V. H. Heywood, D. M. Moore, D. H. Valentine, S. M. Walters, & D. A. Webb (Eds.), *Flora Europaea* (Psilotaceae to Platanaceae, Vol. 1, pp. 68–69). Cambridge: Cambridge University Press.

Chapter 3
The Onset, Course and Intensity of the Pollen Season

Åslög Dahl, Carmen Galán, Lenka Hajkova, Andreas Pauling, Branko Sikoparija, Matt Smith, and Despoina Vokou

Abstract The onset, duration and intensity of the period when pollen is present in the air varies from year to year. Amongst other things, there is an effect upon the quality of life of allergy sufferers. The production and emission of pollens are governed by interacting environmental factors. Any change in these factors may affect the phenology and intensity of the season. Readiness to flower in a plant, and the amount of pollen produced, is the result of conditions during an often long period foregoing flowering. When a plant is ready to flower, temporary ambient circumstances e.g., irradiation and humidity, determine the timing of the actual

Å. Dahl (✉)
Department of Biological and Environmental Sciences, University of Gothenburg,
P.O. Box 461, 405 30 Göteborg, Sweden
e-mail: aslog.dahl@bioenv.gu.se

C. Galán
Department of Botany, Ecology and Plant Physiology, University of Cordoba,
Campus de Rabanales, Edificio Celestino Mutis, 14071 Cordoba, Spain

L. Hajkova
Czech Hydrometeorological Institute/Faculty of Science, Charles University,
Kockovska 2699/18, P.O. Box 2, 400 11 Usti nad Labem, Czech Republic/Albertov 6,
128 43 Prague 2, Czech Republic

A. Pauling
MeteoSwiss, Kraehbuehlstrasse 58, 8044 Zurich, Switzerland

B. Sikoparija
Laboratory for Palynology, Faculty of Sciences, University of Novi Sad,
Trg Dositeja Obradovica 2, 21000 Novi Sad, Serbia

M. Smith
National Pollen and Aerobiology Research Unit, Institute of Health,
University of Worcester, Worcester, WR2 6AJ, United Kingdom

D. Vokou
Department of Ecology, School of Biology, Aristotle University of Thessaloniki,
University Campus, GR-54124 Thessaloniki, Greece

pollen release. In order to understand variation between years and to be able to safely predict future situations, not least due to the ongoing climate change, it is necessary to know the determinants of all related processes and differences between and within species, here reviewed.

Keywords Anemophily • Allergenic plants • Phenology • Readiness to flower • Vernalization • Onset of anthesis • Chilling • Dormancy • Forcing • Pollen release • Pollen emission • Circadian rhythms • Pollen season duration • Pollen index • Flowering intensity • Masting • Pollen season severity • Pollen production • Climate change

3.1 Introduction

Aerobiology studies the behaviour of biological particles in the atmospheric medium. Due to their abundance and allergenic properties, fungal spores and pollen grains, primarily from wind-pollinated (anemophilous plants), are the most important of these particles. Pollen grains from this group of plants are small, very light in mass, aerodynamic, with a relatively thin wall, low sculpturing and a powdery, non-sticky surface.

As "pollen season", we define the period during which pollen is present in the air. The term is used either collectively, referring to pollen from any plant taxon or from each taxon separately. Evidently, the pollen season in a certain area is related to the local flowering season, as for pollen to be present in the air, it has to be previously produced and emitted by mature flowers. However, pollen seasons and flowering seasons usually do not fully coincide; this is the result of intervening winds that allow for mid- and/or long-range transport. The abundance of pollen grains in the air is described by the term pollen season intensity: the more there are, the more intense the season is.

Pollen in the air, which is the result of pollen production and the different aerobiological processes, emission, dispersion and/or transport and deposition, is controlled by factors associated with climate. Any change in these factors may affect the phenological and quantitative features of the season. Monitoring airborne pollen provides substantial information on plants' performance under different environmental conditions and could allow predictions of their response under the ongoing climate change.

The onset, duration and intensity of the pollen season varies from year to year. In order to understand this variation and be able to safely predict future situations, it is necessary to know the determinants of all related processes. At an applied level, this is particularly important because the pollen season and its intensity affects the quality of life of allergic persons, who constitute a considerable part of the human population, particularly in the industrialized countries.

In this chapter, we will attempt to identify the strength and weaknesses in our knowledge and, hence, the requirements for future activity. To this end, we will deal with the factors known to affect pollen-related processes in anemophilous plants giving emphasis to taxa of allergenic importance in both the north and south of Europe.

3.2 Readiness to Flower

The major environmental factors that determine the readiness to flower are photoperiod, temperature and water stress. Water stress can also be important through its effect on growth rate. In some plants, e.g. in many temperate grass species, floral buds are produced only after a sequence of environmental cues that may occur several months apart. The term "readiness" denotes a state of maturity. When a plant is ready to flower, it is temporary ambient circumstances e.g., irradiation and humidity that determine the timing of the actual pollen release (Fig. 3.1). The latter circumstances are described in Sect. 3.3 of this chapter.

3.2.1 *The Influence of Light as Day Length and Light Intensity*

A photoperiod response is a biological response to changes in day length at latitudes where this cue is associated with seasonal progression. Plants discriminate day from night by means of photoreceptors, i.e., pigments that capture different wavelengths that may promote or inhibit flowering. These pigments synchronize biological activities with the day/night cycle, the so-called circadian clock. Plants that are governed by photoperiod may respond to day lengths surpassing a critical threshold in late spring or early summer, securing time for seed maturation, or in contrast, when nights become long so that as much of spring and summer as possible can be used for vegetative growth. For instance, in *Ambrosia artemisiifolia*, floral initiation is not possible before summer (Lewis et al. in Rogers et al. 2006).

Recent studies indicate that plants perceive day length through the degree of coincidence of light with the expression of the gene CONSTANS (CO). This gene encodes a clock-regulated protein that controls the expression of the floral integrator FT, which is ubiquitous in plants, in a light-dependent manner. Although long- and short-day plants contrast in their reactions to light, the differences are due to modifications of the same common basic pathway. Other floral integrator genes react to temperature and/or developmental cues such as hormone levels (Kim et al. 2009). Whereas a response to photoperiod is often overriding in short-lived plants, an interaction between photoperiod and temperature is common in perennials (King and Heide 2009).

Additionally, increasing light intensity during springtime and early summer is likely to influence flowering in summer-flowering plants. This increase can modify the effect of the photoperiod, but also act independently upon the FT-gene in long-day plants (King and Heide 2009).

3.2.2 *Temperature*

Each vital process in living organisms is adjusted to a certain temperature range. The optimal amplitude of this fluctuation range varies between plants adapted to

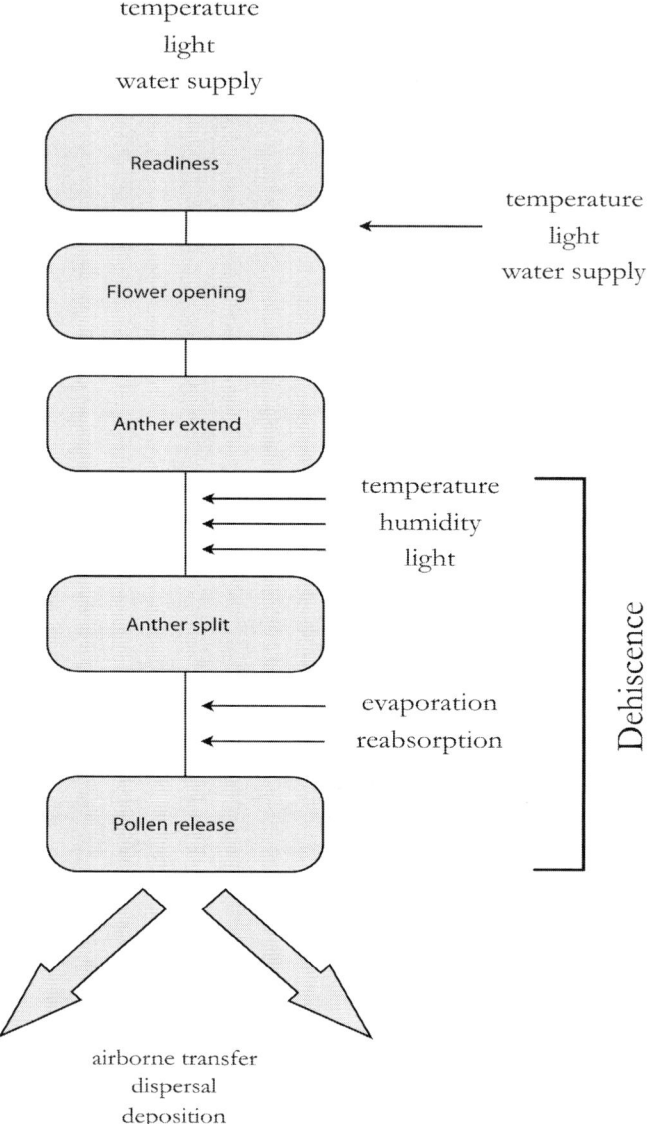

Fig. 3.1 Scheme of the processes involved in the readiness of flowering and anther dehiscence of anemophilous plant species. Modified after Linskens and Cresti (2000)

different climate regimes. For each species/ecotype and process, and for the development of different structures, there is a specific response curve. Below the base temperature, development is zero. Above this threshold, development rate increases linearly up to an optimum temperature at which the rate of development is at a maximum. Above this point, development rate decreases linearly until an upper limit, above which development is again zero. In most temperate woody species and some perennial herbs (Rathcke and Lacey 1985), flowering time is mainly controlled by temperature which usually acts through accumulation of heat above a threshold level.

In temperate regions, many species need a period of exposure to low, but nonfreezing temperatures to acquire reproductive competence, as an adaptation to fit flowering and fruiting to periods when the risks for frost and/or detrimental drought are low. This process is called vernalization (from Latin word *vernus*, meaning "of the spring") and confers the ability of the plant to respond to the correct day length and to initiate flowers. Thus, it is not equal to floral initiation, but must forego this. Many vernalization-requiring species have a winter-annual or biennial habit and many are long-day plants (Kim et al. 2009).

Unlike cold acclimation, which is a relatively rapid process and which follows a separate physiological pathway, the vernalization process must go on for a prolonged period. Otherwise, the plants could be "tricked" into flowering by transient warm conditions before winter is really over. The low temperatures are perceived by the apical meristem, which later will develop into a flower or an inflorescence. Once it has been exposed, the meristem "remembers" that it is vernalized. This memory is stable throughout mitosis, and all daughter cells – also after grafting on a different plant – will retain the reproductive competence as a "memory of winter". In contrast, the effect is not transmissible to tissues that do not originate in the exposed meristem. Although vernalization may have evolved multiple times and thus not necessarily involve identical pathways, it generally appears to be a slow epigenetic process that triggers a series of chromatin modifications, ultimately resulting in silencing of genes that would repress flowering in an active state (King and Heide 2009; Kim et al. 2009).

3.2.3 Resource Accumulation

In plants, there is often a plastic response in flowering time to the availability of water, nutrients and carbon dioxide. Plants of some species are said to reach a level of reproductive maturity only after they have accumulated a threshold level of resources, often measured by plant size. In annuals, large individuals that are large since they have accumulated resources quickly, often flower earlier in the growing season to that of smaller individuals of the same species. The latter are assumed to be slower in resource acquisition (Rathcke and Lacey 1985). In perennial or woody plants, where time between bud formation and flowering is generally longer, one could predict that seasonal resource accumulation determines when floral buds are formed, but that flowering itself is triggered by a more predictable environmental cue. This situation has been documented in three shrubs belonging to the family Ericaceae (Reader in Rathcke and Lacey 1985).

The atmospheric carbon dioxide concentration stimulates leaf-level photosynthesis, which may or may not stimulate growth and, in this way, indirectly influence flowering time, which may be accelerated, delayed or unchanged with increasing CO_2 concentration. The response to elevated concentrations appears to be highly variable both among and within species, and even within populations (Nord and Lynch 2009).

3.2.4 Readiness to Flower in Grasses and Other Herbaceous Plants

In Europe, most indigenous grasses, and with a few exceptions (e.g., the cosmopolitan weed *Cynodon dactylon*), most of those that are important for allergies, belong to the sub-family Pooideae. They are herbaceous plants that may be annual, or more commonly, perennial. A grass plant generally consists of a parent shoot, which has developed from the epicotyl, and a number of lateral shoots or tillers that originate from axillary buds. Each leaf at the base of the parent shoot can subtend such a tiller. Each shoot or tiller is monocarpic, i.e. flowers only once and then dies away after fruiting.

The origin of the subfamily Pooideae is connected with a shift from tropic to temperate regions, and it is postulated that a vernalization requirement originated early within the group, after this shift (Colasanti and Coneva 2009). Many perennial grasses, however, need a dual induction of flowering (King and Heide 2009). In the late autumn, a tiller that is exposed to short day conditions and/or a vernalization period will be able to initiate inflorescence primordia, either directly or after transition to long-day conditions. But these primordia will not develop further, until a secondary induction caused by exposure to long day conditions takes place: hence, the latter is necessary for culm elongation, inflorescence development and anthesis. Each tiller of a perennial species then has a biennial life cycle.

Critical temperatures and day lengths, as well as the critical duration of exposure for primary induction vary greatly among the grass species and may also vary much among ecotypes of different geographic origin within the same species. For a number of temperate grasses of Scandinavian origin, 9°C appears to the most effective temperature to promote flowering during short-days conditions (King and Heide 2009). Generally, ecotypes from high latitudes and especially arctic and alpine ecotypes, have wider ranges of inducing temperatures and photoperiods, and they require fewer cycles of efficient day lengths than their low-latitude and maritime counterparts; for these ecotypes or species, 3–6 weeks can be sufficient, whereas for those at mid-latitudes 16–18 weeks of vernalization could be necessary. The latter is the case for species belonging to the genera *Deschampsia* and *Festuca*. In the Mediterranean grass *Phalaris tuberosa* L., plants from Morocco and Israel have little or no vernalization requirement, but flowering in these ecotypes is accelerated by low temperatures. Populations of the same species from more northern latitudes in Turkey and Greece require up to 8 weeks vernalization to induce complete flowering. Their respective requirements are related to the severity of winters at the collection sites (McWilliam 1968).

Significant variation in vernalization requirements in accordance with different selection regimes is also found in annual grasses like *Bromus tectorum* and *Aegilops tauschii,* which can be found in a variety of habitats that differ in altitude and/or aridity (Meyer et al. 2004; Colasanti and Coneva 2009).

After induction, the morphological changes of the apex by which the various parts of the inflorescence and flower primordia are laid down take place. This occurs during short-day conditions in the autumn in a number of arctic alpine ecotypes and in early-flowering low-latitude species like *Alopecurus pratensis* and *Anthoxanthum odoratum.* In many dual induction grasses, however, morphological changes at the apex take place only after a shift from short to long days. In *Phalaris arundinacea, Agrostis alba, Lolium perenne, Dactylis glomerata* and *Festuca* species, which all have large vernalization requirements, initiation takes place in the spring (King and Heide 2009).

Floral initiation in the non-grass weeds *Ambrosia artemisiifolia* and *Artemisia vulgaris,* belonging to the family Asteraceae, depends on different day-light regimes. In *Ambrosia,* initiation takes place in short-day conditions (Rogers et al. 2006). In *Artemisia vulgaris*, the required photoperiod for flower induction is 4–16 h for 4 weeks (Barney and DiTommaso 2002); hence, it is not a pronounced short-day plant.

After induction and the earliest morphological changes, further development of inflorescences and flowers is a growth process that is much influenced by temperature and day length. In grasses, this development is associated with rapid stem elongation (heading or bolting). Grasses begin growth in spring, when soil and daytime temperatures reach more or less species-specific thresholds that trigger initiation of growth, if soil moisture is available. At this stage, the time to heading decreases with increasing temperature, at least up to about +25°C (King and Heide 2009).

Many different species, with different responses to environmental conditions, contribute to the grass pollen content in the air. Since allergy sufferers react to most grass pollen, and the different taxa are not distinguished in the traditional pollen analysis, it is necessary to find generalized parameters for pollen, if prediction models are to be developed. In spite of the apparent importance of long-day conditions for inflorescence development even to begin, a number of authors have successfully used temperatures even from the pre-equinox part of late winter and early spring (Frenguelli et al. 1989; Spieksma and Nikkels 1998; Emberlin et al. 1993; Smith and Emberlin 2005, Table 3.1). In addition, a correlation between the amount of precipitation and onset of anthesis is found in a number of studies, alone or interacting with temperature; i.e., the more rain, the earlier is the start of flowering (Jones 1995; Ong et al. 1997; Stach et al. 2008).

The December–January–February averages of The North Atlantic Oscillation (NAO) index (i.e. the difference in pressure between Iceland and the Central North Atlantic, see Sect. 3.6.1) was found to be one of the most important predictors of the start of the grass pollen season in Poznan, Poland, as in several other locations in Europe; the strongest associations generally near the Atlantic Coast. A high value of the NAO index is associated with over-average temperatures and precipitation in North Europe. Also, the relationship between the NAO index and terrestrial ecosystems is well-documented (Stach et al. 2008; Smith et al. 2009).

Table 3.1 Environmental variables that have been successfully used to predict the onset of the grass pollen season

Study	Temperature	Precipitation	Other factor	Country
Frenguelli et al. (1989)	Mean and maximum; heat units minus base temperature (5°C) March–April			C. Italy
Spieksma and Nikkels (1998)	Mean temperature April–May			The Netherlands
Emberlin et al. (1993)	Cumulative temperature (>5.5°C) March–April + forecast May forecast			UK
Jones (1995)	10-day aggregated data End of winter, spring	10-day aggregated data Spring		UK
Stach et al. (2008)	Mean temperature May	Mean daily rainfall Days 61–70 from January 1st March	North Atlantic Oscillation Index December–February	Poland
Ong et al. (1997)		Total rainfall July (a winter month)		Australia
Smith and Emberlin (2005)	Minimum temp Days 81–90, from January 1st Maximum temperature Days 101–110, from January 1st			UK

Cleland et al. (2006) investigated responses of phenology to four experimentally simulated changes in warming, elevated CO_2, nitrogen deposition, and increased precipitation in grassland species. Elevated CO_2 delayed flowering in grasses and greening of the canopy at an ecosystem level, whereas N addition dampened the acceleration of greening caused by warming. In this study, increased precipitation had no consistent impacts on phenology.

In the weed *Ambrosia artemisiifolia*, temperatures during early spring seem to have an indirect effect on flowering date through their effect on germination. A difference in the date on which first pollen release was recorded was found among cohorts of transferred-from-cold conditions to climate-controlled green-houses, at 15-day intervals. There was a consistent trend towards a later first date of anthesis in later cohorts. Therefore, in contrast to floral initiation, the onset of anthesis is apparently not controlled by photoperiod (Rogers et al. 2006).

3.2.5 Readiness to Flower in Woody Plants

3.2.5.1 Chilling and Forcing

Many woody plants in temperate areas alternate between a period of active growth in spring and summer, with a period of dormancy in autumn and winter, during which growth and development is temporarily suspended. The ability to survive unfavourable periods and to evade risk for freezing damage of new shoots is essential not only to the plant's fitness, but also for the competence to maximize the duration of the growing season (Repo et al. 1990).

When the environment imposes dormancy, e.g., by low temperatures, the dormant bud is said to be quiescent or ecodormant. When, in contrast, innate agents maintain dormancy, the bud is said to be in rest, or endodormant. Arora et al. (2003) recommend the use of the terms eco- and endodormant (following the Lang et al. 1987 proposal), since they are more physiologically descriptive than the older terms quiescence and rest. During endodormancy, many metabolic and developmental processes go on in the buds and twigs. These processes include respiration, photosynthesis, cell division, enzyme synthesis, production of growth stimulators, and dissipation of growth inhibitors. But growth is arrested: an endodormant bud cannot elongate even under very favourable circumstances. In poplars, the cells in the growing points within buds are cytoplasmically isolated from one another within 5 days after endodormancy induction. In *Betula*, the cell-to-cell communication in the growing points diminishes in association with the start of endodormancy, since the plasmodesmata, the intercellular "communication channels", decrease in number and diameter. They appear to be blocked by 1-3-beta-D-glucan and thus the possibility for the apical meristem to act as an integrated whole is shut down. Endodormant buds have had less free water than ecodormant buds. Bound water status appears to be associated with low temperature stress tolerance rather than directly to dormancy itself. It has been suggested that dormancy induction may be, in part, regulated through dehydration stress via alterations in the quantity or activity of aquaporins, membrane-bound water-channel proteins that are able to increase water transport across membranes 20-fold over diffusion alone. Another class of ubiquitous proteins, the dehydrins, are suggested to be responsible for the binding of free water. Dehydrins appear to be activated by low temperatures and the plant hormone abscisic acid. They also add to cold acclimation (Arora et al. 2003).

The development of endodormancy and cold hardiness, though which a plant can sustain freezing temperatures without being damaged, are from a physiological point of view two separate processes, although they usually occur simultaneously. Both states are usually triggered by decreasing day length and low temperature (Arora et al. 2003). According to prevailing theories and experimental evidence (e.g., Myking 1999), growth-arresting conditions are eliminated when buds are exposed to chilling temperatures for a certain period. For many temperate plants, the optimal temperature for chilling appears to be within the interval 5–7°C (Richardson et al. 1974; Myking 1999).

Winter chilling is associated with a change in the ratio between endogenous gibberellins and some growth inhibitors, such as abscisic acid. When, how and to what degree hormones are involved in dormancy release is still uncertain and evidence both supporting and refuting various growth regulators can be found in the recent literature (Arora et al. 2003). It was suggested that chilling-requirement satisfaction is associated with the gradual release of water into the cytoplasm by the action of gibberellins, and the removal of a permeability barrier that prevents nutrients from entering the bud. Chilling may also lead to restoration of cell-to-cell signalling networks among individual cells of the apical meristem, allowing for symplastic movement of small signalling molecules, hormones, or proteins responsible for dormancy release (Rohde et al. 2000; Arora et al. 2003).

In the northern hemisphere, chilling requirements are generally met in December-February, depending on species and region (Myking 1999). The actual chilling requirement has been related to the risk of freezing damage in different environments (e.g. Chuine and Cour 1999). Plants growing in cold areas could be expected to need a longer chilling period (and a shorter forcing period) than those in a comparatively warmer region (e.g., Jato et al. 2000; Rodriguez-Rajo et al. 2003). However, the important factor for selection of a long period of dormancy may not be the low temperatures *per se*. During endodormancy, growth is prevented during transitory periods of warm temperatures (Arora et al. 2003). Thus, chilling requirements are expected to be inversely related to the length and stability of the winter, rather than to its severity (Table 3.2). Ecotypes of *Betula* from Northern Norway have earlier bud burst than ecotypes from Mid-Norway and Denmark, when they all were kept at the same equal temperature and light regime (Myking 1999). During winter in south Scandinavia, conditions often fluctuate between temperatures above and below the limit for frost damage (−5°C), whereas in the very north, warm temperatures are rare in wintertime. A similar variation in chilling requirements is found between *Betula* ecotypes from areas differing in their degree of continentality. In a transfer experiment, ecotypes native to mild oceanic winters were later released from dormancy than those of lowland continental origin. In contrast, a

Table 3.2 Chilling and forcing base temperatures, and chilling requirements for different tree species, as recorded in literature

Species and area	Chilling base temp	Chilling upper limit	Chill units	Forcing base temp	Source
Alnus glutinosa, Scandinavia	> −9.7°C			−3°C	Myking (1998)
Alnus glutinosa, Denmark, Copenhagen				+2°C	Andersen (1991)
Alnus glutinosa, Spain, Galicia			1,550	+5.5°C	Rodriguez-Rajo et al. (2004)
Alnus glutinosa, Spain, Léon			800	+5.5°C	Gonzalez-Parrado et al. (2006)
Alnus glutinosa, Italy, Perugia			848	+6.5°C	Jato et al. (2000)
Betula pendula, Italy, Perugia (Perugia, Italy)	+7°C		950		Rodriguez-Rajo et al. (2003)
Betula pendula, Norway	< −9.7°C			0 − +1°C	Myking (1998)
Betula pendula, Spain, Vigo, Spain	+5.75°C				Rodriguez-Rajo et al. (2003)
Betula spp., Scandinavia		+12°C			Myking & Heide (1995)
Betula, *Pinus*, Finland	−3.4	+10.4°C			Sarvas (1974)
Buxus sempervirens, *Platanus acerifolia*, Southern and Central Europe	(−5°C)−3.4°C	+10.4°C (+15°C)			Chuine et al. (1999)
Californian plants	0	+7.2°C			Aron (1983)
Cupressaceae, Mediterranean area	+7.1°C				Fuertes-Rodriguez et al. (2007)
Fruit trees		+14°C			Richardson and Anderson (1986)
Northern deciduous trees				0°C	Heide (1993), Myking (1997b)
Olea europaea, Córdoba				12.5°C	Galán et al. (2005)
Olea europaea, Priego				5°C	Galán et al. (2005)
Olea europaea, Granada				6°C	Galán et al. (2005)
Olea europaea, Jaen				7°C	Galán et al. (2005)
Olea europaea, Málaga				10°C	Galán et al. (2005)
Olea, Spain, Mesomediterranean area				5–5.5°C	Alcalá and Barranco (1992)
Olea, cultivar "Picudo", from Southern Spain			997		Orlandi et al. (2004)
Olea, cultivar "Ascolana", from Central Italy			1,848		Orlandi et al. (2002)
Olea, Portuguese cultivars				9–9.7°C	Ribeiro et al. (2006)
Olea, Southern Spain				10–13°C	García-Mozo et al. (2002)
Prunus padus	> −9.7°C				Myking (1998)
Quercus, Southern Spain				8–11°C	García-Mozo et al. (2002)
Tree species, spp., Southern of S. Spain				5–5.5°C	García-Mozo et al. (2002)

continental Mediterranean climate provides a larger freezing damage risk to new *Olea* tissues than a maritime one from an area where frost seldom occurs (Orlandi et al. 2004), thus creating a larger demand for chilling. Myking (1999) suggested a second explanation to the observed differences: that in a region where the growing season is short, early alleviation of dormancy (i.e. a low chilling requirement) secures rapid growth as soon as temperatures are favourable.

Since chilling requirements are usually low in Mediterranean species, thermal time models with temperature accumulation starting around 1 January are often considered to be sufficient for the prediction of flowering start. Any difference in chilling from 1 year to another is considered to be unimportant. However, the importance of chilling is not zero in Mediterranean species, as illustrated in *Olea europea*. Every olive cultivar appears to require its own amount of chilling in order to flower (Table 3.2).

When endodormancy is released, the buds are fully growth competent. But they still need stimulation from the environment before they can burst or flowering can begin. This stimulation has generally been presumed to be the occurrence of forcing temperatures i.e. temperatures above a certain base or threshold temperature. In many studies, 5°C has been considered as the standard threshold (basic) temperature for growth in boreal and temperate species (refs. in Myking 1999; Rodriguez-Rajo et al. 2004), irrespective of origin. However, in several studies, a clinal within-species variation, related to latitude, altitude, and different degrees of continentality, has been demonstrated (Table 3.2). The threshold may also vary with the age and developmental stage of the individual plant, as well as with environmental factors (Wielgolaski 1999).

Although it is possible to experimentally identify the date for the breaking of dormancy, and to approximate it in a model based on phenological and meteorological records, the temporal limitation between endo- and ecodormancy is not necessarily clear-cut. Forcing temperatures may be effective even if chilling requirements have not been fulfilled. In several studies, a dynamic relationship has been observed, i.e. the longer the chilling period, the fewer days of heat accumulation are needed to start the flowering (e.g. Rodriguez-Rajo et al. 2004; Emberlin et al. 2007). However, the range of chilling and forcing intervals is not satisfactorily known (Chuine and Cour 1999), nor is it clarified when and how chilling and forcing temperatures act on bud growth when they alternate. Warm temperatures have been suggested to be able to reverse the effects of previous chilling (e.g. Fuertes-Rodriguez et al. 2007), when they occur before endodormancy release; moreover, temperatures below a certain threshold during the forcing period have been suggested to nullify the effect of heating (Linkosalo et al. 2006). There appears to be a considerable overlap between the intervals when chilling and forcing temperatures are efficient, and their effects are easily confounded. It has been stated (references in Myking 1999) that the most efficient temperature for resumption of growth in dormant buds was the temperature that is both low enough to break endodormancy (through chilling), and high enough to stimulate growth when dormancy diminishes (through forcing). Outdoor fluctuating temperatures have been claimed to be more effective for budburst than artificial constant temperatures in many woody plants. This would imply that the thermal time requirement for budburst is less when the temperature fluctuates than when it is constant. However, in an experiment with *Betula pendula* and *B. pubescens*, no such difference could be detected (Myking 1999).

Results from experiments with tree seedlings of boreal and temperate species (Linkosalo et al. 2006) indicate that there is a gap between dormancy and bud growth. Several analyses of phenological tree series suggest that the starting date of ontogenetic development of buds, usually initiated in previous seasons, falls somewhere in late winter or early spring. The fulfilment of chilling requirements appears usually to be 3–4 months earlier, in late autumn or early winter. The delay is generally explained with prevailing low temperatures, which do not allow growth. However, a number of authors, e.g. Linkosalo et al. (2006), have claimed that something more may be missing from the traditional models. For the ontogenetic development to start, a cue from the diurnal light regime might be necessary. A light dependency would guarantee that development does not take place when temperatures giving frost injury still are likely to occur. While the climate is varying between years, the cyclic changes in light conditions remain stable at certain latitude. Photoperiod has been claimed to be the most important factor to promote flowering in late-successional species like *Fagus sylvatica* and *Quercus robur* (Körner and Basler 2010), but Chuine et al. (2010) argued that no study has shown that it was dominant over temperature even in these genera. Linkosalo et al. (2000) suggested that for *Betula pendula*, the cue could be a combination of day length or night length and twilight conditions. For that species, the initiation of bud development is suggested to occur around spring equinox. After this date, the temperature seldom falls below −5°C, which is the upper limit for freezing damage to occur in *Betula*. Chilling requirement is then a necessary, but not sufficient control mechanism for flowering to take place. In *B. pendula* and *B. pubescens*, short-day conditions stabilize dormancy, whereas long days promote dormancy release (Myking 1999).

3.2.5.2 Similarities and Differences Within and Between Species

The local and regional adaptation of the reproductive phenology of wind-pollinated trees may be surprising, since there generally is a large gene flow, as well as an overlap between generations, that would be expected to counteract selection. Moreover, in north Europe, trees arrived at their present locations fairly recently in the perspective of generation length. A number of quantitative trait loci, and in a few cases single loci that affect induction of endodormancy, chilling requirement and date of bud burst have been identified in *Malus*, *Populus* and *Vaccinium*. In these studies, some of the traits appear to have a large heritability, which would make natural selection possible. On the other hand, heritability of bud burst is very variable in *B. pendula* populations in north Scotland (Arora et al. 2003).

Whereas members of the same genus and species from different climatic regions often differ with regard to reproductive phenology, species of the same genus that grow in the same region often respond in the same way to temperature. There can also be differences within genera, e.g., between *Betula pendula* and

B. pubescens, where the former usually has a lower chilling demand than the latter (Myking 1999).

The similarity is also valid for species from different families, when selection pressures are strong and the successful responses limited. *Salix pulchra* and *Betula nana* from the Arctic had similar mean dates of bud break and could be predicted with the same model parameters (Pop et al. 2000). But there are several other examples where chilling requirements and threshold temperature vary with environmental conditions, ecology, and presumably, with phylogeny. The chilling requirements of *Betula pendula* and *B. pubescens* are in northern Europe often met with already in December, in *Populus tremula* in January, and in *Alnus glutinosa*, in February. *Alnus* has a larger chilling requirement than *Betula*, but flowers much earlier and quite soon after endodormancy alleviation. The low basic temperature may explain its quick growth resumption (Myking 1999).

Although they may have different demands, temperate tree species in the same region come into leafing and flowering in the same sequence from 1 year to another and their flowering dates are highly correlated. The heat accumulation of *Ulmus glabra*, counted from the beginning of flowering of *Alnus glutinosa*, varies little (Frenguelli and Bricchi 1998). The date of alder flowering could thus be used to predict elm flowering.

3.3 Pollen Release and Pollen Traits

3.3.1 *Anther Morphology and Pollen Maturation*

In terrestrial plants, the life cycle is an alternation between a diploid, spore-producing multicellular generation (the sporophyte) and a haploid, also multicellular generation (the gametophyte), producing sperm and/or egg cells. A pollen grain contains the male gametophyte of the seed plant, i.e., the sperm-producing generation. In seedless plants, gametophytes are generally independent from the sporophyte, and consist of hundreds of cells; evolution of the seed-bearing habit involved dependence of the gametophytes upon sporophytes and reduction of their size. Thus, in gymnosperms, the male gametophyte has up to ten cells, whereas in angiosperms, there is only one vegetative cell apart from the two sperm cells (Fig 3.2). The male gametophyte originates from a spore, generally called a microspore, and is still surrounded by this spore's sporopollenin outer wall, the exine (Fig. 3.3c), at maturity. The gametophyte's own wall is called the intine (Fig. 3.3c). Since a microspore is a single cell, it is not correct to use microspore and pollen as synonyms.

In angiosperms, the pollen grains develop within the anthers. An anther generally consists of two "halves" or theca, connected to a filament. Each theca contains two locules, i.e. microsporangia, separated by a septum. There are several deviations from this pattern, and there may be one or more locules in one theca (Endress 1996 in Pacini 2000). The inside of a locule is covered with the tapetum layer

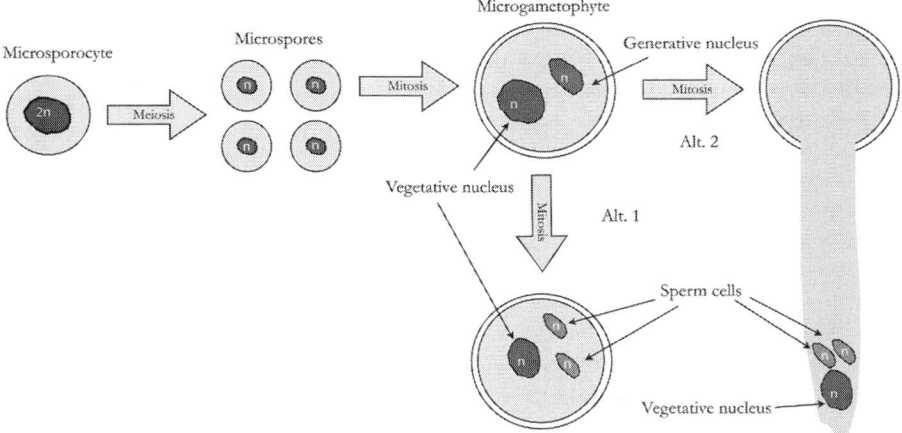

Fig. 3.2 The development of a pollen grain, i. e. a microgametophyte with a resistant outer wall. A gametophyte is the haploid, sexual generation of terrestial plants; a microgametophyte produces sperm cells. A diploid microsporocyte in one of the microsporangia of the anther (see Fig. 3.3a) divides by meiosis into four haploid microspores. Each microspore develops further by mitosis, into a binucleate microgametophyte with a vegetative and a generative cell. During this process, sporopollenin synthesised by the tapetum of the microsporangium is deposited on the microspore to form the exine, the outer pollen wall (see Fig. 3.3c). The generative cell divides to produce two sperm cells, either within the pollen grain before dispersal (Alt. 1) or after germination, within the pollen tube (Alt.2). The alternative locations differ between plant families. (2n = a diploid nucleus, n = a haploid nucleus)

(Fig. 3.3a), which provides the developing spores with nutrients, enzymes and sporopollenin, conveyed through a liquid, the locular fluid. The tapetum disintegrates shortly before anther opening (Fig. 3.3b).

There is also a mechanical layer, which may consist of one or more layers of cells situated in different parts of the anther. They are often dead at the time of pollen maturity, and may have lignified thickenings of different functional types within the same anther. As a result of the drying of all, or part of the anther, and in relation to the disposition of the thickened areas, the cells of the mechanical layer change their shape, causing stretching and folding. This action leads to opening of the anther and pollen presentation or expulsion (Manning 1996; Bianchini and Pacini 1996 in Pacini 2000). In the epidermis of the part of the anther wall that faces the septum, there are one or two thin-walled zones or apertures called stomia (sing, stomium).

At the time of maturation and degeneration, the tapetum also secretes viscous substances, e.g. tryphine (Brassicaceae) or pollenkitt. They cause the pollen to clump together mostly in entomophilous species but have also other functions, e.g. to keep the pollen to remain in the anther until dispersal. In anemophilous species, pollenkitt is usually lacking; this is not the case for *Parietaria judaica* (Fotiou et al. 2010) or *Olea europea* that has relatively "recent" entomophilous ancestors (Pacini 2000). The remaining structures of the disintegrating tapetum are called orbicules (or Ubisch bodies) and consist of sporopollenin, just as the

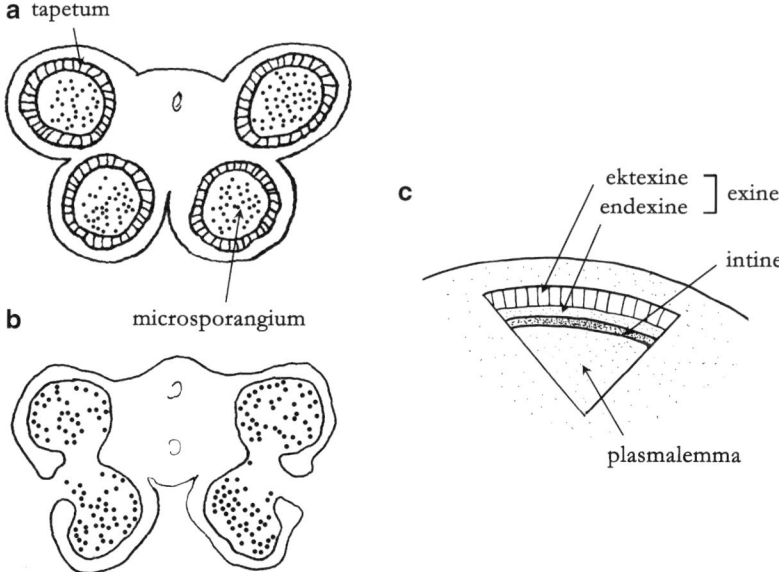

Fig. 3.3 (**a**) A cross section of an immature anther with two microsporangia, each containing numerous microspores and a tapetum layer that provides the microspores with various nutrients and sporopollenin. (**b**) A mature, dehisced anther with pollen grains. The tapetum is now consumed. (**c**) A part of a pollen grain with a cross section of the pollen wall, showing the two layers of the exine (endexine and ectexine) and the largely pectocellulosic intine, which in contrast to the exine is derived from the microgametophyte

pollen exine. Orbicules can be seen on the surface of pollen from some gymnosperms, e.g. *Taxus baccata* and *Juniperus communis*. They may or may not co-occur with pollenkitt, but appear to be absent from the anthers of many predominantly entomophilous angiosperm families, such as Asteraceae. In Fagales, and Poaceae, orbicules are present without pollenkitt. Since they have the same physical properties and electrostatic charge as the pollen grain surface, they have been suggested to facilitate pollen detachment from the anther, which is advantageous in a wind-pollinated species (Pacini and Hesse 2004). They have also been suggested to be allergen carriers (El-Ghazaly et al. 1995).

3.3.2 Anther Dehiscence

Pollen release could be considered to occur in two simple steps: first, the locular fluid disappears and second, the anther wall bursts (Pacini and Hesse 2004; Laursen et al. 2007). The first of these steps may be due to evaporation through the anther wall, or to reabsorbtion through the anther vascular bundles, or to both. The rate of evaporation depends on the ambient relative humidity, although it can be limited by

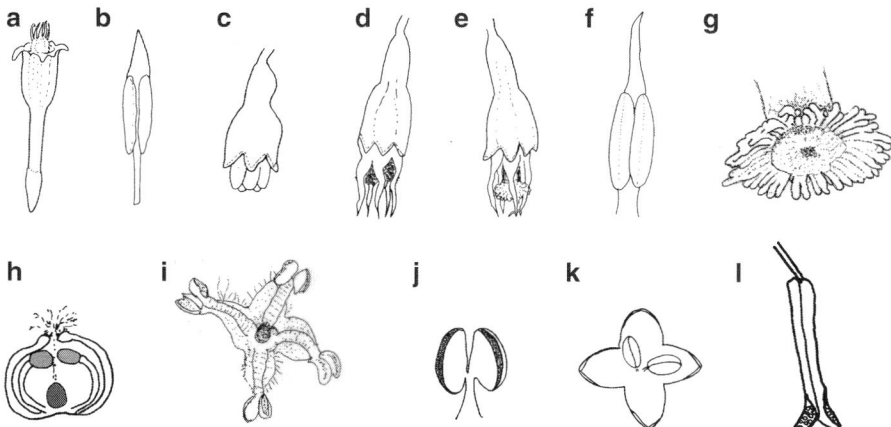

Fig. 3.4 (**a–b**) *Artemisia vulgaris*. (**a**) Male flower in the pollen-presenting stage. (**b**) Stamen. (**c–g**) *Ambrosia artemisiifolia* (**c–e** redrawn after Bianchi et al. (1959); **b** and **g** redrawn after Curtis and Lersten (1995); see also illustrations in Martin et al. (2009, 2010). (**c**) Male flower of with extending anthers. (**d**) Male flower in the pollen-presenting stage. (**e**) Male flower after pollen dispersal. The pistillodium is visible. (**f**) Stamen. (**g**) Pistillodium. (**h–i**) *Parietaria judaica*. (**h**) Bisexual flower in the female stage. The anthers are under tension and pressed towards each other. (**i**) Open flower with straight filaments and released pollen. (**j**) *Betula pendula*. Stamen with cleft anther and open thecas. (**k**) *Olea europaea*, male flower. (**l**) Grass stamen with two apical stomia

specific mechanisms. But reabsorbtion is a programmed process that is regulated by the plant. It enables anther opening at any time of day, whereas evaporation may occur only in the driest hours of the day. The relative importance of the two processes differs among species. The minimum temperature during the 24 h-cycle is also important as it determines whether anthers will dehisce or not. This is corroborated in studies from different taxa, such as Poaceae (Galán et al. 1995), Cupressaceae and *Olea europea* (Cariñanos et al. 2010), Urticaceae (Galán et al. 2000), and *Ambrosia artemisiifolia* (Bianchi et al. 1959).

In anemophilous trees, which flower in early spring, high temperature, low humidity, and moderate wind speed favour a passive dehydration, which then leads to a bursting of the anthers (Pacini and Hesse 2004). There are, however, considerable differences among these species as to whether all pollen will be released simultaneously or if the anther will open and close again according to variations in humidity, e.g. during a rainy day.[1] Apart from generalizing statements, very little is found in the literature on this matter.

In grasses, the anther opens either along a longitudinal slit, or the opening is limited to the apex or both to the apex and the base of this slit (Fig. 3.4l). In *Oryza*, *Hordeum* and probably also in other grass species, dehiscence occurs when the pollen grains swell due to a potassium movement from locular fluid to the grain (Matsui et al. 2000). This is a moisture-requiring process, whereas the widening of

---

[1] Dahl, personal observation.

the splits in both the upper and lower part of the anther wall is caused by desiccation (Matsui et al. 1999). This has been proposed to be the result of evaporation as well as of active retraction (van Hout et al. 2008).

In *Parietaria* and other Urticaceae, pollen is released explosively. All anthers in the flowers are juxtaposed and under tension. The filaments are incurved and form an arch inside the perianth, which compresses the anthers against one another (Fig. 3.4h). Their walls have transverse thickenings (ribs) in their upper two-thirds, whereas there is no mechanical layer in the anther (Franchi et al. 2007). When the anthers dehydrate, they shrink and lose contact with one another. This triggers all the stamen filaments to straighten violently (Fig 3.4i), and the pollen is released as a small cloud through the single stomium (Pacini and Hesse 2004).

3.3.3 Pollen Hydration and Vitality

The pollen wall is perforated by numerous micropores that allow for the transport of water, and it is to a certain degree elastic. At the time of exposure, pollen may contain from 1 to 70% water in its cytoplasm. As a rule, the higher the water content, the more active the metabolism of the pollen. In most species, the pollen grain is partially dehydrated just before or at the time of anther opening, a process that contributes to its longevity. These pollen grains often become ovoid at dispersal due to folding along their colpi (furrow-shaped apertures). Their carbohydrate content contributes to their water capacity and to their ability to stay vital and intact during humidity changes. In general, their longevity is increased at low relative humidities and low temperatures until germination, which generally requires high relative humidity (Gomez-Casero et al. 2004). In contrast, in a number of species that remain partly hydrated at dispersal, e.g. of Urticaceae and Poaceae, there are no adaptations to prevent water loss. Their pollen grains are usually spherical in shape and devoid of colpi, and they do not change in shape when water is lost; instead, they may eventually collapse due to a thin pollen wall, as is usually the case in grass pollen (Pacini 2000). Pollen that is partly hydrated at the time of dispersal is usually short-lived and cannot stand dehydration (Pacini and Hesse 2004). In the context of pollen allergies, it is important to distinguish pollen viability and allergenicity. The pollen allergens, which are glycoproteins produced by the pollen, may retain their impact on the human immune system for a long time, after the pollen grain is essentially dead (e.g. Yli-Panula and Rantio-Lehtimäki 1995).

3.3.4 Diurnal Patterns of Pollen Release

Pollen release may be synchronous or staggered within flowers, inflorescences and entire plants, depending on life form, adaptations or ambient meteorological factors. High temperatures usually speed up maturation and shorten the duration of anthesis.

In many anemophilous trees high temperature, low humidity and moderate wind speed favour a passive dehydration which then leads to bursting of the anthers (Helbig et al. 2004). In grasses, it is well known that each species sheds its pollen at its own time of day, and often with "clock-like regularity". Anthesis starts when filaments elongate and push the anthers outside the florets. In a study of several Indian grass species, Subba Reddi et al. (1988) concluded that each grass species follows the same course of pollen release on successive days with normal weather, but that variation occurs in the initiation, progress and termination of the daily course, when weather changes. They claimed that each course appears to be adapted to a certain "evaporating power", presumably degree of relative humidity and the related vapour pressure deficit. In a study of the circadian rhythm of *Zea mays*, the actual time of pollen release differed from day to day and depended on the time required for the anther tips to dry and open (van Hout et al. 2008), which in turn was affected by the presence or absence of morning dew. In this study, the first pollen concentration peak, detected in Rotorod samplers situated at canopy height, happened shortly before the direct irradiance peak on the anthers suggesting that direct solar irradiation might be important for the desiccation process. The temperature range that prevailed during the study period, 20–30°C, did not seem to have noticeable effects. Jarosz et al. (2003) related the initial morning emission of *Zea* to a decrease in relative humidity below 100 %, corresponding to vapour pressure deficit values of 0.2–0.5 kPa around the anthers.

In a study undertaken in India, 50 of the 52 investigated species shed pollen during only one diurnal period, which lasted for 2–16 h (Subba Reddi et al. 1988). The main period, when more than 50 % of the pollen was shed, could occur at any time, night or day, but always at the same time for the same species and always at the same time from one date to another. Some grass species are reported to exhibit a bimodal diurnal pollen release pattern (Subba Reddi et al. 1988), e.g., *Holcus lanatus*, *Festuca rubra*, and *Lolium perenne*. In all these species as well as in *Cynodon dactylon*, most pollen was released during the second period. In another investigation of *Festuca rubra* and *Lolium perenne* (Liem and Groot 1973 in Subba Reddi et al. 1988), a clear diurnal periodicity with regard to anther emergence was found: most of it took place from noon to midnight, and much less during the reverse period, from midnight to noon. Anthers must protrude from the flowers for pollen to be released, but anther emergence and pollen release are not synchronous. In this investigation, pollen release was not correlated to temperature or light either, but it was to low relative humidity. More pollen was released if the relative humidity fell below 50 %. There was no clear periodicity, and pollen could be released just at any time, depending on when the sufficient minimum humidity was reached. All emerged anthers in a plant were closed or opened simultaneously, which was inferred as a dependence on the same environmental factors. In *Zea mays* (van Hout et al. 2008) the observed bimodal pattern of pollen emission was compared to meteorological factors. The cause of the bimodality was reduced mean wind velocity during mid-day and could not be related to any other meteorological factor. *Imperata cylindrica* and *Pennisetum americanum* studied by Subba Reddi et al. (1988), released pollen during all 24 h, irrespective of weather.

Divergence in daily flowering among grass species has been suggested to act as a regulating mechanism to avoid waste of pollen, stigma sites, and ovules through unwanted "nonsense" pollination (cf. van der Pijl 1978), and also to promote reproductive isolation between closely related species, which otherwise might produce hybrids with low fitness (Grant 1983).

In *Artemisia* and *Ambrosia*, which both belong to Asteraceae, several flowers ("florets") are closely packed together in an inflorescence, a capitulum. The Asteraceae are characterized by secondary pollen presentation, i.e., after anther opening, the pollen is deposited and presented to dispersing agents in other parts of the flower. In *Ambrosia* and some species of *Artemisia* the flowers are unisexual and positioned in likewise unisexual capitulae, whereas in *Artemisia vulgaris*, pollen-producing flowers are hermaphrodite and situated in the centre of the capitulum, surrounded by flowers containing only pistils. In both species, the pollen-producing flower is tube-like and narrow, with five stamens arranged vertically on the inside. In the centre of the flower there is a pistil, which in *Ambrosia* is modified into a columnar structure with a cap of clavate hairs, a pistillodium (Fig. 3.4g). As described above, the anthers open inwardly (Fig. 3.4f). In *Artemisia*, the pollen adheres to the style by means of pollenkitt. When the style subsequently lengthens, it carries the pollen out of the flower (Fig. 3.4a, b). The tips of the stigma lobes are modified into brush-like structures that help in this process (Garnock-Jones 1986; Pacini 2000). In *Ambrosia*, the elongation of the filaments appears to be the main force to cause the complete emergence of the pollen from the mouth of the corolla (Fig. 3.4c, d). But the pistillodium plays a role in the "sweeping out" of the pollen mass also in this species (Fig. 3.4e), at least of the pollen grains that still remain in the anthers after dehiscence, or in the open space between them. The result is a bimodal pollen release (Bianchi et al. 1959; Martin et al. 2010).

In studies of *Artemisia vulgaris* pollen is released during 4.00–8.00 (Finland, Käpylä 1981) and during 6.00–11.00 o'clock (Germany and Poland, von Wahl and Puls 1989; Kazlauskas et al. 2006). However, it is retained by the apical appendages of the anthers. Pollen detachment from the capitulum is therefore likely to be temporarily separate from the actual anther opening within the flower, and thus could take place any time during the day when there is enough wind. Kazlauskas et al. (2006) found a negative correlation between pollen concentration in the air and rain during the morning and forenoon hours, but point out that there is no study on the possibility that anther dehiscence is postponed to the afternoon, if the weather dries up. There is also no study as to the relationship between the role of active reabsorbment and of evaporation for anther opening. In an investigation of *A. tridentata* from sagebrush steppe in Wyoming, USA (Laursen et al. 2007), a baseline diurnal pattern of increasing presentation, release, and transport from late morning to mid- to late afternoon was found. The data suggested that the actual degree and timing of anther development and dehiscence are modified by diurnal variations in temperature and humidity. Evening hours of high humidity were suggested to promote anther development, while early morning hours of decreasing humidity most likely drive anther dehiscence, a process also driven by temperature which in this arid environment generally is inverse to humidity. Wind was responsible for actual pollen detachment.

There was not a linear relationship between wind speed and detachment, but there may be a lower critical limit, which in the Wyoming study was estimated to 2.67 m/s. This fairly mild breeze apparently was strong enough to remove pollen from the capitulae and keep them suspended for subsequent transport as winds continued to increase into the morning. Also Kazlauskas et al. (2006) found a relationship with wind speed and pollen concentration in the air. Finally, Munuera Giner et al. (1999) did not discern any diurnal pattern that could be related to meteorology in the results from their *Artemisia* pollen concentration measurements in Murcia, Spain. Intradiurnal patterns of pollen concentrations were similar for late summer and winter species (*A. campestris* and *A. barrelieri*). During autumn-blooming, the intradiurnal pattern of *A. herba-alba* was particularly erratic. As pointed out by Laursen et al. (2007), will temporal patterns of release from individual plants be obscured when their pollen merge into the well-mixed air streams of the general atmosphere? Moreover, when data are interpreted from pollen traps situated at different elevations, diurnal patterns will appear to differ according to when vertically transported aerosols emitted at ground level reach the trap, as shown by von Wahl and Puls (1989) and others.

In *Ambrosia*, the pollen grains are released from the anthers about or shortly after sunrise, and the pistillodium has swept out the remaining ones about 5 h later (Martin et al. 2010). Pollen grains tend to stick together in clumps, which stay on the capitulae or fall on adjacent vegetation prior to reflotation by wind (Bianchi et al. 1959; Ogden et al. 1969; Martin et al. 2009). Their detachment from the deposition surface is likely to be impeded by the occurrence of morning dew, and the highest concentrations of airborne *Ambrosia* pollen grains are then generally found during midday (Bianchi et al. 1959; Ogden et al. 1969; Solomon and Mathews 1990). Like in *Artemisia*, actual pollen dispersal is also related to wind speed and relative turbulence. Peak releases have been reported to occur in connection with unstable weather with increasing wind forces and shifting wind directions (Barnes et al. 2001).

If populations are gregarious, turbulence will facilitate effective dispersal. But when conspecific plants are separated, turbulence may cause much pollen go to waste (Subba Reddi et al. 1988). Species that tend to appear in low density would be selected to shed pollen when conditions are stable, and pollen concentration at the source can be maintained at high level. Subba Reddi and Reddi (1986) identified 30 such grass species.

3.4 Duration of the Pollen Season and the Shape of the Pollen Curve

The duration of the period, when pollen from a certain taxon is found in the air, and the shape of the curve that describes the change in pollen concentration over time, depend on a number of factors. First, a pollen taxon may comprise one species or an entire family, depending on the taxonomic level where morphological identification is possible. The taxon Poaceae is an amalgam of different species with varying ecology, mating systems and heating requirements, and thus, the curve can have several peaks (e.g. Gonzalez Minero et al. 1998).

The second important factor is the specific schedule for pollen emission, as a result of the proportion of flowers open per day (flowering rate) and of floral or inflorescence longevity (Kudo and Hirao 2006). Flowering may be synchronous or staggered, at the population and/or at individual levels (Rathcke and Lacey 1985; Primack 1985). In wind-pollinated plants, which are the main contributors to the atmospheric pollen load, synchrony among conspecific plants promotes outcrossing and pollen tube competition among a maximum number of potential mates. A high concentration of pollen in the air surrounding a maternal plant is beneficial to the fitness of its progeny. Furthermore, the period for efficient pollination can be limited by predictable environmental changes, selecting for a synchronous behaviour. In temperate regions, many deciduous trees flower before leafing, in the springtime. In the Mediterranean, summer drought is ubiquitous, and at a certain point in time during spring, many annual species allocate their resources away from vegetative growth, into an intense and final period of reproduction. Plants that flower in late summer and autumn, such as *Ambrosia artemisiifolia*, can be limited by the risk for early frost. When flowering is synchronous in a population or even in a region, the pollen curve is often positively skewed (Primack 1985) and the peak of pollen concentration will follow in a fairly short time after flowering onset.

Staggered or asynchronous flowering, on the other hand, can be the case when the outcome of flowering, fruit set or seed germination is uncertain due to the risk of adverse conditions (Rathcke and Lacey 1985; Kudo and Hirao 2006). The result is a platykurtic pollen curve. In contrast to the annuals mentioned above, other annual herbs and grasses, and a number of perennials such as *Parietaria judaica*, are opportunists with indeterminate growth. Their flowers are produced from lateral meristems, i.e., on side shoots, whereas the top of the plant continues to grow vegetatively. This growth pattern permits simultaneous vegetative and reproductive growth and allows for risk-spreading, as well as for the possibility to exploit favourable conditions. This growth pattern allows for pollen dispersal all the year around if there is no frost, although there may be a maximum, e.g., in late spring (Guardia and Belmonte 2004). In winter-flowering trees, such as *Corylus* and *Alnus* (Rodriguez-Rajo et al. 2004), pollen-producing catkins are not injured by frost. In Northwest-Europe, winter temperatures often fluctuate between mild and sub-zero centigrades. In low but frost-free temperatures, only a few anthers open at one time. Moreover, a platykurtic pollen curve can be due to the existence of environmental and genetic variation for flowering time within populations and species, as has been documented for numerous species (Primack 1985; Wielgolaski 1999; Chuine and Belmonte 2004).

Thirdly, apart from the inherent flowering behaviour, flowering rate and therefore also the duration and the timing for the peak of the pollen season, are affected by weather conditions. Temperature affects the rate of growth and development, depending on the species. The main pollination period of *Betula* (mainly *B. pendula*, and in the north, also *B. pubescens*) varies considerably with temperature, as recorded in spring pollen seasons are probably due to the higher dependency on temperature of early spring tree species, such as hazel, alder and birch. This fact could also be explained by the influence of the NAO on pollen season start (D'Odorico et al. 2002; Smith and Emberlin 2006; Smith et al. 2009). The seasonality of the NAO suggests that the phenomenon is more likely to affect ecological mechanisms in winter (Hurrell 1995; Ottersen et al. 2001). Indeed, some relationship was found between grasses and the NAO (Smith et al. 2009), but it should be remembered that several factors other than temperature affect the flowering date, like photoperiod and water availability.

However, it is important to take into consideration that the pollen season advance or delay depends on the period of time selected for the study and the temperature trends during these years, i.e. a moderate increase of temperature resulted in a delay in winter/early spring tree pollen season start (Emberlin et al. 2007; Tedeschini et al. 2006). It also depends on regional characteristics (Orlandi et al. 2009; Recio et al. 2009), what may confound patterns. This is possibly the reason why there are no clear systematic patterns on pollen season dates in South-Eastern Europe, i.e. Greece (Damialis et al. 2007).

An advance in reproductive phenology could give rise to a greater risk of exposure to late frosts, with plants being more exposed to cold stress episodes that can provoke damage on the reproductive structures (Linkosalo et al. 2000; Scheifinger et al. 2002; Garcia-Mozo et al. 2001). In addition, increases on climatic variability and incidences of extreme events are aspects of climate change that can affect less drought-tolerant species (Huynen et al. 2003).

There are also differences in response to climate change depending on the species. Although earlier and longer pollen seasons also occur in some spring flowering herbaceous plants, such as *Artemisia* in Eastern Europe (Stach et al. 2007b), the long-term trends seem to be less remarkable for grass and weeds than for trees (D'Amato and Cecchi 2008). Stable trends, with a slight tendency to delayed weed pollen season start, have been observed in southern Europe (Alcázar et al. 2009; Recio et al. 2009).

3.6.2 *Climate Impacts to Pollen Season Severity and Projections*

As previously mentioned (Sect. 3.5.2), different researchers have observed an increase in annual pollen index in some winter-spring flowering trees (Garcia-Mozo et al. 2006; Damialis et al. 2007; Frei and Gassner 2008; Bonofiglio et al. 2009), being regulated by weather factors together with the system of resource allocation among years (Dahl and Strandhede 1996; Ranta et al. 2005). It should also be noted that there are possible interactions with other components of global change. For instance, it is important to consider that atmospheric increases of CO_2 and the deposition of nitrogenous compounds favour plant activity and photosynthesis, which favours an increase in floral intensity (Garcia-Mozo et al. 2006; Rogers et al. 2006; Ziska et al. 2001, 2008). In general, long-term trends are more evident in the amount et al. 2002). The influence of the NAO is also reduced with increasing distance from the Atlantic coast (Scheifinger et al. 2002; Ahas et al. 2002; Menzel et al. 2006).

Climate change may also influence the behaviour of the NAO. This may enhance or mask climate change on a regional level, which may lead to regionally very different implications of climate change on plant phenology. In this context the influence of the NAO on phenological trends should be mentioned. A recent example is the change of the NAO from predominantly negative to positive states around 1990. This change advanced many phenological dates leading to pronounced trends. Hence recent trends in phenological time series should be interpreted along with changes in the NAO.

3.6.1.1 Impact of Climate on Pollen Content in the Air (Floral Phenology)

Many studies point out the role of phenology as one of the most important bio-indicators to study the direct impact of global warming on different species, both at temporal and spatial level. Climate-related morphological changes in many plants are also good bio-indicators of different conditions that characterize the climes. The pollen content in the air offers a quantitative variable to describe the floral phenology of anemophilous plants. For this reason, pollen season and pollen intensity historical databases have a relevant importance on global warming adaptation studies. For example, a report published by the Minister of Environment in Spain recommends the support of phenological networks, including aerobiology among others, to monitor impacts on plant biodiversity (Fernández-González et al. 2005). In a report produced on an assignment from the Swedish Environmental Agency, it is stated that monitoring of airborne pollen and allergens could be necessary to follow the possible health effects of global warming, and the same agency supports the development of a national phenological network. It should be noted, however, that certain characteristics of the pollen season (such as start or intensity) can be affected by long-distance transport, Therefore, it is necessary to make allowances for this when interpreting the data.

3.6.1.2 Climate Impacts on the Timing of Pollen Seasons

As the World Health Organization (WHO) has concluded in the document on Phenology and human health, allergic disorders, an earlier start and peak of the pollen season is currently more pronounced in species that start flowering earlier in the year, whereas the duration of the season becomes extended in summer flowering species (Huynen et al. 2003).

Different aerobiological studies, based on long historical databases, have shown earlier pollen seasons during recent years as a result of temperature increase, most of them related to spring flowering trees (Garcia-Mozo et al. 2006; Galán et al. 2005; Emberlin et al. 2007; Damialis et al. 2007; Frei and Gassner 2008; Bonofiglio et al. 2009; Orlandi et al. 2009). The greater advances observed in earlier rather than later community has pointed out the value of historical phenological databases for the climate change research. A meta-analysis, comprising different species in European countries, has shown that the response of spring phenology to temperature is unquestionable (Menzel et al. 2006). Changes in phenology depend on the season when they take place, and also on climatic trends (Gordo and Sanz 2005). In general it has been demonstrated that greater phenological changes are recorded in those events that occur earlier in the year, from early spring to summer. This emphasizes the fact that changes in temperature are more pronounced in winter and early spring (Scheifinger et al. 2002; Peñuelas et al. 2002; Menzel et al. 2006). However, for phenological events that occur in autumn, it is more difficult to define a pattern since greater interannual oscillations occur (Menzel et al. 2006; Gordo and Sanz 2005).

In Europe, an increase of annual mean temperature in the last 30 years has been associated with an advance of the growing season start and increase of its duration; the end of the growing season showed a lower variability in all regions of Europe (Chmielewski and Rötzer 2001). Studies at regional scale have shown that the rate of change is higher in Western Europe and Scandinavia, and that different phenological rhythm and trends occur in the Eastern part of Europe (Ahas et al. 2002; Chmielewski and Rötzer 2001). Most long-term phenological studies to date have been focused on Northern Europe, while comparatively few have addressed the Mediterranean region (Peñuelas et al. 2002; Gordo and Sanz 2005). For instance, it is important to consider that in the Mediterranean region other variables related with changes in rainfall and water availability are also important (Peñuelas et al. 2004). However, Gordo and Sanz (2005) suggested that the relationship between water availability and different phenological phases is difficult to quantify.

Furthermore, at regional level in Europe, it is important to consider the role of North Atlantic Oscillation (NAO) in governing the temporal variability of the lower atmosphere, and thus phenological dates in Europe (Scheifinger et al. 2002; Chmielewski and Rötzer 2001). The NAO is the dominant mode controlling (above all winter and spring) climate over Europe. It is characterised by the longitudinal oscillation of atmospheric mass between the two dominant pressure systems over the North Atlantic sector, namely the Azores high and the Icelandic low. Its positive state (high pressure difference between the Azores high and the Icelandic low) is connected with increased zonal flow and causes northern Europe to experience wet and mild winters while over southwestern Europe, winters are anomalously dry (Xoplaki et al. 2004). During its negative state, northern Europe experiences dry and cold winters while the winters in the western Mediterranean are wetter than normal (e.g. Wanner et al. 2001). For central Europe, however, the NAO is of minor importance as this region is located at a transition zone of the main centers of action (Azores high, Icelandic low and Russian high). For a detailed discussion of this phenomenon see Wanner et al. (2001). The NAO accounts for a significant amount of the interannual variation in temperatures in the North Atlantic (Menzel et al. 2006; Post and Stenseth 1999). Plant phenology in this region is therefore influenced by year-to-year variations in the NAO (D'Odorico et al. 2002; Stenseth et al. 2002). The maximum influence of the NAO on phenological dates is found on early phases and decreases with an increasing number of days after January 1st (Scheifinger

Due to projected rates of future atmospheric CO_2 increase given by current climate change scenarios, scientists became interested in the impacts of CO_2 on growth and pollen production, especially in plants with C3-type photosynthesis that are currently carbon-limited. The increase of CO_2 is suspected to stimulate photosynthesis, resulting in increased vegetative growth and pollen production (Skjoth et al. 2007). For ragweed, it was shown that large plants produce more pollen grains and seeds than small ones (Fumanal et al. 2007). As is experimentally confirmed any ecological factor resulting into biomass increase inevitably leads to increased pollen production per plant (references in Rogers et al. 2006).

Ziska and Caulfield (2000, in Rogers et al. 2006) performed a highly controlled experiment involving ragweed reared in growth chambers. Pollen production per plant increased significantly with increasing CO_2 concentration: 132 % more pollen was produced by plants living in the current CO_2-concentration as compared to plants that were exposed to pre-industrial CO_2-levels. Another study (Rogers et al. 2006) confirmed these results. It also showed that earlier germination induces higher pollen production. Similar responses of ragweed plants to increased CO_2 concentrations were also found in field studies. This has implications (increasing biomass and pollen production) for plants that live in areas enriched with greenhouse gases. In places like roadsides, it was experimentally proved that an enriched CO_2 environment induces increased levels of *Amb a* 1 allergen in the pollen grains, although their total protein content remained unchanged (references in Rogers et al. 2006)

Elevated CO_2 concentrations resulted in greater biomass and/or seasonal pollen production in loblolly pine (*Pinus taeda*) at the Duke University forest Free-Air CO_2 Enrichment (FACE) site (reference in Rogers et al. 2006) and in field studies of soybean (*Glycine max*), velvetleaf (*Abutilon theophrasti*) and tomato (*Lycopersicon esculentum*) (Ziska et al. 2001).

3.6 Impacts of Climate Change on Plants

During the twentieth century, the average annual surface temperature has increased by 0.8°C in most of Europe. Results of the Intergovernmental Panel on Climate Change (IPCC), Working Group II on Impact, Adaptation and Vulnerability (Alcamo et al. 2007) showed that from 1901 to 2005, average temperature in Europe increased by 0.9°C, with a larger increase in winter than in summer. During the most recent period (1979–2005), the trend is considerably higher (+0.4°C/decade). Precipitation trends in the twentieth century showed an increase (10–40 %) in northern Europe and a decrease in southern Europe (up to 20 % in some parts).

3.6.1 Impact of Climate on Phenological Events

Climate change affects the physiology, phenology and distribution of European plant and animal species. Phenological data are simple to record, and the scientific

Such a pattern may be more evident in areas where environmental conditions are limiting factors to reproduction. In *Olea europaea*, winter chilling (below 9°C) was suggested to cause such synchronization in for flowering bud differentiation in areas where relatively warm winters are common (Hartmann and Prolingis 1957 in Lavee 2007) A correlation between the amount of chilling and the pollen index has been reported; rainfall during the vegetative period exerted the greatest influence on olive flowering intensity in Spain (Galán et al. 2001). In oak, rainfall and temperature prior to the pollen season are the most important independent variables for predicting the pollen season intensity for *Quercus* in the Iberian peninsula (Garcia-Mozo et al. 2006). Similarly, the amount of rainfall during the rainy season of the year previous to flowering is positively correlated to the annual pollen sum for oak in northern California (reference in Antepara et al. 1995). In North Europe, a number of studies show the importance of the weather conditions during the previous growing season for the intensity of birch flowering (Emberlin et al. 1993; Dahl and Strandhede 1996; Rasmussen 2002; Ranta et al. 2005; Ranta and Satri 2007). Emberlin et al. (1993) used pre-seasonal meteorological parameters as independent variables in a multiple regression model. Among the tested parameters, precipitation and temperature values until the end of April, performed best.

A number of authors emphasize the importance of combining biological and meteorological factors when predicting pollen intensity in anemophilous trees. Lavee (2007) described the intricate interaction between endogenous processes and the environment in *Olea*. Dahl and Strandhede (1996) found that favourable conditions could compensate for the innate competition between leaf expansion and flowering in *Betula*. If late spring and early summer are warm enough, assimilation will be promoted and hence the trade-off effects caused by competition between the catkins from the previous year and those from the current year decreased. If a number of favourable years follow one another, the endogenous flowering rhythm, in the case of *Betula* a biannual cycle, will not be apparent. Ranta et al. (2008) tested a combined resource-budget and environmental effect model developed in Japan by Masaka and Maguchi (2001) for birch. This model is based on the assumption that abundant flowering and seeding in a mast year depend on resource accumulation over several years, affecting the amount of catkin initiation in springtime, as well as weather factors stimulating catkin growth and development during late spring and early summer during the year prior to flowering. They suggested that the model is not suitable for data sets longer than 10 years, if there is some long-term trend in environmental conditions affecting the amount of flowering such as increase in mean annual temperatures induced by climate change. However, for shorter time-spans, the annual fluctuations are well explained.

3.5.4 *The Effects of Atmospheric Carbon Dioxide*

Besides temperature and water, atmospheric CO_2 could be considered as an important environmental factor influencing plant physiology, principally through its direct effects on photosynthesis, photorespiration, dark respiration and stomatal physiology.

especially in wind-pollinated trees, abundant fruit set, satiating seed predators (Janzen 1971), and vigorous progeny (Ranta et al. 2008). In contrast, the resource matching hypothesis states that reproductive output will vary in response to environmental variation in the absence of selection for (or against) masting (Kelly 1994).

Many species are claimed to have an innate flowering rhythm, where mast years recur every second or third year, depending on specific characteristics and processes within the tree. A biannual cycle has been described for *Betula* (Jäger et al. 1991; Jato et al. 2002; Latałowa et al. 2002), *Olea europaea* (Lavee 2007), *Alnus glutinosa* (Rodriguez-Rajo et al. 2006) and *Quercus ilex* (Latorre 1999 in Weryszko-Chmielewska et al. 2006). In *Fraxinus*, there is a tendency to a 3-year cycle, corresponding to the number of seasons between floral initiation and pollen emission (Dahl and Wallander, personal data).

The specific cycle length appears to be affected by the shoot architecture, the schedule of development of different organs and their relative demands, and possible competition for energy (Dahl and Strandhede 1996). These properties could actually be interpreted as evolutionary constraints for a constant reproductive output. For various fruit species, the developing fruit has been suggested to act as a sink competing for metabolites (Monselise and Goldschmidt 1982), but the negative effect upon future reproduction is not universal (Dahl and Strandhede 1996; Lavee 2007). In contrast, Dahl and Strandhede (1996) found evidence for competition for resources between elongating inflorescences and expanding leaves in the same shoot in *Betula pendula*, resulting in a diminished assimilation capacity and a negative effect on the development of new inflorescence initials and the following pollen-production. In a study of *Styrax obassia*, Miyazaki et al. (2002) found a lower amount of starch in reproductive than in non-reproductive shoots during the growing season of a mast year. Leaves of reproductive shoots had a significantly smaller area, a lower mass per area and lower concentrations of nitrogen than leaves of non-reproductive shoots, indicating a similar competition between reproductive and assimilating organs.

In *Olea*, differences in growth regulators in leaves and buds between the so-called "on" (heavy flowering) and "off" (lean flowering) years have been reported (Lavee 2007). They indicate that masting indeed appears to be the result of selection in this species, and not just the result of morphological and physiological constraints. The growth regulators are supposed to balance the development of vegetative and reproductive shoots and, at the same time, act as vectors to initiate the specific metabolic activity controlling the flowering and fruit potential for the next year. According to resource-budget hypotheses (Isagi et al. 1997), a tree will reproduce when accumulated resources exceed a certain threshold, which can be determined by the cost of reproduction in the species in question, e.g. according to fruit size, but not necessarily by environmental factors.

As a support for the resource-matching hypothesis, several studies show the effects of varying environmental conditions, causing a high degree of synchronization between conspecific trees in the same region, and even, as found by Ranta et al. (2008) for *Betula pendula* and *B. pubescens*, over wide geographical areas.

Agrostis castellana (Chastain and Young 1998). Nitrogenous fertilizers applied before the onset of primary induction increase the number of inflorescences in several dual induction grasses (Heide 1994); again, the effect is mainly indirect through an increase in the number and vigour of inducible tillers in autumn.

Grass pollen counts derive from many different species, which may result in multiple peaks. The long-term (26 year) dataset of daily airborne grass pollen concentrations in Leiden was analysed (Spieksma and Nikkels 1998) in order to identify trends and variations. Analysis showed that individual years had their own characteristic periods of high and low grass pollen concentrations, very often with multiple peaks. Average daily grass pollen concentrations over the 26-year period showed that the data had a tendency towards a bimodal distribution, as observed by Norris-Hill (1995), with a main peak in mid-June followed by a secondary peak in the first week of July.

In the Mediterranean, Poaceae anthesis may continue throughout the year. It is generally most intense in late spring and early summer, but peaks may appear also in early spring and later, during summer, the latter depending on rainfall during May (Gonzalez Minero et al. 1998). In North Europe, the grass pollen period continues from late spring until early autumn. Models of the course of grass pollen anthesis can be strengthened when the grass pollen season is divided into three periods (Smith and Emberlin 2006). Pollen counts during the earliest period (end of May) can be used to predict the counts in July and August.

The intensity of the grass pollen season has a positive correlation with duration, because both rely on common variables. Davies and Smith (1973) used cumulated temperature divided with the number of days during two months prior to anthesis to predict the peak date. In a study from Bilbao, both season length, as well as the time to the peak date, was related to cumulated temperatures (800–900°C). The results of Emberlin and her colleagues (Emberlin et al. 1993, 1999) show that an early start date for the pollen season will lead to a higher total pollen catch. A similar relationship to phenology was confirmed by a model based on the dates when the cumulated pollen counts reach a particular value (Davies and Smith 1973). The earlier these cumulated counts are obtained, the greater is the likelihood of a high pollen index.

3.5.3 Flowering Intensity in Trees

In trees and other perennial plants, it is common that reproductive effort varies among years, e.g. in *Betula* (Emberlin et al. 1993; Dahl and Strandhede 1996; Mahura et al. 2007; Ranta et al. 2008), *Olea* (Galán et al. 2001), *Alnus* (Rodriguez-Rajo et al. 2006) and *Quercus* (Jato et al. 2007). During one year, many flowers and fruits are produced in a population and in an entire region, whereas in other years, reproduction is moderate or sparse. The phenomenon is called masting, and a year of high abundance is called a mast year.

A number of non-mutually exclusive hypotheses as to the evolutionary advantages of masting have been proposed. The result would be efficient pollination,

In the Mediterranean parts of South Europe, precipitation is an especially strong limiting factor to determine flowering intensity, not least because of the fact that it is irregular. In Sevilla, Southern Spain, the annual pollen index of grass pollen appeared to decrease over a period with recurring drought (Gonzalez Minero et al. 1998). A lack of water will have a general negative effect on vigour and growth potential in all grasses since they have shallow roots, and especially in annual species, which constitute a relatively large proportion of the Mediterranean flora. Galán et al. (1995) observed that the number of rainy days was even more important than the amount of precipitation collected in these areas, since rain tends to be torrential and almost all is lost as run-off. In the Mediterranean area, the grass pollen season may continue throughout the year, but is most intense in late spring and early summer. In contrast to the results from Sevilla, Recio et al. (2010) found increasing grass pollen counts and more days with numbers over 100 grains/m^3 in Malaga during the springs 1992–2007, which they related to increased temperature and minimum temperatures during early spring, i.e., at seed germination and vegetative growth. Also the study from Sevilla, Spain (Gonzalez Minero et al. 1998), the total pollen index increased with increasing rainfall between 1 January and 2 months onward. However, neither temperature prior to anthesis nor autumn precipitation was found to be important. In Melbourne, Australia, with a semiarid climate, the seasonal total of grass pollen was significantly correlated to the rainfall sum of the preceding year (Schappi et al. 1998). Craine et al. (2010) found an interspecific difference in the response to precipitation during the growing season at different times in three American tallgrass prairie grass species, which they related to differences in their phenology of resource uptake.

In a study of *Artemisia vulgaris*, the highest annual pollen indexes were recorded during years when the maximum temperature was moderately high (between 22 and 25°C) and there also was a high amount of precipitation during the 4 months preceding flowering. Higher maximum temperatures combined with lower precipitation, i.e., a tendency to drought, were correlated to a lower annual total. Especially, rainfall recorded in the first fortnight of July had a positive effect on pollen season intensity (Stach et al. 2007b).

Studies on yield in grass species used for forage indicate that in northern temperate areas, growth conditions in the year previous to flowering could affect the annual pollen index. Thus, the carbohydrate status at the time of floral induction can influence the magnitude of reproductive output during the next summer. In several cool-season perennial grasses, as much as 92 % of the seed yield is said to be set before the onset of conditions favouring vernalization, i.e. during the growth season previous to flowering. But there is a variation among species. In *Poa pratensis* and *Festuca rubra*, the tiller height at the end of post-harvest regrowth period was consistently related to flowering and seed yield. A connection between the basal diameter of vegetative tillers in autumn and flowering and seed yield was found in young stands of the same species, but not in older ones. Also, in *Dactylis glomerata* and *Festuca arundinacea*, there was an age-related correlation between the number and size of vegetative tillers before induction with flowering and seed yield in young stands. Vegetative characteristics were not related to flowering and seed production in *Lolium perenne* and crops, such as *Olea* (Galán et al. 2001) and *Vitis* (Cristofolini and Gottardini 2000). The annual variations in the amount of registered pollen indicate variations in abundance of the source. Thus, changes in the amount of registered pollen have been used in studying changes in vegetation (Pidek 2007; Kobzar 1999) and in environmental conditions (e.g. Damialis et al. 2007), which could be of particular interest for forecasts. However, the pollen index can be a biased measure of the reproductive output of the plants, as it is influenced by factors such as the atmospheric transport and deposition of pollen, and weather factors affecting the emission process (Ranta et al. 2008). Episodes of long distance transport from remote sources occur at a regular basis, and can contribute significantly to the counts registered at a certain site (Mahura et al. 2007). Although the reproductive output of tree species could be synchronized at a large geographical scale (Ranta et al. 2008), the local pollen index may be influenced by factors acting far away from the immediate environment.

3.5.2 *Flowering Intensity in Grasses and Other Herbaceous Species*

Pollen concentrations experienced by allergy sufferers depends on the weather during anthesis. Nevertheless, in temperate areas, the potential severity of the pollen season of grasses and other herbaceous species is mainly determined earlier, by the accumulated temperature and precipitation from the beginning of the growth period and onwards during the spring, when the culms elongate and pollen is formed in the anthers. Thus, the important factors are about the same that influence the start date. They are difficult to foresee far in advance, but forecast is possible when some time into the development period has passed. As emphasized in Emberlin et al. (1999), specific forecast models must be developed for different sites, since the absolute time periods that are important differ. Emberlin et al. (1993) found that for London, they could predict the likely severity of the season at the end of May or alternatively when 75 pollen grains were recorded, by using cumulative temperatures for March and April, combined with forecasts of temperature and precipitation in June in multiple regression analyses based on pollen data over a 20-year period. Predictions could be made even earlier, at the end of April, replacing the latter variables with long-term weather forecasts for May. The severity of the Poaceae pollen seasons was also modelled at a network of sites in the UK (Jones 1995). Total monthly rainfall, daily average temperatures compiled to monthly mean and cumulated temperatures as day degrees above 5.5°C were used as variables in a multiple regression analysis. It was, however, found that 10-day aggregated variables often gave the best results. Three models were developed, enabling the prediction of the grass pollen season within 4 days of the actual start dates, but with varying success at London, Cardiff and Derby, respectively. Emberlin et al. (1999), employed 10-day aggregates of cumulative daily temperatures above 5.5°C and precipitation at the same locations, and achieved a high level of explanation for them all.

several studies from different parts of Europe (Mendez et al. 2005). The length and timing of the peak of the grass pollen season in Bilbao (Spain) differ between years, according to temperature (Antepara et al. 1995). When nights are not hot enough as to favour anther dehiscence (i.e. >10–15°C) in mid-autumn, the pollen curve tails off in *Ambrosia artemisiifolia* (Bianchi 1959; Stark et al. 1997), although there may not have been any night frost. The effect of heating and of related meteorological variables can alter as the season progresses, especially as a result of changes in the amounts of remaining pollen to be dispersed or of pollen production (Smith and Emberlin 2005; Vázquez et al. 2003). Heat favours anther dehiscence in *Olea europaea* during the pre-peak period but makes flowering decrease during the post-peak period (reference in Galán et al. 2001).

Rainfall before anthesis can have an effect on the grass pollen season: in southern Spain, an abundant rainfall during development leads to a short and intense peak, whereas drier conditions will cause the season to be long and without a pronounced maximum (Gonzalez Minero et al. 1998). In contrast, rainfall and a high relative humidity during anthesis will prolong the season in Poaceae (Antepara et al. 1995), in *Olea* (Gonzalez Minero and Fernandez-Mensaque 1996) and probably also in other species since they hamper anther dehiscence.

García-Mozo et al. (1999) investigated variations in the *Quercus* sp. pollen season at selected sites in Spain and detected that no relationship was observed between the amount of pollen produced (the pollen index, see subchapter 3.5.1) and season duration: years with higher pollen counts did not have a longer pollen season and vice-versa. However, there was a connection between the number of days on which more than 50 grains/m^3 were recorded and the pollen index, in years with lower pollen production, fewer days were registered despite the season maintaining its normal duration.

3.5 Factors That Influence the Magnitude of the Pollen Index/Pollen Sum

3.5.1 The Pollen Index Parameter

Ever since Blackley constructed the first pollen trap in the late 1860s, the amount of registered airborne pollen during defined period of time has become a basic aerobiological parameter. The amount of pollen registered during a year is described as the "annual pollen sum" or the "pollen index". This has become the standard parameter both for quantitative descriptions of the registered airborne pollen within the same monitoring site.

The magnitude of the pollen sum/index reflects the flowering intensity of the pollen-producing plants. The availability of pollen can be important for successful pollination and fertilization, and is of interest in studies of plant reproduction behaviour (McClanahan 1986) or of the production of agricultural anemophilous of pollen produced by trees and shrubs than by herbs (Damialis et al. 2007). Precipitation trends in Europe may also affect floral intensity. This is because of the generally low water availability in the Mediterranean region; especially for herbs and grasses, declining precipitation can provoke declining annual pollen counts (Recio et al. 2010). However, relating changes of pollen season intensity to climate change cannot be straightforward. Relationships between climate change and pollen production increase are not always evident and depend on the site and the period of time studied (Clot 2003). Among the factors that may influence pollen counts are changes to land use and urban green space designs (like the planting of ornamental plants in urban areas).

The NAO may have an influence on the intensity of the pollen season. There is a significant positive relationship between total pollen count and averages of the NAO the previous year in Poznan (Poland). The city has a continental climate, but it is located in an area where winds arrive predominantly from the west and southwest (Woś 1994; Stach et al. 2008; Corden et al. 2000). In the likewise Polish city Krakow, the intensity of *Betula* pollen seasons do not appear to be influenced by the NAO, which is probably the result of a more continental climate. Also, in contrast to the situation in Poznan, in Worcester and London, the temperatures and averages of the NAO during the year previous to pollen dispersal were negatively related to the pollen counts. Instead, the intensity of the birch pollen season at these locations was positively correlated to temperatures and averages of the NAO during winter and spring in the current year (Stach et al. 2008).

Predictive modelling based on climate projections is an important tool to allow the adoption of mitigation measures for the anticipated impacts. However, only few studies in aerobiology have focused on this topic. Future meteorological data modelled by the Hadley Meteorological Centre, included in the Intergovernmental Panel on Climate Change (IPCC), have been used for estimating future projections on pollen season start dates and intensity for different trees in the Iberian Peninsula (Garcia-Mozo et al. 2006; Galán et al. 2005) and the UK (Emberlin et al. 2007). Studies on some spring flowering trees in southern Europe indicated that, under a double CO_2 scenario, at the end of the twenty-first century pollen seasons for these species will start from 1 week to 1 month earlier, depending on the plant and the study area, with stronger effects in inland areas.

3.6.3 *Climate Impacts on Plant Migration*

With climate change, non-native species from adjacent areas may cross over frontiers and become new elements of the biota. When such movements cover long distances, they can usually be attributed to human activity (Walther et al. 2002). Either plant invasion or long-range transport of foreign allergenic pollen exposes citizens to new allergenic pollen grains.

Ragweed (*Ambrosia*) is an important cause of pollen allergy in Eastern and Central Europe and can be presented as a recent example of biological invasion.

It has been introduced in Europe with cereal grain from North America (Smith et al. 2008; Stach et al. 2007a). Ragweed populations do not tend to thrive under a maritime climate and in northern Europe the growing season is often too short for seed maturation (Comtois 1998; Dahl et al. 1999; Saar et al. 2000). *Ambrosia* has been recorded as far north as Poland, the Baltic States and even Sweden, but populations are often ephemeral and scattered and reliant on the regular introduction of seeds from outside sources (Dahl et al. 1999; Saar et al. 2000; Stepalska et al. 2002). Often, ragweed does not reach maturity in northern latitudes and so the occurrence of *Ambrosia* pollen may be reliant on conditions being appropriate for both germination and flowering or long-range transport (Stach et al. 2007a). However, when seed set does occur, it may contribute to a persistent seed bank. Climate change may therefore allow ragweed to spread to areas where populations have not previously been able to establish.

3.6.4 Climate Impacts on Public Health

Climate change can prolong or shorten pollen seasons, increase or decrease pollen season intensity and aid the spread of species with allergenic interest. In addition, the higher occurrence of heavy precipitation events and an increase of episodes of urban pollution can provoke or aggravate health problems (Ziska et al. 2008; D'Amato and Cecchi 2008) including negative effects on respiratory allergic diseases. However, the data and tools currently available make predictions quite uncertain. To face this limitation, a number of activities should be undertaken. As also recommended by Cecchi et al. (2010), (a) aerobiological data should be collected in a structured way at the European level, (b) multidisciplinary research teams should be created, promoted and supported, (c) there should be more attention to and emphasis on the pollen/spore exposure in the guidelines for the diagnosis and treatment of respiratory and allergic diseases, and (d) communicating the importance of aerobiological research and lobbying the European Union and other funders to finance this research should become priority activities.

References

Ahas, R. A., Aasa, A., Menzel, A., Fedotova, V. G., & Scheifinger, H. (2002). Changes in European spring phenology. *International Journal of Climatology, 22*, 1727–1738.

Alcalá, A. R., & Barranco, D. (1992). Prediction of flowering time in olive for the Cordoba olive collection. *Hortscience, 27*, 1205–1207.

Alcamo, J., Moreno, J. M., Nováky, B., Bindi, M., Corobov, R., Devoy, R. J. N., Giannakopoulos, C., Martin, E., Olesen, J. E., & Shvidenko, A. (2007). Europe climate change 2007: Impacts, adaptation and vulnerability. In M. L. Parry, O. Canziani, F. J. P. Palutikof, P. J. van der Linden, & C. E. Hanson (Eds.), *Contribution of working group II to the fourth assessment report of the intergovernmental panel on climate change* (pp. 541–580). Cambridge: Cambridge University Press.

Alcázar, P., Stach, A., Nowak, M., & Galán, C. (2009). Comparison of airborne herb pollen types in Córdoba (Southwestern Spain) and Poznan (Western Poland). *Aerobiologia, 25*, 55–63.

Andersen, T. B. (1991). A model to predict the beginning of the pollen season. *Grana, 30*, 269–275.
Antepara, I., Fernandez, J. C., Gamboa, P., Jauregui, I., & Miguel, F. (1995). Pollen allergy in the Bilbao area (European Atlantic seaboard climate): Pollination forecasting methods. *Clinical and Experimental Allergy, 25*, 133–140.
Aron, R. (1983). Availability of chilling temperatures in California. *Agricultural Meteorology, 28*, 351–363.
Arora, R., Rowland, L. J., & Tanino, K. (2003). Induction and release of bud dormancy in woody perennials: A science comes of age. *HortScience, 38*, 911–921.
Barnes, C., Pacheco, F., Landuyt, J., Hu, F., & Portnoy, J. (2001). The effect of temperature, relative humidity and rainfall on airborne ragweed pollen concentrations. *Aerobiologia, 17*, 61–68.
Barney, J. N., & DiTommaso, A. (2003). The biology of Canadian weeds. 118. *Artemisia vulgaris* L. *Canadian Journal of Plant Science, 83*, 205–215.
Bianchi, D. E., Schwemmin, D. J., & Wagner, W. H., Jr. (1959). Pollen release in the common ragweed (*Ambrosia artemisiifolia*). *Botanical Gazette, 120*, 235–243.
Bonofiglio, T., Orlandi, F., Sgromo, C., Romano, B., & Fornaciari, M. (2009). Evidences of olive pollination date variations in relation to spring temperature trends. *Aerobiologia, 25*, 227–237.
Cariñanos, P., Galán, C., Alcazar, P., & Domìnguez, E. (2010). Airborne pollen records and status of the anemophilous flora in arid areas of the Iberian Peninsula. *Journal of Arid Environments, 74*, 1102–1105.
Cecchi, L., D'Amato, G., Ayres, J. G., Galán, C., Forastiere, F., Forsberg, B., Gerritsen, J., Nunes, C., Behrendt, H., Akdis, C., Dahl, R., & Annesi-Maesano, I. (2010). Projections of the effects of climate change on allergic asthma: The contribution of aerobiology. *Allergy, 65*, 1073–1081.
Chastain, T. G., & Young, W. C. (1998). Vegetative plant development and seed production in cool-season perennial grasses. *Seed Science Research, 8*, 295–301.
Chmielewski, F. M., & Rötzer, T. (2001). Response of tree phenology to climate change across Europe. *Agricultural and Forestry Meteorology, 108*, 101–112.
Chuine, I., & Belmonte, J. (2004). Improving prophylaxis for pollen allergies: Predicting the time of course of the pollen load of the atmosphere of major allergenic plants in France and Spain. *Grana, 43*, 65–80.
Chuine, I., & Cour, P. (1999). Climatic determinants of budburst seasonality of temperate-zone trees. *The New Phytologist, 143*, 339–349.
Chuine, I., Cour, P., & Rousseau, D. D. (1999). Selecting models to predict the timing of flowering of temperate trees: Implications for tree phenology modelling. *Plant, Cell & Environment, 22*, 1–13.
Chuine, I., Morin, X., & Bugmann, H. (2010). Warming, photoperiods, and tree phenology. *Science, 329*(5989), 277–278.
Cleland, E. E., Chiariello, N. R., Loarie, S. R., Mooney, H. A., & Field, C. B. (2006). Diverse responses of phenology to global changes in a grassland ecosystem. *Proceedings of the National Academy of Sciences, USA, 103*, 13740–13744.
Clot, B. (2003). Trends in airborne pollen: An overview of 21 years of data in Neuchatel (Switzerland). *Aerobiologia, 19*, 227–234.
Colasanti, J., & Coneva, V. (2009). Mechanisms of floral induction in grasses: something borrowed, something new. *Plant Physiology, 149*, 56–62.
Comtois, P. (1998). Ragweed (*Ambrosia* sp.): The Phoenix of allergophytes. In: 6th international congress on aerobiology. Satellite symposium proceedings: *Ragweed in Europe*. Perugia, Italy: ALK Abelló.
Corden, J., Millington, W., Bailey, J., Brookes, M., Caulton, E., Emberlin, J., Mullins, J., Simpson, C., & Wood, A. (2000). UK regional variations in *Betula* pollen (1993–1997). *Aerobiologia, 16*, 227–32.
Craine, J. M., Towne, E. G., & Nippert, J. B. (2010). Climate controls on grass culm production over a quarter century in a tallgrass prairie. *Ecology, 91*, 2132–2140.
Cristofolini, F., & Gottardini, E. (2000). Concentration of airborne pollen of *Vitis vinifera* L. and yield forecast: A case study at S. Michele all'Adige, Trento, Italy. *Aerobiologia, 16*, 125–129.
Curtis, J. D., & Lersten, N. R. (1995). Anatomical aspects of pollen release from staminate flowers of *Ambrosia trifida* (Asteraceae). *International Journal of Plant Sciences, 165*, 29–36.

D'Amato, G., & Cecchi, L. (2008). Effects of climate change on environmental factors in respiratory allergic diseases. *Clinical and Experimental Allergy, 38*, 1264–1274.

Dahl, Å., & Strandhede, S.-O. (1996). Predicting the intensity of the birch pollen season. *Aerobiologia, 12*, 97–106.

Dahl, Å., Strandhede, S.-O., & Wihl, J.-A. (1999). Ragweed, an allergy risk in Sweden? *Aerobiologia, 15*, 293–297.

Damialis, A., Halley, J. M., Gioulekas, D., & Vokou, D. (2007). Long-term trends in atmospheric pollen levels in the city of Thessaloniki, Greece. *Atmospheric Environment, 41*, 7011–7021.

Davies, R. R., & Smith, L. P. (1973). Forecasting the start and severity of the hay fever season. *Clinical Allergy, 32*, 263–267.

D'Odorico, P., Yoo, J. C., & Jager, S. (2002). Changing seasons: An effect of the North Atlantic Oscillation? *Journal of Climate, 15*, 435–445.

El-Ghazaly, G., Takahashi, Y., Nilsson, S., Grafström, E., & Berggren, B. (1995). Orbicules in *Betula pendula* and their possible role in allergy. *Grana, 34*, 300–304.

Emberlin, J., Savage, M., & Jones, S. (1993). Annual variations in grass pollen seasons in London 1961–1990: Trends and forecast models. *Clinical and Experimental Allergy, 23*, 911–918.

Emberlin, J., Mullins, J., Cordon, J., Jones, S., Millington, W., Brooke, M., & Savage, M. (1999). Regional variations in grass pollen seasons in the UK, long term trends and forecast models. *Clinical and Experimental Allergy, 29*, 347–356.

Emberlin, J., Smith, M., Close, R., & Adams-Groom, B. (2007). Changes in the pollen seasons of the early flowering trees *Alnus* spp. and *Corylus* spp. in Worcester, United Kingdom, 1996–2005. *International Journal of Biometeorology, 51*, 181–191.

Fairley, D., & Batchelder, G. L. (1986). A study of oak-pollen production and phenology in northern California: Prediction of annual variation in pollen counts based on geographic and meteorologic factors. *The Journal of Allergy and Clinical Immunology, 78*, 300–307.

Fernández-González, F., Loidi, J., Moreno, J. C., del Arco, M., Férnández-Cancio, A., Galán, C., García-Mozo, H., Muñoz, J., Pérez-Badia, R., Sardinero, S., & Tellería, M. (2005). Impact on plant biodiversity. In J. M. Moreno (Ed.), *Impacts on Climatic Change in Spain* (pp. 183–248). Madrid: OCCE, Ministerio de Medio Ambiente.

Fotiou, C., Damialis, A., Krigas, N., Halley, J. M., & Vokou, D. (2010). *Parietaria judaica* flowering phenology, pollen production, viability and atmospheric circulation, and expansive ability in the urban environment: Impacts of environmental factors. *International Journal of Biometeorology, 55*, 35–50.

Franchi, G. G., Nepi, M., Matthews, M. L., & Pacini, E. (2007). Anther opening, pollen biology and stigma receptivity in the long blooming species, *Parietaria judaica* L. (Urticaceae). *Flora, 202*, 118–127.

Frei, T., & Gassner, E. (2008). Climate change and its impact on birch pollen quantities and the start of the pollen season, an example from Switzerland for the period 1969–2006. *International Journal of Biometeorology, 52*, 667–674.

Frenguelli, G., & Bricchi, E. (1998). The use of the pheno-climatic model for forecasting the pollination of some arboreal taxa. *Aerobiologia, 14*, 39–44.

Frenguelli, G., Bricchi, E., Romano, B., Mincigrucci, G., & Spieksma, F.Th.M. (1989). A predictive study on the beginning of the pollen season for Gramineae and Olea europaea L. *Aerobiologia, 5*, 64–70.

Fuertes-Rodriguez, C. R., Gonzalez-Parrado, Z., Vega-Maray, A. M., Valencia-Barrera, R. M., & Fernandez-Gonzalez, D. (2007). Effect of air temperature on forecasting the start of Cupressaceae pollen type in Ponferrada (León, Spain). *Annals of Agricultural and Environmental Medicine, 14*, 237–242.

Fumanal, B., Chauvel, B., & Bretagnolle, F. (2007). Estimation of pollen and seed production of common ragweed in France. *Annals of Agricultural and Environmental Medicine, 14*, 233–236.

Galán, C., Emberlin, J., Dominguez, E., Bryant, R. H., & Villamandos, F. (1995). A comparative analysis of daily variations in the Gramineae pollen counts at Cordoba, Spain and London, UK. *Grana, 34*, 189–198.

Galán, C., Alcázar, P., Cariñanos, P., Garcia, H., & Dominguez-Vilches, E. (2000). Meteorological factors affecting daily Urticaceae pollen counts in southwest Spain. *International Journal of Biometeorology, 43*, 191–195.

Galan, C., Garcia-Mozo, H., Carinanos, P., Alcazar, P., Dominguez-Vilches, E. (2001). The role of temperature in the onset of the *Olea europaea* L. pollen season in southwestern Spain. *International Journal of Plant Biometeorology, 45*, 8–12.

Galán, C., Garcia-Mozo, H., Vazquez, L., Ruiz, L., Diaz de la Guardia, C., & Trigo, M. M. (2005). Heat requirement for the onset of the *Olea europaea* L. pollen season in several sites in Andalusia and the effect of the expected future climate change. *International Journal of Biometeorology, 49*, 184–188.

Garcia-Mozo, H., Galán, C., Cariñanos, P., Alcazár, P., Mendez, J., Vendrell, M., Alba, F., Saenz, C., Fernandez, D., Cabezudo, B., & Dominguez, E. (1999). Variations in the *Quercus* sp. pollen season at selected sites in Spain. *Polen, 10*, 59–69.

Garcia-Mozo, H., Galán, C., Aira, M. J., Belmonte, J., Diaz de la Guardia, C., Fernandez, D., Gutierrez, M., Gutierrez, M., Rodriguez, F. J., Trigo, M. M., & Dominguez-Vilches, E. (2001). Model for forecasting *Olea europaea* L. airborne pollen in South-West Andalusia, Spain. *Agricultural and Forest Meteorology, 110*, 247–257.

García-Mozo, H., Galán, C., Aira, M. J., Belmonte, J., Díaz de la Guardia, C., Fernández, D., Gutierrez, F. J., Trigo, M. M., & Domínguez, E. (2002). Modelling start of oak pollen season in different climatic zones in Spain. *Agricultural and Forest Meteorology, 110*, 247–257.

Garcia-Mozo, H., Galán, C., Jato, V., Belmonte, J., Diaz de la Guardia, C., Fernandez, D., Gutierrez, M., Aira, M. J., Roure, J. M., Ruiz, L., Trigo, M. M., & Dominguez-Vilches, E. (2006). *Quercus* pollen season dynamics in the Iberian Peninsula: Response to meteorological parameters and possible consequences of climate change. *Annals of Agricultural and Environmental Medicine, 13*, 209–224.

Garnock-Jones, P. J. (1986). Floret specialization, seed production and gender in *Artemisia vulgaris* L. (Asteraceae, Anthemideae). *Botanical Journal of the Linnean Society, 92*, 286–302.

Gomez-Casero, M. T., Hidalgo, P. J., Garcia-Mozo, H., Dominguez, E., & Galán, C. (2004). Pollen biology in four Mediterranean *Quercus* species. *Grana, 43*, 22–30.

Gonzalez Minero, F. J., & Fernandez-Mensaque, P. C. (1996). Prediction of the beginning of the olive full pollen season in south-west Spain. *Aerobiologia, 12*, 91–96.

Gonzalez Minero, F. J., Candau, P., Tomas, C., & Morales, J. (1998). Airborne grass (Poaceae) pollen in southern Spain. Results of a 10-year study (1987-96). *Allergy, 53*, 266–274.

Gordo, O., & Sanz, J. J. (2005). Phenology and climate change: A long-term study in a Mediterranean locality. *Oecologia, 146*, 484–495.

Grant, V. (1983). *Plant speciation*. New York: Columbia University Press.

Guardia, R., & Belmonte, J. (2004). Phenology and pollen production of *Parietaria judaica* L in Catalonia (NE Spain). *Grana, 43*, 57–64.

Heide, O. M. (1993). Daylength and thermal time responses of budburst during dormancy release in some northern deciduous trees. *Physiologia Plantarum, 88*, 531–540.

Heide, O. M. (1994). Control of flowering and reproduction in temperate grasses. *New Phytologist, 128*, 431–462.

Helbig, N., Vogel, B., Vogel, H., & Fiedler, F. (2004). Numerical modelling of pollen dispersion on the regional scale. *Aerobiologia, 20*, 3–19.

Hurrell, J. W. (1995). Decadal trends in the North Atlantic Oscillation: regional temperatures and precipitation. *Science, 269*, 676–679.

Huynen, M., Menne, B., Behrendt, H., Bertollini, R., Bonini, S., Brandao, R., Brown-Fahrländer, C., Clot, B., D'Ambrosio, C., De Nuntiis, P., Ebi, K. L., Emberlin, J., Erdei Orbanne, E., Galán, C., Jäger, S., Kovats, S., Mandrioli, P., Martens, P., Menzel, A., Nyenzi, B., Rantio-Lehtimäki, A., Ring, J., Rybnicek, O., Traidl-Hoffmann, C., van Vliet, A. J. H., Voigt, T., Weiland, S., & Wickman, M. (2003). *Phenology and human health: Allergic disorders*. Rome: Health and global environmental change.

Isagi, Y., Sugimura, K., Sumida, A., & Ito, H. (1997). How does masting happen and synchronize? *Journal of Theoretical Biology, 187*, 231–239.

Jäger, S., Spieksma, F. Th. M., & Nolard, N. (1991). Fluctuation and trends in airborne concentrations of some abundant pollen types, monitored at Vienna, Leiden and Brussels. *Grana, 30*, 309–312.

Janzen, D. H. (1971). Seed predation by animals. *Annual Review of Ecology and Systematics, 2*, 465–492.

Jato, V., Rodriguez Rajo, F. J., & Aira, M. J. (2007). Use of *Quercus ilex* subsp. ballota phenological and pollen-production data for interpreting *Quercus* pollen curves. *Aerobiologia, 23*, 91–105.

Jarosz, N., Loubet, B., Durand, B., McCartney, A., Fouellassar, X., & Huber, L. (2003). Field measurements of airborne concentration and deposition rate of maize pollen. *Agricultural and Forest Meteorology, 119*, 37–51.

Jato, V., Frenguelli, G., Rodriguez, F. J., & Aira, M. J. (2000). Temperature requirements of *Alnus* pollen in Spain and Italy (1994–1998). *Grana, 39*, 240–245.

Jato, V., Dopazo, A., & Aira, M. J. (2002). Influence of precipitation and temperature on atmospheric pollen concentration in Santiago de Compostela (Spain). *Grana, 41*, 232–241.

Jones, S. (1995). Allergenic pollen concentrations in the United Kingdom. PhD Thesis. London: University of North London.

Käpylä, M. (1981). Diurnal variation of non-arboreal pollen in the air in Finland. *Grana, 20*, 55–59.

Kazlauskas, M., Sauliene, I., & Lankauskas, A. (2006). Airborne *Artemisia* pollen in Siauliai (Lithuania) atmosphere with reference to meteorological factors during 2003–2005. *Acta Biologica Universitatis, 6*, 1–2.

Kelly, D. (1994). The evolutionary ecology of mast seeding. *Trends in Ecology & Evolution, 9*, 465–470.

Kim, D. H., Doyle, M. R., Sung, S., & Amasino, R. M. (2009). Vernalization: Winter and the timing of flowering in plants. *Annual Review of Cell and Developmental Biology, 25*, 277–299.

King, R. W., & Heide, O. M. (2009). Seasonal flowering and evolution: The heritage from Charles Darwin. *Functional Plant Biology, 36*, 1027–1036.

Kobzar, V. N. (1999). Aeropalynological monitoring in Bishkek, Kyrgyzstan. *Aerobiologia, 15*, 149–153.

Körner, C., & Basler, D. (2010). Phenology under global warming. *Science, 327*(5972), 1461–1462.

Kudo, G., & Hirao, A. S. (2006). Habitat-specific responses in the flowering phenology and seed set of alpine plants to climate variation: Implications for global-change impacts. *Population Ecology, 48*, 49–58.

Lang, G. A., Earl, J. D., Martin, G. C., & Darnell, R. L. (1987). Endo-, para- and ecodormancy: Physiological terminology and classification for dormancy research. *HortScience, 22*, 371–377.

Latałowa, M., Miętus, M., & Uruska, A. (2002). Seasonal variations in the atmospheric *Betula* pollen count in Gdańsk (southern Baltic coast) in relation to meteorological parameters. *Aerobiologia, 18*, 33–43.

Latorre, F. (1999). Differences between airborne pollen and flowering phenology of urban trees with reference to production, dispersal and interannual climate variability. *Aerobiologia, 15*, 131–141.

Laursen, S. C., Reiners, W. A., Kelly, R. D., & Gerow, K. G. (2007). Pollen dispersal by *Artemisia* tridentata (Asteraceae). *International Journal of Biometeorology, 51*, 465–481.

Lavee, S. (2007). Biennal bearing in olive (*Olea europaea*). *Annales Series Historia Naturalis, 17*, 101–112.

Linkosalo, T., Carter, T. R., Hakkinen, R., & Hari, P. (2000). Predicting spring phenology and frost damage risk of *Betula* spp. under climatic warming: A comparison of two models. *Tree Physiology, 20*, 1176–1182.

Linkosalo, T., Hakkinen, R., & Hanninen, H. (2006). Models of the spring phenology of boreal and temperate trees: Is there something missing? *Tree Physiology, 26*, 1165–1172.

Linskens, H. F., & Cresti, M. (2000). Pollen allergy as an ecological phenomenon; A review. *Plant Biosystems, 134*, 341–352.

Mahura, A. G., Korsholm, U. S., Baklanov, A. A., & Rasmussen, A. (2007). Elevated birch pollen episodes in Denmark: Contributions from remote sources. *Aerobiologia, 23*, 171–179.

Martin, M. D., Chamecki, M., Brush, G. S., Meneveau, C., & Parlange, M. B. (2009). Pollen clumping and wind dispersal in an invasive angiosperm. *American Journal of Botany, 96*, 1703–1711.

Martin, M. D., Chamecki, M., & Brush, G. S. (2010). Anthesis synchronization and floral morphology determine diurnal patterns of ragweed pollen dispersal. *Agricultural and Forest Meteorology, 150*, 1307–1317.

Masaka, K., & Maguchi, S. (2001). Modelling the masting behaviour of *Betula platyphylla* var *japonica* using the resource budget model. *Annals of Botany, 88*, 1049–1055.

Matsui, T., Omasa, K., & Horie, T. (1999). Rapid swelling of pollen grains in response to floret opening unfolds anther locules in rice (*Oryza sativa* L.). *Plant Production Science, 2*, 196–199.

Matsui, T., Omasa, K., & Horie, T. (2000). Anther dehiscence in two-rowed barley (*Hordeum distichum*) triggered by mechanical stimulation. *Journal of Experimental Botany, 51*, 1319–1321.

McClanahan, T. R. (1986). Pollen dispersal and intensity as criteria for the minimum viable population and species reserves. *Environmental Management, 10*, 381–383.

McWilliam, J. R. (1968). Nature and genetic control of the vernalization response in *Phalaris tuberosa* L. *Australian Journal of Biological Sciences, 21*, 359–408.

Mendez, J., Comptois, P., & Iglesias, I. (2005). *Betula* pollen: One of the most important aeroallergens in Ourense, Spain. Aerobiological studies from 1993 to 2000. *Aerobiologia, 21*, 115–123.

Menzel, A., Sparks, T. H., Estrella, N., Koch, E., Aasa, A., Ahas, R., Alm-Kubler, K., Bissolli, P., Braslavska, O., Briede, A., Chmielewski, F. M., Crepinsek, Z., Curnel, Y., Dahl, A., Defila, C., Donnelly, A., Filella, Y., Jatczak, K., Mage, F., Mestre, A., Nordli, O., Peñuelas, J., Pirinen, P., Remisova, V., Scheifinger, H., Striz, M., Susnik, A., Van Viet, A. J. H., Wielgolaski, F., Zach, S., & Zust, A. (2006). European phenological response to climate change matches the warming pattern. *Global Change Biology, 12*, 1969–1976.

Meyer, S. E., Nelson, D. L., & Carlson, S. L. (2004). Ecological genetics of vernalization response in *Bromus tectorum* L. *Annals of Botany, 93*, 653–666.

Miyazaki, Y., Hiura, T., Kato, E., & Funada, R. (2002). Allocation of resources to reproduction in *Styrax obassia* in a masting year. *Annals of Botany, 89*, 767–772.

Monselise, S. P., & Goldschmidt, E. E. (1982). Alternate bearing in fruit trees. *Horticultural Reviews, 4*, 128–173.

Munuera Giner, M., Carrión García, J. S., & García Sellés, J. (1999). Aerobiology of *Artemisia* airborne pollen in Murcia (SE Spain) and its relationship with weather variables: Annual and intradiurnal variations for three different species. Wind vectors as a tool in determining pollen origin. *International Journal of Biometeorology, 43*, 51–63.

Murray, M. B., Cannell, M. G. R., & Smith, R. I. (1989). Date of budburst of fifteen tree species in Britain following climatic warming. *Journal of Applied Ecology, 26*, 693–700.

Myking, T. (1997a). Dormancy, budburst and impacts of climatic warming in coastal-inland and altitudinal *Betula pendula* and *B. pubescens* ecotypes. In H. Lieth & M. D. Schwarz (Eds.), *Phenology in seasonal climates I* (pp. 51–66). Leiden: Backhuys.

Myking, T. (1997b). Effects of constant and fluctuating temperature on time to budburst in *Betula pubescens* and its relation to bud respiration. *Trees, 12*, 107–112.

Myking, T. (1998). Interrelations between respiration and dormancy in buds of three hardwood species with different chilling requirements for dormancy release. *Trees, 12*, 224–229.

Myking, T. (1999). Winter dormancy release and budburst in *Betula pendula* Roth. and *B. pubescens* Ehrh. ecotypes. *Phyton, 39*, 139–146.

Myking, T., & Heide, O. M. (1995). Dormancy release and chilling requirement of buds of latitudinal ecotypes of *Betula pendula* and *B. pubescens*. *Tree Physiology, 15*, 697–704.

Nord, E. A., & Lynch, J. P. (2009). Plant phenology: A critical controller of soil resource acquisition. *Journal of Experimental Botany, 60*, 1927–1937.

Norris-Hill, J. (1995). The modelling of daily Poaceae pollen concentrations. *Grana, 34*, 182–188.

Norris-Hill, J., & Emberlin, J. (1991). Diurnal variation of pollen concentration in the air of north-central London. *Grana, 30*, 229–234.

Ogden, E., Hayes, J., & Raynor, G. (1969). Diurnal patterns of pollen emission in *Ambrosia*, *Phleum*, *Zea* and *Ricinus*. *American Journal of Botany, 56*, 16–21.

Ong, E. K., Taylor, P. E., & Knox, R. B. (1997). Forecasting the onset of the grass pollen season in Melbourne (Australia). *Aerobiologia, 13*, 43–48.

Orlandi, F., Fornaciari, M., & Romano, B. (2002). The use of phenological data to calculate chilling units in *Olea europaea* L. in relation to the onset of reproduction. *International Journal of Biometeorology, 46*, 2–8.

Orlandi, F., Garcia-Mozo, H., Vazquez Ezquerra, L., Romano, B., Dominguez, E., Galán, C., & Fornaciari, M. (2004). Phenological olive chilling requirements in Umbria (Italy) and Andalusia (Spain). *Plant Biosystems, 138*, 111–116.

Orlandi, F., Sgromo, C., Bonofiglio, T., Ruga, L., Romano, B., & Fornaciari, M. (2009). A comparison among olive flowering trends in different Mediterranean areas (south-central Italy) in relation to meteorological variations. *Theoretical and Applied Climatology, 97*, 339–347.

Orlandi, F., Garcia-Mozo, H., Galán, C., Romano, B., de la Guardia, C. D., Ruiz, L., del Mar Trigo, M., Dominguez-Vilches, E., & Fornaciari, M. (2010). Olive flowering trends in a large Mediterranean area (Italy and Spain). *International Journal of Biometeorology, 54*, 151–163.

Ottersen, G., Planque, B., Belgrano, A., Post, E., Reid, P. C., & Stenseth, N. C. (2001). Ecological effects of the North Atlantic Oscillation. *Oecologia, 12*, 1–14.

Pacini, E. (2000). From anther and pollen ripening to pollen presentation. *Plant Systematics and Evolution, 22*, 219–243.

Pacini, E., & Hesse, M. (2004). Cytophysiology of pollen presentation and dispersal. *Flora, 199*, 273–285.

Peñuelas, J., Fillela, I., & Comas, P. (2002). Changed plant and animal life cycles from 1952 to 2000 in the Mediterranean region. *Global Change Biology, 8*, 531–544.

Peñuelas, J., Fillela, I., Zhang, X., Llorens, L., Ogaya, R., Lloret, F., Comas, P., Estiarte, M., & Terradas, J. (2004). Complex spatiotemporal phenological shifts as a response to rainfall changes. *The New Phytologist, 161*, 837–846.

Pidek, I. A. (2007). Nine-year record of *Alnus* pollen deposition in the Roztocze region (SE Poland) with relation to vegetation data. *Acta Agrobotanica, 60*, 57–64.

Pop, E. W., Oberbauer, S. F., & Starr, G. (2000). Predicting vegetative bud break in two arctic deciduous shrub species, *Salix pulchra* and *Betula nana*. *Oecologia, 124*, 176–184.

Post, E., & Stenseth, N. C. (1999). Climatic variability, plant phenology and northern ungulates. *Ecology, 80*, 1322–1339.

Primack, R. B. (1985). Patterns of flowering phenology in communities, populations, individuals and single flowers. In J. White (Ed.), *The Population Structure of Vegetation* (pp. 571–593). Dordrecht: Junk.

Ranta, H., & Satri, P. (2007). Synchronised inter-annual fluctuation of flowering intensity affects the exposure to allergenic tree pollen in North Europe. *Grana, 46*, 274–284.

Ranta, H., Oksanen, A., Hokkanen, T., Bondestam, K., & Heino, S. (2005). Masting by *Betula*-species; Applying the resource budget model to north European data sets. *International Journal of Biometeorology, 49*, 146–151.

Ranta, H., Hokkanen, T., Linkosalo, T., Laukkanen, L., Bondestam, K., & Oksanen, A. (2008). Male flowering of birch: Spatial synchronization, year-to-year variation and relation of catkin numbers and airborne pollen counts. *Forest Ecology and Management, 255*, 643–650.

Rasmussen, A. (2002). The effects of climate change on the birch pollen season in Denmark. *Aerobiologia, 18*, 253–265.

Rathcke, B., & Lacey, E. P. (1985). Phenological patterns of terrestrial plants. *Annual Review of Ecology and Systematics, 16*, 179–214.

Recio, M., Rodriguez-Rajo, F. J., Jato, V., Mar Trigo, M., & Cabezudo, B. (2009). The effect of recent climatic trends on Urticaceae pollination in two bioclimatically different areas in the Iberian Peninsula: Malaga and Vigo. *Climatic Change, 97*, 215–228.

Recio, M., Docampo, S., García-Sanchez, J., Trigo, M., Melgar, M., & Cabezudo, M. (2010). Influence of temperature, rainfall and wind trends on grass pollination in Malaga (western Mediterranean coast). *Agricultural and Forest Meteorology, 160*, 931–940.

Repo, T., Mäkälä, A., & Hänninen, H. (1990). Modeling frost resistance of trees. In H. Jozefek (Ed.), *Modeling to understand forest functions* (Silva Carelia, Vol. 15, pp. 61–74). Joensuu: University of Joensuu.

Ribeiro, H., Cunha, M., & Abreu, I. (2006). Comparison of classical models for evaluating the heat requirements of olive (Olea europaea L.) in Portugal. *Journal of Integrative Plant Biology, 48*, 664–671.

Richardson, E. A., & Anderson, J. L. (1986). The omnidata biophenometer (TA45-p). A chill unit and growing degree hour accumulator. *Acta Horticulturae, 184*, 95–99.

Richardson, E. A., Seeley, S. D., & Walker, D. R. (1974). A model for estimating the completion of rest for "Redhaven" and "Elberta" peach trees. *HortScience, 9*(4), 331–332.

Rodriguez-Rajo, F. J., Frenguelli, G., & Jato, M. V. (2003). Effect of air temperature on forecasting the start of the *Betula* pollen season at two contrasting sites in the south of Europe (1995–2001). *International Journal of Biometeorology, 47*, 117–125.

Rodriguez-Rajo, F. J., Dopazo, A., & Jato, V. (2004). Environmental factors affecting the start of the pollen season and concentrations of airborne *Alnus* pollen in two localities of Galicia (NW Spain). *Annals of Agricultural and Environmental Medicine, 11*, 35–44.

Rodriguez-Rajo, F. J., Valencia-Barrera, R. M., Vega-Maray, A. M., Suarez, F. J., Fernandez-Gonzalez, D., & Jato, M. V. (2006). Prediction of airborne *Alnus* pollen concentration by using ARIMA models. *Annals of Agricultural and Environmental Medicine, 13*, 25–32.

Rogers, C. A., Wayne, P. M., Macklin, E. A., Muilenberg, M. L., Wagner, C. J., Epstein, P. R., & Bazzaz, F. A. (2006). Interaction of the onset of spring and elevated atmospheric CO_2 on ragweed (*Ambrosia artemisiifolia* L.) pollen production. *Environmental Health Perspectives, 114*, 865–869.

Rohde, A., Howe, G. T., Olsen, J. E., Moritz, T., van Montagu, M., Junttila, O., & Boerjan, W. (2000). Molecular aspects of bud dormancy in trees. In S. M. Jain & S. C. Minocha (Eds.), *Molecular Biology of Woody Plants* (Vol. 1, pp. 89–134). Dordrecht: Kluwer Academic Publishers.

Saar, M., Gudiskas, Z., Plompuu, T., Linno, E., Minkien, Z., & Motiekaityt, V. (2000). Ragweed plants and airborne pollen in the Baltic states. *Aerobiologia, 16*, 101–106.

Sanchez Mesa, J. A., Smith, M., Emberlin, J., Allitt, U., Caulton, E., & Galán, C. (2003). Characteristics of grass pollen seasons in areas of southern Spain and the United Kingdom. *Aerobiologia, 19*, 243–250.

Sarvas, R. (1974). Investigations of the annual cycle of development of forest trees. II. Autumn dormancy and winter dormancy. *Communicationes Instituti Forestalis Fennia, 84*. 101p.

Schappi, G. F., Taylor, P. E., Kenrick, J., Staff, I. A., & Suphioglu, C. (1998). Predicting the grass pollen count from meteorological data with regard to estimating the severity of hayfever symptoms in Melbourne (Australia). *Aerobiologia, 14*, 29–37.

Scheifinger, H., Menzel, A., Koch, E., & Peter, C. (2002). Atmospheric mechanisms governing the spatial a temporal variability of phenological phases in central Europe. *International Journal of Climatology, 22*, 1739–1755.

Scheifinger, H., Menzel, A., Koch, E., & Peter, C. (2003). Trends of spring frost events and phenological dates in Central Europe. *Theoretical and Applied Climatology, 74*, 41–51.

Skjoth, C. A., Sommer, J., Stach, A., Smith, M., & Brandt, J. (2007). The long-range transport of birch (*Betula*) pollen from Poland and Germany causes significant pre-season concentrations in Denmark. *Clinical and Experimental Allergy, 37*, 1204–1212.

Smith, M., & Emberlin, J. (2005). Constructing a 7-day ahead forecast model for grass pollen at north London, United Kingdom. *Clinical and Experimental Allergy, 35*, 1400–1406.

Smith, M., & Emberlin, J. (2006). A 30-day-ahead forecast model for grass pollen in north London, United Kingdom. *International Journal of Biometeorology, 50*, 233–242.

Smith, M., Skjøth, C. A., Myszkowska, D., Uruska, A., Puc, M., Stach, A., Balwierz, Z., Chlopek, K., Piotrowska, K., Kasprzyk, I., & Brandt, J. (2008). Long-range transport of Ambrosia pollen to Poland. *Agricultural and Forest Meteorology, 148*, 1402–1411.

Smith, M., Emberlin, J., Stach, A., Rantio-Lehtimäki, A., Caulton, E., Thibaudon, M., Sindt, C., Jäger, S., Gehrig, R., Frenguelli, G., Jato, V., Rajo, F., Alcázar, P., & Galán, C. (2009). Influence of the North Atlantic Oscillation on grass pollen counts in Europe. *Aerobiologia, 25*(4), 321–332.

Solomon, W., & Mathews, K. (1990). Aerobiology and inhalant allergens. In E. Middleton, C. E. Reed, E. F. Ellis, N. F. Adkinson, & J. W. Yunginger (Eds.), *Allergy principles and practice* (Vol. 1, pp. 312–372). St Louis: Mosby.

Spieksma, F. T., & Nikkels, A. H. (1998). Airborne grass pollen in Leiden, the Netherlands: Annual variations and trends in quantities and season starts over 26 years. *Aerobiologia, 14*, 347–358.

Stach, A., Smith, M., Skjøth, C. A., & Brandt, J. (2007a). Examining *Ambrosia* pollen episodes at Poznan (Poland) using back-trajectory analysis. *International Journal of Biometeorology, 51*, 275–286.

Stach, A., Garcia-Mozo, H., Prieto-Baena, J. C., Czarnecka-Operacz, M., Jenerowicz, D., Silny, W., & Galán, C. (2007b). Prevalence of *Artemisia* species pollinosis in Western Poland: Impact of Climate Change on aerobiological trends, (1995–2004). *Journal of Investigational Allergy and Clinical Immunology, 17*, 39–47.

Stach, A., Emberlin, J., Smith, M., Adams-Groom, B., & Myszkowska, D. (2008). Factors that determine the severity of *Betula* spp. pollen seasons in Poland (Poznan and Krakow) and the United Kingdom (Worcester and London). *International Journal of Biometeorology, 52*, 311–321.

Stark, P. C., Ryan, L. M., McDonald, J. L., & Burge, H. A. (1997). Using meteorologic data to predict daily ragweed pollen levels. *Aerobiologia, 13*, 177–184.

Stenseth, N. C., Mysterud, A., Ottersen, G., Hurrell, J. W., Chan, K. S., & Lima, M. (2002). Ecological effects of climate fluctuations. *Science, 297*, 1292–1296.

Stepalska, D., Szczepanek, K., & Myszkowska, D. (2002). Variation in *Ambrosia* pollen concentration in Southern and Central Poland in 1982–1999. *Aerobiologia, 18*, 13–22.

Subba Reddi, C., & Reddi, N. S. (1986). Pollen production in some anemophilous angiosperms. *Grana, 25*, 55–61.

Subba Reddi, C., Reddi, N. S., & Atluri Janaki, B. (1988). Circadian patterns of pollen release in some species of Poaceae. *Review of Palaeobotany and Palynology, 54*, 11–42.

Tedeschini, E., Rodriguez-Rajo, F. J., Caramiello, R., Jato, V., & Frenguelli, G. (2006). The influence of climate changes in *Platanus* spp. pollination in Spain and Italy. *Grana, 45*, 222–229.

Van der Pijl, L. (1978). Reproductive integration and sexual disharmony in floral functions. In A. J. Richards (Ed.), *The pollination of flowers by insects* (pp. 79–88). London: Academic.

van Hout, R., Chamecki, M., Brush, G., Katz, J., & Parlange, M. B. (2008). The influence of local meteorological conditions on the circadian rhythm of corn (*Zea mays* L.) pollen emission. *Agricultural and Forest Meteorology, 148*, 1078–1092.

Vázquez, L. M., Galán, C., & Domínguez, E. (2003). Influence of meteorological parameters in *Olea* pollen. *International Journal of Biometeorology, 48*, 83–90.

von Wahl, P.-G., & Puls, K. E. (1989). The emission of mugwort pollen (*Artemisia vulgaris* L.) and its flight in the air. *Aerobiologia, 5*, 55–63.

Walther, G. R., Post, E., Convey, P., Menzel, A., Parmesan, C., Beebee, T. J., Fromentin, J. M., Hoegh-Guldberg, O., & Bairlein, F. (2002). Ecological responses to recent climate change. *Nature, 416*(6879), 389–395.

Wanner, H., Bronnimann, C., Casty, D., Gyalistras, J., Luterbacher, C., Schmultz, D., Stephenson, B., & Xoplaki, E. (2001). North Atlantic Oscillation – concepts and studies. *Surveys in Geophysics, 22*, 321–382.

Wielgolaski, F. E. (1999). Starting dates and basic temperatures in phenological observations of plants. *International Journal of Biometeorology, 42*, 158–168.

Woś, A. (1994). *Klimat Niziny Wielkopolskiej* (p. 192). Poznan: Wyd. 1 Adam Mickiewicz University. written in polish.

Xoplaki, E. J., Gonzalez-Rouco, F., Luterbacher, H., & Wanner, H. (2004). Wet season Mediterranean precipitation variability: influence of large-scale dynamics and predictability. *Climate Dynamics, 23*, 63–79.

Yli-Panula, E., & Rantio-Lehtimäki, A. (1995). Birch pollen antigenic activity of settled dust in rural and urban homes. *Allergy, 50*, 303–307.

Ziska, L. H., Ghannoum, O., Baker, J. T., Conroy, J., Bunce, J. A., Kobayashi, K., & Okada, M. (2001). A global perspective of ground level, 'ambient' carbon dioxide for assessing the response of plants to atmospheric CO_2. *Global Change Biology, 7*, 789–796.

Ziska, L. H., Epstein, P. R., & Rogers, C. A. (2008). Climate change, aerobiology, and public health in the Northeast United States. *Mitigation and Adaption Strategies for Global Change, 13*, 607–613.

Chapter 4
Monitoring, Modelling and Forecasting of the Pollen Season

Helfried Scheifinger, Jordina Belmonte, Jeroen Buters, Sevcan Celenk, Athanasios Damialis, Chantal Dechamp, Herminia García-Mozo, Regula Gehrig, Lukasz Grewling, John M. Halley, Kjell-Arild Hogda, Siegfried Jäger, Kostas Karatzas, Stein-Rune Karlsen, Elisabeth Koch, Andreas Pauling, Roz Peel, Branko Sikoparija, Matt Smith, Carmen Galán-Soldevilla, Michel Thibaudon, Despina Vokou, and Letty A. de Weger

Abstract The section about monitoring covers the development of phenological networks, remote sensing of the season cycle of the vegetation, the emergence of the science of aerobiology and, more specifically, aeropalynology, pollen sampling instruments, pollen counting techniques, applications of aeropalynology in agriculture and the European Pollen Information System. Three data sources are directly related with aeropalynology: phenological observations, pollen counts and remote sensing of the vegetation activity. The main future challenge is the assimilation of these data streams into numerical pollen forecast systems. Over the last decades

---

H. Scheifinger (✉) • E. Koch
Zentralanstalt für Meteorologie und Geodynamik, Hohe Warte 38, 1190 Wien, Austria
e-mail: helfried.scheifinger@zamg.ac.at

J. Belmonte
CREAF, Universidad Autónoma de Barcelona, 08193 Bellaterra, Spain

J. Buters
ZAUM - Center of Allergy & Environment, a joint institute of the Technische Universität München and Helmholtz Zentrum München, Biedersteiner Str. 29, 80802, Munich, Germany

S. Celenk
Science Faculty, Biology Department, Uludag University, Görükle, 16059 Bursa, Turkey

A. Damialis • D. Vokou
Department of Ecology, School of Biology, Aristotle University,
GR-54124, Thessaloniki, Greece

C. Dechamp
Association Française d'Etude des Ambroisies: A F E D A, 25 Rue Ambroise Paré,
F 69 800, Saint-Priest, France

H. García-Mozo • C. Galán-Soldevilla
Departamento de Biologia Vegetal, Campus de Rabanales,
Universidad de Cordoba, 14071, Cordoba, Spain

R. Gehrig • A. Pauling
Bio- and Environmental Meteorology, Climate Division, MeteoSchweiz,
Krähbühlstr. 58, CH-8044, Zürich, Switzerland

M. Sofiev and K-C. Bergmann (eds.), *Allergenic Pollen: A Review of the Production, Release, Distribution and Health Impacts*, DOI 10.1007/978-94-007-4881-1_4,
© Springer Science+Business Media Dordrecht 2013

consistent monitoring efforts of various national networks have created a wealth of pollen concentration time series. These constitute a nearly untouched treasure, which is still to be exploited to investigate questions concerning pollen emission, transport and deposition. New monitoring methods allow measuring the allergen content in pollen. Results from research on the allergen content in pollen are expected to increase the quality of the operational pollen forecasts.

In the modelling section the concepts of a variety of process-based phenological models are sketched. Process-based models appear to exhaust the noisy information contained in commonly available observational phenological and pollen data sets. Any additional parameterisations do not to improve model quality substantially. Observation-based models, like regression models, time series models and computational intelligence methods are also briefly described. Numerical pollen forecast systems are especially challenging. The question, which of the models, regression or process-based models is superior, cannot yet be answered.

Keywords Aerobiology • Aeropalynology • Phenology • Pollen modelling • Phenological modelling

---

L. Grewling
Laboratory of Aeropalynology, Faculty of Biology, Adam Mickiewicz University, Umultowska 89, 61-614, Poznan, Poland

J.M. Halley
Department of Biological Applications and Technology, University of Ioannina, GR-45110, Ioannina, Greece

K. Hogda • S. Karlsen
NORUT ITEK, Postboks 6434, Forskningsparken, N-9294, Tromsø, Norway

S. Jäger • M. Smith
HNO Klinik der Medizinischen Universität Wien, Währinger Gürtel
18-20, A-1090, Wien, Austria

K. Karatzas
Informatics Applications and Systems Group, Faculty of Engineering, Aristotle University,
Egnatia Str., Box 483 54124, Thessaloniki, Greece

R. Peel
National Pollen and Aerobiology Research Unit, Institute of Health, University of Worcester,
WR2 6AJ, Worcester, UK

B. Sikoparija
Laboratory for Palynology, Faculty of Sciences, University of Novi Sad,
Trg Dositeja, Obradovica 2, 21000, Novi Sad, Serbia

M. Thibaudon
Réseau National de Surveillance Aérobilogique,
Chemin des Gardes BP 8, F-69610, Saint Genis L'Argentière, France

L.A. deWeger
Department of Pulmonology, Leiden University Medical Centre,
P.O. Box 9600, 2300 RC, Leiden, The Netherlands

List of Acronyms

AFEDA	French Association for Ragweed Study
ANN	Artificial Neural Networks
ARIMA	Autoregressive Integrated Moving Average
AVHRR	Advanced Very High Resolution Radiometer
CART	Classification and Regression Trees
COST725	COST Action 725: Establishing a European Phenological Data Platform for Climatological Applications
CAgM	Commission for Agrometeorology
DEM	Digital Elevation Model
DWD	Deutscher Wetterdienst
EAN	European Aeroallergen Network
ELISA	Enzyme-linked Immunosorbent Assay
EUMETNET	The Network of European Meteorological Services
GIMMS	Global Inventory Modeling and Mapping Studies
IC	Computational Intelligence
IAA	International Association for Aerobiology
IBP	International Biological Programme
INERIS	Industrial Environment and Risks National Institute
IPG	International Phenological Garden
Landsat TM	Landsat Thematic Mapper, Satellite
LUT	Look Up Table
MODIS	Moderate Resolution Imaging Spectroradiometer
NDVI	Normalised Difference Vegetation Index
NHMS	National Hydrometeorological Services
NOAA	National Oceanic and Atmospheric Administration
PCR	Polymerase Chain Reaction
PEP725	Pan European Phenological Database
PM	Particulate Matter
RMSE	Root Mean Square Error
SPOT	Satellite Pour l'Observation de la Terre
SOM	Self-Organising Maps
SVMs	Support Vector Machines
TSM	Temperature Sum Model
UM	Use and Management of Biological Resources
WCDMP	World Climate Data and Monitoring Programme
WCP	World Climate Programme
WIBS	Wide-Issue Bioaerosol Spectrometer
WMO	World Meteorological Organisation

4.1 Introduction

Input for the aeropalynological core topics of monitoring, modelling and forecasting of the pollen season have been drawn from an array of disciplines and cast into this review chapter. History, current state, recent developments and future prospects of phenological and pollen counting networks have been reviewed in the first section. Both, phenological observations and pollen counts collected by various networks form the observational basis of any quantitative description of the relationship between the seasonal cycle of plants and their atmospheric environment. The various modelling strategies and their applications are extensively elucidated in the second section.

Although phenology and aeropalynology experienced separate historical developments, they meet here and share the same models, which forecast the beginning of flowering and the beginning of pollen shedding, respectively. Links between aeropalynology and phenology are scattered throughout this review, but are explicitly summarised under the headings of "Phenological observations" and "Process-based phenological models":

- The natural relationship between phenology and aeropalynology may be expressed in the assumption that the beginning of flowering equals the beginning of pollen shedding into the atmosphere. Pollen emission modelling can benefit much from the knowledge, observations and modelling of flowering phenology.
- The effort to maintain a phenological network is less than to maintain a pollen observing network. Therefore in many regions the spatial density of phenological networks is higher than that of pollen traps and phenological time series are longer than pollen concentration time series. Thus it is possible to infer something about the pollen problem from phenology with a higher spatial density and/or further back in time than it would be possible based on pollen data alone.
- Phenology has made substantial progress during the last decade in various aspects like phenological modelling, satellite observation of the vegetation cycle, relation with climate variability and others, so that the problem of pollen allergenicity now can benefit from that progress in phenological research.

The recent boost in the interest in phenology as climate impact factor has been motivated by the discussion about human influence on climate, which became manifest in an increasing flood of publications with phenological background and an extended chapter of the 4AR about the role of phenology in climate impact research (Rosenzweig et al. 2007). Aeropalynology benefits a great deal from the enhanced interest in phenology within the frame of the climate impact discussion. Both fields of interest have more in common than it appears at first glance, a factor, which has still to be exploited.

4.2 Monitoring

4.2.1 Phenological Observations

4.2.1.1 Monitoring Networks

Systematic phenological observation can now look back on a history, which reaches back as far as the eighteenth century, when Carolus Linnaeus started the first phenological network in Sweden and Finland 1750–1752 (Nekovar et al. 2008). A few decades later phenological observations were also included in the first pan European meteorological network of the Societas Meteorologicae Palatinae (1781–1792). In the mid-nineteenth century the first national networks began their operation in the USA and Europe, although in most cases only for a limited time period. Ihne and Hoffmann managed to run their phenological network in a number of European countries over 1883–1941 (Fig. 4.1). During the 1950s the idea of International Phenological Gardens with a cloned set of plants was born, which resulted in a still operating and expanding phenological network in Europe. During the same period most national phenological networks began collecting phenological observations systematically, as recommended by the Commission for Agrometeorology (CAgM) of the World Meteorological Organisation (WMO).

A detailed global overview about phenological networks can be found in Schwartz (2003) and Koch (2010), whereas Nekovar et al. (2008) summarise the current situation in Europe.

National Monitoring Networks

Phenological research relies on phenological observations, collected mostly by national meteorological and hydro-meteorological services (NMHS). Phenological data collection with its rather small data volume has been usually running unobtrusively alongside the main stream collection of meteorological and climate data and thus survived in many NMHSs the ups and downs of the interest in phenological science through time. Another advantage of NMHSs is their experience in running station networks, quality controlling the incoming data, digitising and storing them on appropriate devices. Due to the efforts of COST Action 725 and the growing concern about climate change impacts, the Commission for Climatology (CCl) of the WMO now recommends the NHMS to organise phenological observations, whereas the World Climate Data and Monitoring Programme (WCDMP) and World Climate Programme (WCP) are working to stimulate and coordinate climate and climate impact monitoring activities around the world, which include phenological observations (www.omm.urv.cat/media/documents/WMO.pdf).

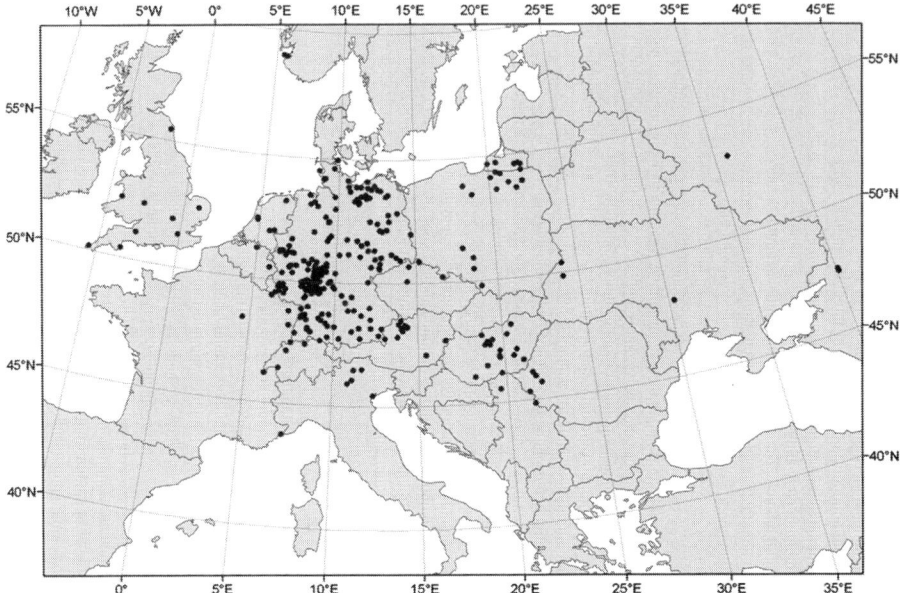

Fig. 4.1 Stations of the Hoffmann–Ihne phenological network from 1883 to 1941. Only stations with a minimum number of five observations are being displayed. The historical phenological database HPDB is maintained by the DWD (after Koch et al. 2008)

The recently published report of the COST Action 725 summarises the information about national European phenological networks (Nekovar et al. 2008). A few "phenophases", which are relevant for pollen allergies, entered the COST Action 725 data base. Here is a list of plants shedding allergenic pollen with a useful number of observations of phenological events (beginning of flowering) in this data base: Norway maple (*Acer platanoides*), horse chestnut (*Aesculus hippocastanum*), black alder (*Alnus glutinosa*), meadow foxtail (*Alopecurus pratensis*), mugwort (*Artemisia vulgaris*), birch (*Betula pendula*), hazel (*Corylus avellana*), forsythia (*Forsythia suspensa*), ash (*Fraxinus excelsior*) and goat willow *(Salix caprea)*.

An increasing number of national weather services and other organisations interested in nature observation have been creating internet-based observation networks, where volunteers can enter their georeferenced phenological observations (Table 4.1). Some weather services merge the data from their traditional network and the entries from the web.

International Monitoring Networks and Data Collection Initiatives

Contrary to national networks, the distribution of plants is not influenced by national borders. There exist two basic strategies to overcome the problem of national differences in observational methods, either by an international network in the first place or by merging national networks.

Table 4.1 List of web-based phenological observational networks

Number	Country	Name	Web address
1	The Netherlands	Natuurkalender	http://www.natuurkalender.nl
2	UK	Nature's Calendar	http://www.naturescalendar.org.uk/
3	Ireland	Nature's Calendar	http://www.biology.ie
4	USA	National Phenology Network	http://www.usanpn.org/?q=home http://www.usanpn.org/participate/observe
5	USA	Appalachian Mountain Club	http://www.outdoors.org/conservation/mountainwatch/index.cfm
6	USA	Project Budburst	http://www.windows.ucar.edu/citizen_science/budburst/results.php
7	Canada	Alberta Plantwatch	http://plantwatch.fanweb.ca/
8	Canada	PlantWatch	http://www.naturewatch.ca/english/plantwatch/intro.html
9	Austria	Phänologie	http://zacost.zamg.ac.at/phaeno_portal/
10	Sweden	Svenska fenologinätverket	http://www.blommar.nu/index.php
11	Australia	ClimateWatch	http://www.climatewatch.org.au/

The International Phenological Gardens (IPGs), for instance, were thought to obtain comparable and standardised large-scale phenological observations across Europe (Chmielewski 2008). With the same idea in mind, the Global Phenological Monitoring has been launched (http://www.agrar.hu-berlin.de/struktur/institute/pfb/struktur/agrarmet/phaenologie/gpm).

National data collection initiatives achieve their final value only after a number of national networks get merged over larger areas. Unfortunately, such merging efforts are only few because of a number of difficulties. COST Action 725 "Establishing a European Phenological data Platform for Climatological Applications" (2004–2009) aimed at creating a European reference data set of phenological observations that can be used for climatological purposes (Koch et al. 2005; Nekovar et al. 2008). The proposal for a follow up of COST725 so called "PEP725" (Pan European Phenological Database) was accepted by EUMETNET, has started in 2010 and will run over 5 years. PEP725 will maintain and update the COST initiated phenological database. Additionally, it will incorporate phenological data from before 1951 and develop better quality checking procedures. PEP725 will ensure an open access to the database for research and education. An attractive webpage will make phenology and climate impacts on vegetation more visible for the public, enabling a real time monitoring of vegetation development.

The European Phenology Network (http://www.pik-potsdam.de/~rachimow/epn/html/frameok.html) represents a broad based compilation of meta-information on phenological and related networks across the world.

Monitoring for Special Scientific Studies

For some research projects, special phenological observational data sets are required, because observations from ordinary networks are insufficient, not applicable or not available. Special networks are operated at a limited number of observational points and over a short period of time. For instance, this was the case for the larch phenological study in the Western Alpine Aosta valley by Migliavacca et al. (2008), where the influence of elevation and topography on the phenology of larch (*Larix decidua*) was studied. Ziello et al. (2010) linked phenological, meteorological and palynological data along an altitudinal gradient in the German Alps. The study of the flowering phenology of herbaceous plants in a lawn community required a special observational setup (Marletto et al. 1992) as did the observation of the beginning of male flowering of trees of the cypress family (*Cupressaceae*) for pollen modelling purposes (Torrigiani Malaspina et al. 2007).

Monitoring for Pollen Forecasting Purposes

In the traditional phenological monitoring setup the observational sheets are returned to the network operator at the end of each year. For more immediate information on the state of the vegetation, some network operators introduced a rapid information system (e.g. Sofortmeldenetz of the German and Swiss weather services). Such immediately transmitted phenological information supports the pollen forecast system of the German weather service, for instance. Assuming that the observers enter their observed entry dates immediately, information on the current state of the vegetation can be derived from the web-based networks (Table 4.1). Remote sensing systems, like satellites and real time digital cameras, can also serve the same purpose, but are still to be included into the operational procedures. Likewise, assimilation systems, which consistently merge the observational real time data into phenological and pollen dispersion models, still have to be developed (Stöckli et al. 2008).

4.2.1.2 Remote Sensing

Normalised Difference Vegetation Index (NDVI)

Live green plants absorb solar radiation in the photosynthetically active spectral region (400–700 nm), which they use as a source of energy for photosynthesis. At the same time leaf cells do not absorb but reflect and transmit solar radiation in the near-infrared spectral region. This large contrast in reflectance properties between red and near-infrared spectral regions is unique for photosynthetically active plants, and can be used by remote sensing sensors to distinguish them from other land cover types as soil, bare rock and snow.

Accordingly, phenological changes during the growing season can be studied by examining changes in the remote sensing-based Normalised Difference Vegetation index (NDVI) value. The NDVI is defined as:

$$\mathrm{NDVI} = (\mathrm{Ch2} - \mathrm{Ch1}) / (\mathrm{Ch2} + \mathrm{Ch1}), \tag{4.1}$$

where Ch1 and Ch2 represent reflectance measured in the near infrared and red channels, respectively (Lillesand and Kiefer 1994). NDVI quantifies the contrast between red surface reflectance, which decreases with chlorophyll content, and near infrared surface reflectance, which increases with leaf area index and crown density. NDVI of an area containing a dense green vegetation canopy will tend to have high positive values (typically 0.6–0.8); more sparsely vegetated areas will have lower values while clouds and snow fields will be characterised by negative values of the NDVI index.

Atmospheric noise in the NDVI caused by clouds, dust and aerosols is generally negatively biased. This is due to the additive path radiance, which causes an increase in red reflectance, while lower atmospheric transmission reduces near infrared reflectance (Guyot et al. 1989). Maximum value compositing (Holben 1986) is a common method of minimising such noise. In this method, only the highest NDVI value in a predefined compositing period (typically 15–16 days) is retained. This results in fewer but more reliable NDVI values representing the time series.

Satellite Sensors

Maximum value composite NDVI datasets with global coverage and bi-monthly compositing period have been created using data from sensors with large swath widths as the National Oceanic and Atmospheric Administration (NOAA) Advanced Very High Resolution Radiometer (AVHRR) and the VEGETATION sensor aboard Satellite Probatoire d'Observation de la Terre (SPOT). The starting dates of these time series are 1981 and 1998, respectively, and the products are available at spatial resolutions of 8 and 1 km, respectively.

Since 2000, NDVI products have been available from the Moderate Resolution Imaging Spectroradiometer (MODIS) aboard the Terra and Aqua satellites. Compared with the NOAA AVHRR GIMMS dataset, MODIS NDVI data have improved calibration and atmospheric correction, a spatial resolution of 236 m and compositing intervals of 16 days. This medium-resolution dataset opens new possibilities within global phenological monitoring.

Recently, satellite series as Formosat-2, Komsat-2, and RapidEye were launched. This new generation of satellite sensors, with both high temporal and spatial resolution (<10 m), opens new possibilities for local phenological monitoring.

Phenological Observations by Satellites

Satellite image-aided analysis of phenology of natural vegetation provides spatially complete coverage that can be used to interpolate traditional ground-based phenological observations, and NDVI has evolved as the primary tool for monitoring changes in vegetation activity. Probably the most commonly used long-time dataset is the Global Inventory Modeling and Mapping Studies (GIMMS) dataset based on the AVHRR instrument (Tucker et al. 2005). It has been used by many researchers to study the effect of global climate change on phenological timing and primary production (e.g. Myneni et al. 1997; Walker et al. 2003; Stöckli and Vidale 2004).

Karlsen et al. (2006, 2007) mapped the onset of the growing season of the whole of Fennoscandia by applying the GIMMS-NDVI dataset, where the onset of season was well correlated with the phenophase "onset of leafing of birch". They compared the NDVI-defined start of growing season with registrations of the onset of leafing of birch at 15 phenological registration sites across Fennoscandia. Most of the stations (13 out of 15) showed a moderately high correlation ($r^2 = 0.22$–0.65) between field data and the NDVI-defined start of growing season. Four of the stations had 20- or 21-year-long time-series. For these stations, the mean coefficient of determination (r^2) between the start of growing season and the onset of leafing of birch was 0.39 ($p < 0.05$). For all stations except one, the mean time span between the NDVI-defined start of growing season and the onset of leafing of birch was less than 1 week, and the root mean square error between field data and NDVI data was less than 10 days for all stations. For bi-monthly maximum value composited NDVI time series this is probably as close as it is possible to get. To decrease the difference it is necessary to use daily NDVI data.

Birch Flowering Measured by Satellites

Onset of flowering of birch and leaf-bud burst of birch are well correlated. Linkosalo (1999, 2000) found in southern Finland that the difference in time from male flowering to the first date of bud burst is only 1.1 days with male flowering occurring first. This indicates that the phenophase observed as leaf-bud burst could be used to determine the timing of local birch pollen release. Also, since bud burst of birch is accurately measured by remote sensing, measurements of NDVI could be used to determine the timing of local birch pollen release. Høgda et al. (2003) used the GIMMS-NDVI dataset, correlated it with birch pollen measurements from five stations, and found correlation values (r) in the range from 0.55 to 0.85. They also found trends in the timing of the start of pollen seasons, consistent with effects of climate change.

Because of its mountainous topography, deep fjords, and long distance from north to south, Norway is climatically and ecologically very diverse. The number of pollen traps is also relatively low so developing pollen forecasts in Norway is a challenging task. Karlsen et al. (2009a) used MODIS-NDVI satellite data with 16-days time resolution and 236 m spatial resolution to map the average onset of birch flowering in Norway for the 2000 to 2007 period (Fig. 4.2).

In those studies, they found high correlation with phenological field data of onset of leafing of birch, as well as with the date when the annual birch pollen sum reaches 2.5% of the annual total from the ten Burkard traps across Norway. Accordingly, the satellite data can be used to determine the best location of the pollen traps and define the area with similar timing for start of birch pollen seasons as each trap.

Karlsen et al. (2009a) also identified the NDVI threshold for each pixel when the onset of birch flowering occurred. On this basis they developed a model for

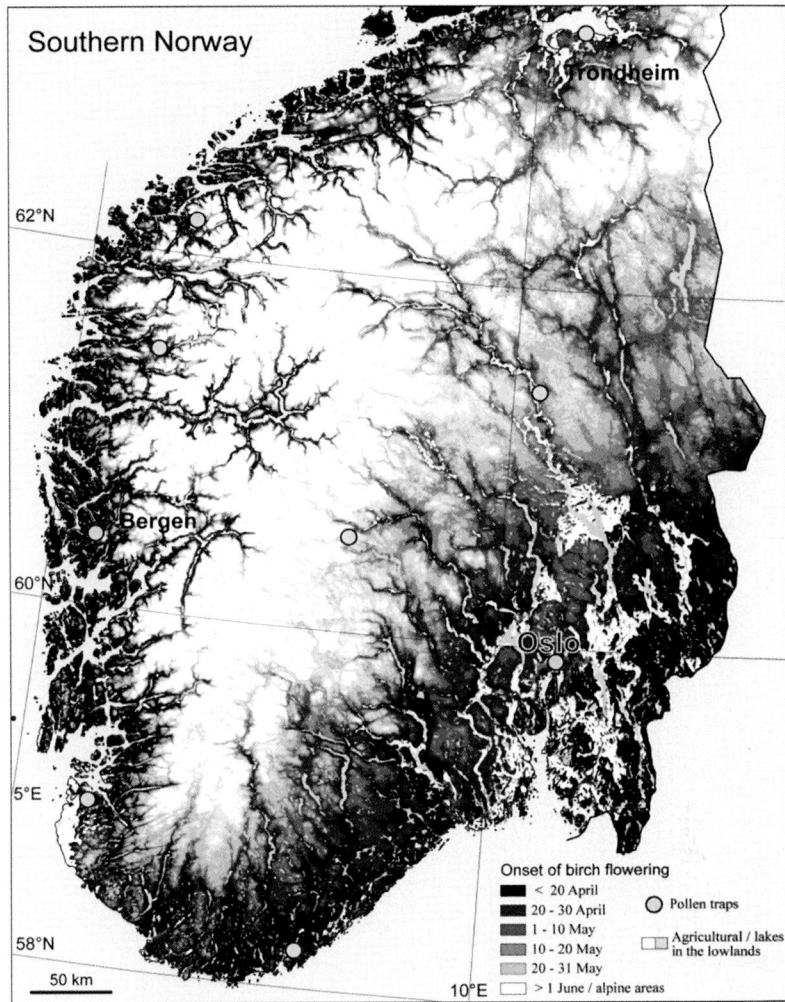

Fig. 4.2 Onset of birch flowering in southern Norway, based on mean values from the MODIS-NDVI dataset for the period 2000–2007, after Karlsen et al. (2009a). The map also shows the position of the pollen traps used in the study

real-time monitoring of the birch flowering. In the model they used additional geo-data, where the most important was a Landsat TM based vegetation map of all of Norway used to identify vegetation types where birch occurs. The model was applied to monitor the onset of birch pollen season in near real-time during spring 2009 (Karlsen et al. 2009b). The method showed in most cases good agreement with data from the pollen traps. However, the model does not give any information about the amount of birch pollen or about long-distance transported pollen. This will be a limitation for forecasting in areas where long-distance transport is an important factor.

Future Prospects

The next step to consider would be to apply the model, which was developed for the start of birch pollens season in Norway, at other places. In areas with complex topography, as the Alps, where atmosphere-based models are less reliable, remote sensing-based methods would be particularly useful. The same procedure would have to be followed as was applied in Norway, where at first the average onset of birch flowering had to be mapped and afterwards a model for real-time measuring was installed. It is also believed to be possible to further develop the model to map the onset of the grass pollen seasons as well, but with slightly less accuracy. Since the release of e.g. grass pollen occurs after the peak of the seasonal NDVI curve, simple threshold based methods as used for birch pollen will not be sufficient. Comparing time integrated NDVI and Growing Degree Days Karlsen et al. (2006) found a very high correlation. Accordingly, one method to apply satellite monitoring for estimating release dates of different pollen types could be to use time integrated NDVI as an additional data source.

The new generation of satellite sensors, with both daily data and high resolution (as Formosat-2, Komsat-2, and RapidEye), provide an opportunity to monitor the onset of the pollen season at a local scale. Due to the high data costs, it is only realistic to use these data for relatively small areas ($<\sim$1,000 km^2) and only some years. However, this new scale of observation creates a link between field observation of phenology/data from pollen traps and medium resolution sensors as MODIS. This would increase our understanding of the seasonal dynamics of the vegetation and improve the up-scaling from pollen traps to large regions, for instance, by the use of MODIS-NDVI data.

In the years to come there will be an increasing number of satellites with a range of scales in spatial resolution suitable to be used for near real time phenological monitoring (Ward 2008). Our conclusion is that satellite based monitoring of phenology is rapidly developing and observations will be assimilated into phenological models.

4.2.2 Aerobiological Observations

There is a wide spectrum of micro-organisms and biological particulate matter present in the atmosphere, which is investigated with a similarly wide range of methods and instruments (Cox and Wathes 1995). Biological aerosol sources are located in any place where biological activity exists. Many small organisms use the atmosphere as a means to be transported or to transport their own propagules. Bacteria, microalgae, microfungi, protozoa, insects and sometimes viruses are among the organisms that change their geographic location along their life cycle through the air. Fungal spores, lower plant spores or higher plant pollen grains are propagules transported by air. Fragments of fungi, animal or vegetal origin are also present in the atmosphere.

Aerobiology studies the release of biological particulate matter into the atmosphere, its transport through the atmosphere, and its deposition and re-suspension. In order to be taken up and transported, biological particles have to be released into the atmosphere, which is achieved by different mechanisms: (a) active, explosive or turgid; (b) passive, from an external agent. For example, in the case of most lower plants, i.e. bryophytes and pteridophytes, one finds active mechanisms for spore emission. In some cases, special sporangia and other devices enable an active emission by means of a catapult like discharge of the spores. However, these spores are usually large and heavy, which limits their dispersion through the atmosphere. In contrast, most of higher plants possess passive mechanisms for pollen emission. The role of pollen grains is to transfer the male gametophytes to the female reproductive organs. This is called pollination and is achieved via different mechanisms depending on the plant, which may be classified as anemophilous, entomophilous or hygrophilous. In order to guarantee an effective pollination, the entire structure of pollen grain is subject to selection pressure. However, this review will concentrate mainly on airborne pollen from wind-pollinated plants that are primarily responsible for seasonal allergies, the prevalence of which has been increasing substantially during the last decades.

4.2.2.1 Pollen Monitoring History

Aerobiology is a young scientific discipline that made great advances in the second half of the twentieth century, largely due to the introduction of advanced methods of monitoring. This brought a larger number of devotees to the subject and witnessed the rise of networks monitoring pollen and fungal spores on a national scale. Though aerobiology is related par excellence to ecology, it grew up following the major advancements in "allergology". The term "Aerobiology" was defined in the 1930s by Fred Campbell Meier (1893–1938), but Aerobiology did not become a recognised discipline until the 11 September 1974, when the International Association for Aerobiology (IAA) was founded at the 1st International Congress of Ecology, which was held at The Hague, in The Netherlands. Prior to this, in 1964, aerobiology had become a theme when the International Biological Program (IBP) was established. The major objective of the IBP was to study the biological basis for productivity of the world's ecosystems. NASA supported the Atmospheric Biology Conference, with the idea that the atmospheric dispersion of biological materials might be given attention by the IBP. An aerobiological programme was subsequently established in 1968, through the efforts of Benninghoff & Gregory, under the IBP section Use and Management of Biological Resources (UM). The IBP officially finished in 1974, when it was recognised that the studies on Aerobiology at international level should continue, and the IAA was formed 4 years later. Aerobiology is currently considered an experimental and multidisciplinary science that includes workers from botany, palynology, mycology, agronomy, microbiology, acarology, bioclimatology, meteorology, allergology and ecology. Aerobiology is made up of many different scientific disciplines and so it is not easy to trace the most significant milestones in its history.

The Origin of the Aerobiology

In ancient times, we find several references to the idea of micro-organisms and to the hypothesis that air can be a vector for diseases. For example, the Greek physician Hippocrates (460–377 BC) argued in "De Flatibus Corpus Hippocraticum" that people fall ill with fever after having inhaled infected air, although he was unaware of the nature of the infection. The author M. Terentius Varro (116–27 BC) cited invisible animals, which penetrated the body through mouth and nostrils thus causing disease. This concept was taken up in 1546 by Girolamo Fracastoro (1478–1553), who realised that some diseases were caused by "life seeds" that contaminate man, assuming that the body is reached by these particles by direct contact or by breathing in infected air. The Latin poet and philosopher Titus Lucretius Caro (98–55 BC) also mentioned small particles that can infect man.

Precursors of Aerobiology

Another important step was accomplished at the end of the fifteenth century by the invention of the microscope, the instrument that gave green light to explore the unknown world of the infinitely small and that allowed the investigation of aerobiological particles. The natural philosopher Robert Hooke (1635–1703) made a number of accurate microscopic observations in the book Micrographia (1665). This work inspired the Dutch biologist Antonie van Leeuwenhoek (1632–1723), who between 1673 and 1683 first described bacteria, the animalcules (protozoa) or diatomaceous, as well as some yeasts and moulds, and assumed that they were transported by the wind along with floating dust. In 1682 Nehemiah Grew published his book "Anatomy of Plants", which contains the first known description of pollen (Fig. 4.3).

Birth of the Experimental Aerobiology

Around 1860, French biologist Louis Pasteur (1822–1895) began to study the bio-aerosols in the atmosphere. He built a series of specially designed glass bottles with a long curved neck ("a swan neck") with a spout at their end that could be sealed. The bottle was positioned in a way that the dust containing spores and germs were deposited on infusions that were sterilised by boiling in the bottle. When the air had been filtered or heated at a temperature high enough to kill all germs, the infusion remained sterile, but exposure to dust instead of air caused the deposition and growth of microorganisms on the infusion. Thanks to these experiments, Pasteur was able to demonstrate the heterogeneity of aerospora and the dispersion of germs in the atmosphere. Therefore, Pasteur is also considered a pioneer in aerobiology, the one who designed the first aerobiological experiments to examine the biological contents of dust in the air of Paris.

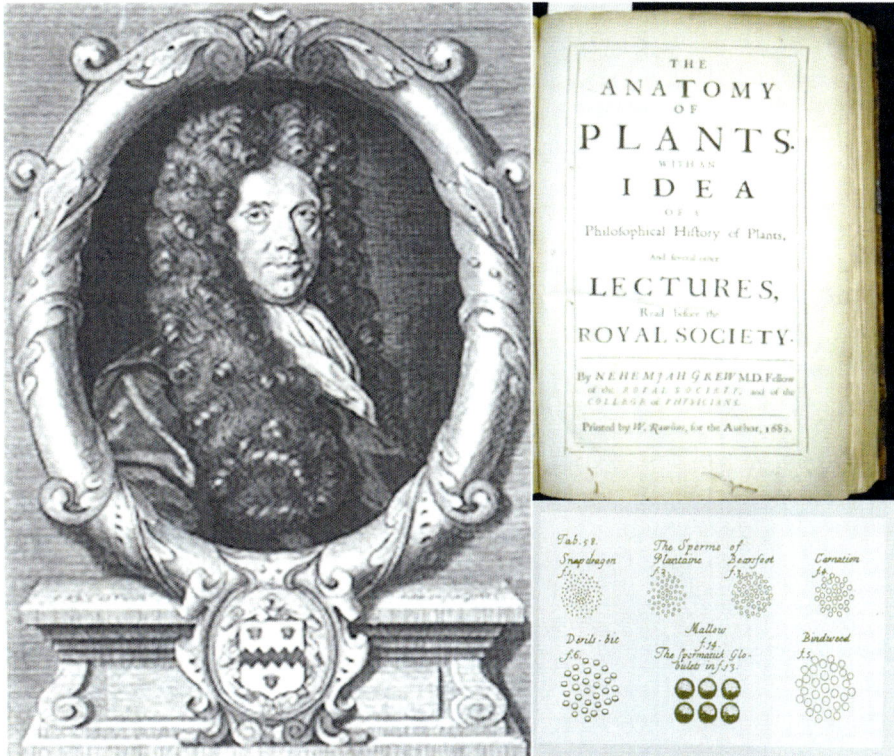

Fig. 4.3 Nehemiah Grew (*left*), who published the "Anatomy of plants" in 1682 (*top right*). His book contains a plate depicting the first description of pollen (*bottom right*)

His most direct descendant was without doubt the French physician Pierre Miquel (1850–1922), who continually monitored pollen and fungal spores in the atmosphere for years with various types of samplers of his own design. The results of his research were published in "Les organismes vivants de l'atmosphère" (1883), which presents interesting data and many graphs on the amount of fungal spores and bacteria in the air correlated with some meteorological parameters such as temperature, rainfall, humidity and wind speed.

During the nineteenth century, the German physician and naturalist Christian Gottfried Ehrenberg (1795–1876) worked as one of the founders of the science dealing with micro-organisms transported by the atmosphere. He examined samples of soil, water, sediments, atmospheric dust and rocks, describing hundreds of new species of bacteria, protozoa, diatoms, rotifers and fossils in about 400 publications. Irish physicist John Tyndall (1820–1893) became famous for his studies on light (the Tyndall effect) and sterilisation (Tindalizzazione). He also conducted aerobiological studies investigating the physical aspects of atmospheric particles and physiological growth of micro-organisms. Indeed, in essays on the floating matter of the air in relation to putrefaction and infection (1881), which represents his most important contribution to aerobiology, Tyndall studied the organic nature of dust in the atmosphere and claimed that epidemic diseases and putrefaction were caused by germs, thus refuting the abiogenesis. He was among the first to observe that seeds are abundantly present in and transported by the atmosphere. In the same century, Florentine Giorgio Roster (1843–1927), professor of biological chemistry and hygiene at the Royal Institute for Advanced Studies in Florence, can certainly be called one of the first experts on urban air pollution. He was the first true Italian aerobiologist.

Birth of "Aeropalynology" from "Aerobiology"

During the second half of the nineteenth century the study of micro-organisms expanded greatly. This stimulated an interest in studying also the pollen grains in the air, which led to the development of the field of aeropalynology. During this time, a number of hypotheses emerged, which tried to explain the seasonal appearance of respiratory allergy in humans. It should be recalled that in 1819 the English physician John Bostock (1773–1846) set forth in detail the clinical picture of the disease.

Origins of Pollen Monitoring

In 1873, another English physician, Charles H. Blackley (1820–1900), went down in history as the father of aerobiology. He was the first to show that pollen was present in large quantities in summer and was the cause of breathing problems (described as Catarrhus æstivus or summer catarrh), demonstrating the direct relationship between the amount of pollen in the air and the severity of symptoms. From 1866, he compiled pollen calendars of Manchester, UK, having counted the pollen that he collected with a self-made sampler under a microscope.

During the same period, Morrill Wyman (1812–1903) described the autumnal catarrh in the United States of America, which appeared each year in August and September. He attributed it to the flourishing of ragweed (*Ambrosia*). Elias Marsh (1835–1908), who created the first pollen calendars for ragweed pollen in 1875 for Paterson (New Jersey), must also be mentioned.

The factor that contributed most to the increase of knowledge about aerospora, after the invention of the microscope, was the introduction of air sampling devices. Starting from the first sampler of Pasteur (1860) and Blackley (circa 1866), during the second half of the nineteenth century, many scientists devised their own equipment to conduct aerobiological investigations. Most famous among them were the aeroscopes of Maddox (1870), Cunningham (1873) and Miquel (1878), the aeroscope recorders for bacteria and moulds (France, circa 1884) and the electric suction pump of Roster (1885). These instruments relied on the state of the art technology of their time, but more rudimentary samplers were in operation between the late 1800s and the early decades of the twentieth century. For instance, samplers were built with clothes pegs and a glass slide covered with glycerine fixed to a backing of wood and often covered with a small roof to shield them from the elements. These low precision instruments enjoyed some success primarily for economic reasons and they aroused the interest of many scholars for aerobiology and pollen monitoring. In 1946, O.C. Durham introduced his gravimetrically standardised sampler, which became the most frequently used instrument around the world for many years. It was also adopted by the "Pollen and Mold Committee of the American Academy of Allergy" as standard equipment. In 1967, the American Botanists Ogden and Raynor devised the rotary impact sampler "Rotoslide". In 1952, the Englishman Jim Hirst designed a sampler with a suction pump, which was the ancestor of the modern Burkard and Lanzoni traps. Hirst-type volumetric samplers are still operated by the majority of monitoring centres throughout the world.

Aerobiology in 1800–1900

Without doubt the discovery of numerous micro-organisms responsible for infectious diseases have to be listed among the most significant achievements of biomedical science between the second half of the 1800s and the first decades of the 1900s. German bacteriologist and hygienist Carl Flügge (1847–1923), a colleague of Robert Koch who devised many bacteriological techniques and established the bacterial causes of a number of infectious diseases, proved around 1890 that coughing and sneezing releases small droplets, defined as "droplets of Fluggi", which contain numerous pathogenic micro-organisms that remain suspended in the air and are responsible for the transmission of infectious diseases to new hosts. Around 1912, the Czech-Austrian botanist and chemist Hans Molisch (1856–1937) coined the term "Aeroplankton" by including it in all sorts of particles and especially those of biological origin such as pollen, fungal spores, algae, etc.

Birth of Pollen Monitoring Networks

Pollen monitoring at national level started for the first time in the US by O.C. Durham (1889–1967) in an attempt to correlate patient and pollen data of *Ambrosia* in 1928. Within a few years, the American network had expanded to more than 50 stations scattered throughout the country and the measurements were extended to all types of pollen. The network spread in a short time over much of the continent including Canada, Mexico and Cuba. Durham continued to coordinate this pollen recording network until the 1950s. In 1955, Durham supervised the publication of the first report of the monitoring network, which constituted the first of its kind. The first conference to deal exclusively with aerobiological topics was held in 1942 under the auspices of the "American Association for the Advancement of Science". After 1970, national monitoring networks were established in most European countries and the "European Aerobiology Society" was founded in 2008.

Aerobiology is still developing as a discipline and considerable advances are expected in the coming years. For example, many aerobiologists still use sampling equipment that is based on a design from the 1950s (Hirst 1952) and the analysis of samples is by light microscopy, which is labour intensive and extremely time consuming. One of the main areas of future development is expected to be in the monitoring of airborne organic particles with automated detection and analysis technologies (Rogers 2006), such as monitoring the allergen content of the air, DNA analysis or image analysis. There are also likely to be improvements in atmospheric modelling. Many of these advances are expected to be fuelled by an increasing need for aerobiological data due to the effect of climate change on human health (Huynen et al. 2003).

4.2.2.2 Monitoring Instruments and Sampling Methods in Aerobiology

Conventional Pollen Monitoring Instruments

Mullins and Emberlin (1997) reviewed strategies applied in sampling airborne pollen. The authors stressed that airborne pollen samplers, if they are to be effective, should be able to obtain volumetric record of all particles (5–50 μm), irrespective of the wind velocity. Numerous strategies including cylinder traps, sedimentation traps, impactors and filters have been used in trapping pollen.

Volumetric pollen samplers based on the Hirst design (Hirst 1952) are used as the standard method in many national networks for measuring the pollen concentration in the atmosphere. Air is sucked in through a 2 mm × 14 mm nozzle at a rate of 10 l/min. The rate of 10 l/min was chosen for operation in the field because efficiency varied less at different wind speeds than at an alternative rate of 17.5 l/min. At a suction rate of 10 l/min there was also less danger of obscuring the spore deposit with fine dust particles, which are less efficiently impacted with a lower velocity in the orifice (Hirst 1952). Airborne particles are deposited on a tape mounted on a drum, which is slowly turned by clockwork (Levetin et al. 2000). The sticky tape is then mostly stained with suitable dye and analysed under an optical light microscope, whereby the pollen and spores of different taxa are determined and their numbers per surface area counted according to standardised procedures (Makinen 1981; BAF 1995; Galán et al. 2007). The Hirst type volumetric pollen trap supplies pollen concentration data at a temporal resolution of up to 2 h. Relating the pollen counts with the exposure time, the number of pollen grains per cubic meter and time can be calculated. In order to avoid the distortion of the pollen count by local emissions, the traps are located on the roof of buildings, often at 12 m above street level (Winkler et al. 2001).

In the US, whirling arm samplers, such as the Rotorod, are preferred. In whirling arm samplers, airborne particles impact on one side of translucent 6 cm long square rods (1.6 × 1.6 mm) that whirl through the air at 2,400 rpm resulting at a sampling rate of about 120 l/min. The duration of whirling period determines the sampling period.

The Cour's method (Cour 1974) is based on the principle of passive sampling, which has the advantage that it does not require any power supply. The traps are mounted in 3 m above the surface. The pollen is collected passively on a filter. It is a volumetric method, because the results are expressed as the number of pollen grains per m^3 of air. The method has been applied for agronomical questions and has exclusively been used by the French Association for Ragweed Study (AFEDA) for the last 29 years. A pair of 20×20 cm filters (400 cm^2) composed of six sterile cellulose gauzes impregnated with silicone oil are exposed over a week. The trap functions with a weather vane; it is always oriented perpendicular to the wind direction and retains pollen grains transported by air currents. Wind speed, measured by an anemometer at the level of the trap, is used to compute the volumetric quantity filtered and the number of pollen grains per cubic meter of air. In the laboratory, the gauze with the collected material is dissolved in hydrochloric acid, hydrofluoric acid, acetone and potassium hydroxide. Chemical treatment empties pollen grains of the nucleus and the cytoplasm. The pollen grains are concentrated in the residue which is diluted in glycerine and homogenised. A volume of 50 μl of this dilution is deposited between a slide and a 22×50 mm cover slip and examined using light microscopy.

Pollen Counting Methods for Conventional Instruments

Whatever method for airborne pollen sampling is used, further analysis requires identification and quantification of registered pollen types. The identification of pollen requires knowledge on basic palynology (primarily pollen morphology), and it is performed either based on a comparison with reference microscopic slides or by using pollen identification keys and atlases. Due to their small size, pollen grains are commonly analysed under light microscope. The magnification is chosen so that pollen can be safely identified according to the characteristics specific for each taxon. The most widely used magnification in aerobiological monitoring is ×400.

Because the Hirst-type pollen samplers are the most common ones today, the various quantification methods will be briefly described below. When performing the quantitative analysis of a sample collected by the Hirst-type volumetric sampling procedure, the most accurate method would be to count pollen on the entire surface of the 24 h sample. However, from a routine pollen monitoring point of view and in the context of producing data for forecasting and informing public on the prevalent allergy risk, this would be unacceptably time-consuming. Therefore, three sub-sampling methods, which analyse only a fraction of 24-h slide, are proposed:

a. The random field method (Makinen 1981) considers the examination of a certain number of fields chosen at random from the entire daily surface, and counting the pollen present in each single field. This is probably the quickest method for slide analysis but, although it is good at estimating the daily mean concentration, it is unable to estimate short term concentrations (bi-hourly) (Kapyla and Penttinen 1981). Furthermore, the application of the random field method can result in underestimates or overestimates of the pollen concentration, because their depositing is not uniform on the tape, but depends on the particular biological cycle, environmental conditions and the type of pollen (Tormo et al. 1996).

b. The transverse traverses method (Emberlin et al. 1994) considers either the examination of successive tangent fields in 12 transversal lines or the examination of complete 12 transversal lines, separated by 4 mm distance from one another. In this way a line is read for every 2 h, enabling an estimation of both daily and bi-hourly pollen concentrations. The choice of the position of the lines could influence the final result obtained by this sub-sampling method, because pollen deposited within a very short time on the tape might be missed.
c. The longitudinal traverses method considers either the examination of successive tangent fields (Mandrioli 1990) positioned on 3 or 4 or 5 horizontal lines or the examination of 3 or 4 or 5 complete horizontal (Dominguez et al. 1991) lines separated by a space of about 2 mm. Although this method enables the estimation of both daily and bi-hourly pollen concentrations, it was noted that overestimates can arise from counting only the central regions of the slide, where most of the pollen is deposited (Tormo et al. 1996).

All of these sampling methods produce the pollen count expressed as concentration in pollen grains/m^3, which is calculated having in mind the suction rate of the used sampler and the ratio between the total sample surface and the sub-sampled surface of the slide based on the formula: pollen grains/m^3 = (pollen count*total sample surface)/(sub-sample surface on which pollen are counted*total volume of air sampled) (BAF 1995).

Since the main disadvantage of sub-sampling in the airborne pollen monitoring is the analysis of only a small proportion of the daily sample, Comtois and his colleagues (1999) checked the effect of sub-sampling on the accuracy of the quantitative analysis. They found that, when comparing concentrations obtained by counting the total slide surface versus counting only a fraction of it, none of the sub-sampling methods was able to reproduce the counting result of the total slide nor did the fractional counting give exactly the same result. Furthermore, the sub-sampling error was much higher than what is commonly believed, and it was significantly correlated with the abundance of pollen taxa on the sampled slide. Although each method has its advantages and disadvantages, all proposed methods enable a fairly good estimation of the whole biological population contained in a certain volume of air (Comtois et al. 1999; Sterling et al. 1999; Carinanos et al. 2000).

Automated Pollen Counting Techniques

For forecasting purposes, a continuous delivery of pollen counts and most suitably in an hourly time resolution would be very valuable. This cannot be achieved by manual counting systems, but could be obtained with automated pollen counting systems. In recent literature several different methods for automated pollen detection have been described:

1. Systems that make use of multifocal optical microscopic images of air samples collected by a conventional Hirst-type pollen sampler. A first step in automated counting of the pollen is the discrimination of the pollen grains from other airborne material in the images (Landsmeer et al. 2009; Bonton et al. 2001).

For subsequent identification of the pollen grain, several characteristic pollen features, including shape, statistical grey-level and specific pore/colpus features are extracted from the images by pattern recognition software tools (Boucher et al. 2002; Chen et al. 2006). These methods report various levels of success in identification of specific pollen types: 77% in samples from airborne pollen (Boucher et al. 2002) or 97.2% in samples containing three allergenic pollen taxa (Chen et al. 2006).
2. A fully integrated pollen sampling system that automatically collects, prepares and records by making use of a conventional light microscope (Ronneberger 2007). The method developed for the recognition of the pollen employs digitised images, using the grey-value of each pixel (Ronneberger et al. 2002). This system reached a recognition rate in "real world" samples of 84.3% (Ronneberger 2007). Up until now, it was not developed beyond a stage of a prototype and it did not reach the stage of becoming commercially available.
3. Other systems do not make use of digitised images of pollen, but are based on the technology of particle counters by laser light.
   - In the system described by Kawashima et al. (2007), pollen is characterised by the sideways and forward scattering of laser light. Air, containing the airborne particles, is passed through the optical system and irradiated by a laser beam. The scattering of light signals caused by the pollen grains is recorded in real time and processed by a computer. During a sampling period in late summer, pollen from nettle (*Urticaceae*), ragweed and grass (*Poaceae*) could be separated well by different scattering patterns. For other European pollen taxa, the system has not been tested yet.
   - In Japan, another real-time airborne pollen counter was developed by the company Kowa. The technology is based on a laser particle counter and on the characteristic distribution of pollen on the scattered diagram according to the grain size versus the fluorescent hue. In Japan, this counter is used by the Tokyo pollen information network systems (Suzuki et al. 2008).
   - Recently, a new methodology was presented on the 9th International Congress of aerobiology: the WIBS 4 (Wide-Issue Bioaerosol Spectrometer). This instrument combined information from laser light scattering with 2D-spectroscopic measurements. The instrument was successfully used in an area with a low diversity of pollen (Sodeau et al. 2010).
4. Another method is based on the Coulter counting principle (Zhang et al. 2005). Pollen was suspended in a KCl aqueous suspension and passed through a microchannel. The changes in conductance, due to the passing of the pollen, were recorded and analysed. In this system juniper (*Juniperus*) and grass pollen could be discriminated.

Airborne Allergen Monitoring Instruments

Allergologists have become increasingly interested in questions concerning the allergenic potency of pollen, which cannot be answered by the conventional pollen counts. For the purpose of allergen measurement in pollen new types of pollen traps needed to be developed, which now allow the application of immunological analysis methods like ELISA or immuno-fluorescence. Spieksma et al. (1990, 1999) used a high volume sampler that operated at 1,130 l/min equipped with a five-stage cascade impactor (HSV), sampling onto glass fibre impaction sheets to fractionate particles by size class. The HVS had a high capture efficiency (96–99 % for particles > 0.3 μm diameter) but sampling was not isokinetic or unidirectional. Also, Rantio-Lehtimaki et al. (1994) used a static, size selective bio-aerosol sampler (virtual impactor) with a flow rate of 18.5 l/min that collected samples onto three filters. Alternatively, Emberlin and Baboonian (1995) collected particles over a wide range of sizes in Eppendorf tubes for immunological analysis using a cyclone sampler operating at 16.6 l/min (Mullins and Emberlin 1997).

More recent approaches include the Coriolis® Delta of Bertin that works at a flow rate of 300 l/min and transfers pollen into a liquid collection media. It has an efficiency of 90–100% for particles with a diameter of 3 μm upwards (Bertin Technologies 2007).

Other samplers have been designed that are able to separate particulate matter from ambient air according to its size. The Andersen sampler aspirates 28 l/min of air (Andersen 1958). This sampler may contain up to six impaction stages, capturing particles with an aerodynamic size of PM > 8.2 μm, 10.4 μm > PM > 5.0 μm, 6.0 μm > PM > 3.0 μm, 3.5 μm > PM > 2.0 μm, 2.0 μm > PM > 1.0 μm and < 1 μm, enabling the study of the distribution of the allergenic particles according to their sizes (De Linares et al. 2007, 2010). The Chemvol® high-volume cascade impactor represents a more recent development (Demokritou et al. 2002). Ambient air is aspirated at 800 l/min and split into three identical airstreams, each impacting on a porous polyurethane impacting substrate. A cascade of stages with cut-offs at 10, 2.5, 1, 0.12 μm and an absolute stage can be mounted. Pollen and allergen are mostly detected in the stage PM > 10 μm with an additional 10–15% of allergen in the stage 10 μm > PM > 2.5 μm. The smallest stage (particles with a diameter of 2.5 μm > PM > 0.12 μm) is seldom used when working with pollen or allergen in ambient air as no allergen was found in this stage. The authors postulate that this could be due to concomitant collection of diesel particles at this stage, absorbing the possibly available allergen (Buters et al. 2010). This phenomenon could be shared by all samplers.

Personal Samplers

The Burkard Personal Volumetric Air Sampler (Burkard Manufacturing Co., Rickmansworth, UK) is a portable battery-powered device similar in operation to the Hirst trap. Air is drawn through a vertically orientated slot-shaped intake and impacted directly onto an adhesive covered microscope slide. The sample may thus be examined under the light microscope with little additional effort (Aizenberg et al. 2000). Whilst the term "personal sampler" is often applied to such portable devices, true personal samplers that sample from the breathing zone are designed to be worn by one person. For instance, the CIP 10, available from Arelco, was developed by

INERIS (Industrial Environment and Risks National Institute) to measure the exposure of workers to dusts in coal mines (Arelco 2004). The sampler can be set up to select either the respirable alveolar, thoracic or inhalable particle size fraction, collected in a foam filter. The Button Aerosol Sampler (SKC, Eighty Four, PA, USA) is a filter device that targets the inhalable particle fraction. Air is aspirated at a rate of 4 l/min through a curved, porous inlet designed to minimise wind sensitivity and promote uniform particle distribution. Sample analysis is by microscopy, immunoassay or PCR (SKC 2010). The Nasal Air Sampler is a passive impaction device worn inside the nasal cavity, thus truly measuring personal exposure rather than particle concentration in the breathing zone. Inhaled air is drawn past a specially designed adhesive strip onto which particles with sufficient inertia are impacted. Samples may be analysed using ELISA, or be mounted for microscopic examination (Graham et al. 2000).

Quality Standard

In order to produce comparable aerobiological data a Quality Control working group has been established within the European Aerobiology Society (EAS) in 2008, which intends to create an internationally recognized standard. As a first step towards such a quality standard a preliminary list of "Minimum requirements" for all monitoring stations involved in the European Aeroallergen Network (EAN) has been compiled. More details of the Quality Control working group discussions have been published in the IAA Newsletters (beginning with number 67: http://www.isac.cnr.it/aerobio/iaa/IAABULL.html).

4.2.2.3 Applications of Aerobiological Monitoring

Mandrioli and Ariatti (2001) stated that aerobiology must be considered as a discipline by itself as well as a tool for other disciplines. Because pollen grains of a number of plant species induce allergic reactions, aerobiology and in particular airborne pollen research developed as a discipline in close relation to medical research. In addition to that, allergenic airborne pollen is the only object of aerobiological research for which routine monitoring on a daily basis is widely accepted and often implemented in country, regional or continental networks, producing long time series of data available from numerous regions worldwide. Bryant (1989) identified pollen as "fingerprints of plants", which are closely related with flowering, reproduction and distribution of the vegetation. Such relationships provide the potential for airborne pollen observations to be used in wide spectra of studies dealing with anemophilous plant species.

Frenguelli (1998) reviewed the potential contribution of aerobiological observations to agriculture. There it can for instance be applied to yield forecasting. Two approaches have been followed: (1) the pollen index indicates the number of developed flowers, which correlates with the number of fruits in monoecious plants; (2) the pollen index indicates the amount of pollen available for fertilisation in anemophilous plants (a higher number of successful fertilisations is linked with a higher fruit productivity). These approaches were successfully implemented in forecasting the production of olives (*Olea Europeae*, Moriondo et al. 2001; Galán et al. 2004, 2008; García-Mozo et al. 2007a), grapes (*Vitis vinifera*, Cristofolini and Gottardini 2000) and even forest species, such us oaks (*Quercus*, García-Mozo et al. 2007b) or birch (Litschauer 2003).

On the other hand, with the development of Genetically Modified (GM) crops, there emerged the need for monitoring the potential gene flow (Stokstad 2002). In that context, aerobiological observations could offer data on pollen dispersal patterns primarily originating from wind pollinated plants, such as oak (Schueler et al. 2005), which show an intermediate to high potential of gene flow (Govindaraju 1988). In the case of oilseed rape (*Brassica*), although it is primarily insect-pollinated, aerobiological studies prove transport of airborne pollen (Fiorina et al. 2003). These findings address the potential risk of gene flow in GM oilseed rape threatening the surrounding crops. Therefore, aerobiology should support the development of methods to predict the concentration of viable pollen as a function of distance from the source. Such predictions should be a guideline for policy makers when defining distances needed between crop fields in order to prevent gene flow.

Aerobiological observations enable indirect analysis of plant responses (in particular linked to male reproductive systems) to environmental factors. For example, stress situations caused by frost could lead to male sterility, which would result in a lower pollen index, as it was observed in the case of the Mediterranean cork-oak (*Quercus suber*, García-Mozo et al. 2001). Also, climate change induced differences in the timing and duration of the flowering phenophase of anemophilous plants that could be observed by analysis of the duration, start and end dates of the airborne pollen season (Tedeschini et al. 2006; Frei and Gassner 2008). Such changes were also observed by Fotiou et al. (2011) who modelled the flowering process of north- and south-facing populations of spreading pellitory (*Parietaria judaica*) and compared the duration of the flowering and pollen seasons locally.

Pollen analysis has a great potential for providing a continuous record of pollen production going back thousands of years, due to the fact that pollen is produced in large quantities, dispersed widely and remains well preserved in wet anaerobic environments. In order to be able to reconstruct past climate, Autio and Hicks (2004) suggested deducing an empirical relation between pollen production and meteorological conditions. Applying sedimentational Tauber traps, these authors analysed the annual variation in pollen production in the studied area in relation to meteorological parameters. The network of stations for pollen monitoring using Hirst type volumetric traps is well distributed all over Europe. Furthermore significant correlations exist between data obtained by volumetric Burkard and sedimentation Tauber traps (Levetin et al. 2000). This allows the description of the influence of climate variability on pollen production and deposition, which supports the quantitative reconstruction of past climate.

4.2.2.4 The European Pollen Information System

At present, the Europe-wide pollen information system consists of two coherent units:

- the European aeroallergen database EAN
- the public web portal www.polleninfo.org

EAN Database

Both units form the basis of the aerobiological unit. Access to the database is restricted to a defined user group: those who contribute data have read access to all available data for internal use. By agreement, the use of data for publication or for commercial purposes without the consent of the data owners is prohibited. Founded in 1988, the database has got a new structure in 1999 and 2009.

The majority of data sets cover the last decade, but some time series extend back as far as to 1974. In total, over 700 monitoring stations from 38 countries are incorporated in the database, collecting pollen and spore counts of over 200 different taxa. This results in about 1 million annual reports (pollen types per station and year). The main goal concerning pollen information services is to assist in forecasting and to help in developing and testing forecast models. Another frequently used feature is providing data for multicentre studies in allergology, forestry, and climatology.

The Public Web Portal (www.polleninfo.org)

Since 1997, a European platform for pollen information has been provided for the public (www.cat.at/pollen/). A hierarchic structure allows navigation from general overviews down to highly specific local information contents. The up-to date information county by country are available both in English and in the country language(s). The main goal is to provide links to pollen information services on a common European pollen information portal – in particular for travellers and for vacation planning. A new platform www.polleninfo.org has been launched in mid April 2003 with financial support from epi Ltd. replacing the old cat.at/pollen/site.

4.3 Modelling and Forecasting of the Pollen Season

Although very different in the way of being observed and measured, phenological events and pollen counts can be traced back to the same phenomenon, the flowering of plants. Similarly, both kinds of data can in many respects be modelled with a similar set of observation-based models. Simple regression models can predict entry dates of phenological phases and likewise the start, peak and end of the pollen season or, given a greater number of independent variables, the day to day variability of the pollen counts. Phenological models will equally well predict the entry dates of phenological phases as well as the start, peak and end of the pollen season.

Phenological models are sometimes grouped into the class of process-based models, because they are built on assumptions rooted in experimental results on plant physiological responses to various environmental variables (Chuine et al. 2000). The other modelling approaches presented in this section are summarised under the term observation-based models, because no a-priori assumptions are involved. This includes regression models, time-series modelling and a survey about applications of artificial intelligence methods to pollen data. A discussion of special problems related to numerical pollen forecast completes the literature survey.

4.3.1 Observation-Based Models

Observation-orientated models relate pollen records (dependent variable) to one or more variables (independent variables) that can be measured or predicted, and are constructed without knowledge of the sources, emission or calculations of diffusion (Norris-Hill 1995). On the contrary, everything starts from the pollen counts being recorded at the pollen traps (the receptors). Pollen data usually produce mean daily values for the studied area and in some cases hourly or 2-h values. These data can be used for producing forecasts of day-to-day variations in pollen concentrations, or for predicting characteristics of the pollen season, such as start dates and severity.

All methods use certain mathematical tools in order to describe and imitate the behaviour of pollen count (its temporal and in some cases spatial variations); they may be applied for better understanding, description and knowledge concerning pollen season problems. The most rudimentary method for pollen forecasting is the pollen calendar. Recording seasonal variations in the timing and abundance of different pollen types is the first task undertaken by operators of pollen-monitoring sites. When sufficient data has been recorded, it is possible to relate temporal variations in pollen records (diurnal variations, daily average values or seasonal characteristics) to meteorological data, such as temperature and rainfall. This is achieved using a variety of statistical techniques that include correlation analysis, parametric Pearson correlations and non-parametric Spearman Rank correlation analysis (e.g. Stach et al. 2008; Smith and Emberlin 2006; Galán et al. 2000; Rodriguez Rajo et al. 2005), Factor Analysis (Makra et al. 2004) and hierarchical multiple regression analysis (Emberlin et al. 2007). The results from these analyses can be used to improve qualitative predictions or provide the theoretical rationale on which quantitative forecast models are built.

4.3.1.1 Regression Models

Regression analysis is an empirical technique that is often used in aerobiological studies. It is used to predict a score on one variable from a score on the other and as a result is often referred to as a causal method of statistical modelling. Causal models predict the future by modelling the past relationships between a dependent variable and one or more other variables called either independent or predictor variables. The goal of regression analysis is to arrive at the set of B values for independent variables, called regression coefficients, that bring predicted Y values from the equation as close as possible to the Y values obtained by measurement. The ability to find potentially causal relationships that not only predict but also explain the dependent variable makes regression analysis a very powerful technique (DeLurgio 1998; Tabachnick and Fidell 2001).

In simple bivariate (two-variable) linear regression, a straight line between the two variables is found. The best fitting straight line goes through the means of X and Y and minimises the sum of the squared distances between the data points and the line (Kinnear and Gray 1999; Tabachnick and Fidell 2001). This technique is useful when the dataset is not large enough for multiple regression analysis (Stach et al. 2008) as studies based on relatively small datasets are inappropriate for the multiple regression process because they result in inflated regression coefficients of determination (R^2 values) and inaccurately estimated coefficients. In such cases, it may be more suitable to use simple linear regression (Ong et al. 1997).

In multiple regression analysis, the values of the dependent variable are estimated from those of two or more independent variables (Kinnear and Gray 1999). Multiple regression analysis makes a number of assumptions about the data (such as normality, linearity and homoscedasticity) and is not forgiving if they are violated (Kinnear and Gray 1999; Pallant 2001). Many of the problems associated with these factors can be addressed by transforming the data prior to analysis. Different methods of data transformation have been used by a variety of authors, mainly when predicting daily variations in pollen counts; these include square root (Smith and Emberlin 2005), lognormal (Alcazar et al. 2004) and log10 (Stach et al. 2008). Toro et al. (1998) investigated the use of different methods of data transformation. The authors transformed daily mean grass pollen data (x_t) expressed as the number of pollen grains/m³ into different scales before attempting to construct regression models: $\log(x_t+1)$, $\sqrt{x_t}$, and $\ln((x_t*1,000, x)+1)$. The latter method is the transformation proposed by Moseholm et al. (1987). Toro et al. (1998) found that a regression equation obtained from data transformed by square root usually resulted in a better prediction because substantial errors can be introduced in de-transforming the data to the usual scale (pollen grains/m³), although its R^2 value was lower than equations obtained with other transformations. Furthermore, it was suggested that the transformation proposed by Moseholm et al. (1987) should not be used to construct short-term predictive models because the margin of error is further increased by the inclusion of an additional factor $\sum x$ (total of pollen grains for the whole season), which has to be predicted before it can be transformed (Toro et al. 1998). In addition to data transformations, datasets also have to be cleaned prior to analysis in order to remove or alter univariate and multivariate outliers that regression models are sensitive to, as they can distort statistics (DeLurgio 1998; Tabachnick and Fidell 2001).

There are several methods of multiple regression analysis commonly used in observation based modelling. Step-wise multiple regression analysis has often been used in aerobiological studies (Galán et al. 1995; Goldberg et al. 1988; Bringfelt et al. 1982; Moseholm et al. 1987). An alternative method used in aerobiological research is standard multiple regression analysis (Stach et al. 2008; Smith and Emberlin 2005; Makra et al. 2004), where all the independent variables are entered into the equation simultaneously and each independent variable is evaluated in terms of its predictive power compared to that of the other independent variables (Kinnear and Gray 1999; Pallant 2001). However, non-linear statistics should be used, if the data do not show a normal distribution (Toro et al. 1998), such as polynomic regression (Antepara et al. 1995) or semi-parametric Poisson regression models, where the variance of the data is proportional to its mean (Stark et al. 1997; Erbas et al. 2007).

Regression analysis has been used for predicting daily values (Stach et al. 2008; Smith and Emberlin 2005), as well as the start (Emberlin et al. 1993a; Laaidi 2001a, b; Galán et al. 1998, 2001a; Frenguelli et al. 1989; Davies and Smith 1973), peak (Orlandi et al. 2006), duration (Laaidi et al. 2003) and severity (Emberlin et al. 1993b; Galán et al. 2001b) of pollen seasons and the beginning of flowering (Crepinsek et al. 2006). A number of different independent variables were used in these analyses. Variables that affect the timing of pollen release from allergenic plants are used to predict the start of pollen seasons and the beginning of flowering, such as monthly (Stach et al. 2008; Emberlin et al. 1993a; Galán et al. 1998, 2001a; Frenguelli et al. 1989; Davies and Smith 1973; Crepinsek et al. 2006) or 10-day periods (sometimes referred to as decades-of-days; Stach et al. 2008; Smith and Emberlin 2005; Spieksma and Nikkels 1998) of meteorological data, as well as winter averages of the North Atlantic Oscillation (Stach et al. 2008; Spieksma and Nikkels 1998). Similar variables are also used when attempting to predict the severity of seasons (Emberlin et al. 1999; Laaidi 2001b; Galán et al. 1998). Observed pollen season starting dates, which occur before the start dates of the pollen season to be modelled, can also be considered as only or additional independent regression parameter (Norris-Hill 1998).

A variety of different independent variables have also been used to predict daily average pollen counts, and include minimum (Toro et al. 1998), maximum (Iglesias et al. 2007; Rodriguez Rajo et al. 2004, 2005; Mendez et al. 2005; Toro et al. 1998) and mean (Toro et al. 1998; Rodriguez Rajo et al. 2004; Goldberg et al. 1988) daily temperatures, rainfall (Stark et al. 1997; Rodriguez Rajo et al. 2004; Toro et al. 1998), relative humidity (Stach et al. 2008; Smith and Emberlin 2005; Toro et al. 1998), sunshine hours (Stach et al. 2008; Toro et al. 1998), wind speed (Bringfelt et al. 1982) and also direction and persistence (Damialis et al. 2005), and the amount of pollen recorded in the previous days (Stach et al. 2008; Smith and Emberlin 2005; Iglesias et al. 2007; Sánchez-Mesa et al. 2002; Rodriguez Rajo et al. 2004, 2005; Mendez et al. 2005). The division of grass pollen seasons prior to analysis has become an accepted methodology when attempting to predict daily pollen counts, because the relationship between pollen counts and environmental data tends to change during the pre-peak and post-peak periods (Sánchez-Mesa et al. 2003; Galán et al. 1995; Toro et al. 1998). Examples include splitting the grass pollen season into two and defined the pre-peak period as dating from the beginning of the main pollen season to the peak day itself (Toro et al. 1998), or dividing the grass pollen seasons into three (pre-peak, peak and post-peak periods of pollen release) because pollen counts around the peak day behave as one population, whereas pollen counts from the pre-peak and post-peak periods can be treated separately (Smith and Emberlin 2005).

When selecting the independent variables to enter into regression analysis, it is important to avoid multicollinearity, which exists when the independent variables are highly correlated. For instance, highly correlated variables (such as maximum, minimum and mean daily air temperatures) should not be included in the same multiple regression. It is also important not to attempt to assess prediction accuracy with the same data used to construct the model (Stark et al. 1997).

4.3.1.2 Time Series Modelling

Modelling and forecasting of pollen counts based on regression equations is simple and straightforward and can be carried out with any statistical package providing multivariate regression procedures. However, it suffers from a number of disadvantages. The most important one is that usually time is kept fixed and different time periods are handled as different variables, which prevents any exploration of the role of timescales. In order to do this, it is usually preferable to use time-series approaches.

The classical time-series method for the analysis and forecasting of pollen levels is the Box-Jenkins approach (Box et al. 1994). This is based on the successive refinement of the model by fitting different deterministic and stochastic components of variability. First, the average value is found and then subtracted from the series. Then, a trend is fitted to these centred residuals and the values of this trend-line are subtracted to get the detrended values. Next, successive years are stacked on a daily, weekly or monthly basis and we find the average over all years for each day, week or month. For example, if we use a monthly basis, the model $S(m)$ for $m=$ June is the average of the detrended values for June averaged over all years. Finally, after this cyclic model has been subtracted from the data, the autocorrelation structure is fitted using ARIMA techniques (Box et al. 1994). Ideally, at the end of this stage, the residuals should be free of correlation. This is a widely-used approach, which can be regarded as standard, and features in several aerobiological studies (Moseholm et al. 1987; Stephen et al. 1990; Rodriguez-Rajo et al. 2006; Aznarte et al. 2007). The model may thus be described as:

$$X(t) = M + Tr(t) + S(m_t) + \varepsilon(t) \qquad (4.2)$$

Here, $Tr(t)$ is the trend component at time t, a steadily rising or falling background that is sometimes observed in pollen records (Damialis et al. 2007), and M is the mean value of detrended series. The seasonal component is $S(m_t)$, where m_t is the month (or week or day) of the year. Finally $\varepsilon(t)$ is the residual noise, whose structure can be described by an ARIMA model. It is often desirable to transform data, for example using the logarithm or square-root, before carrying out a Box-Jenkins analysis.

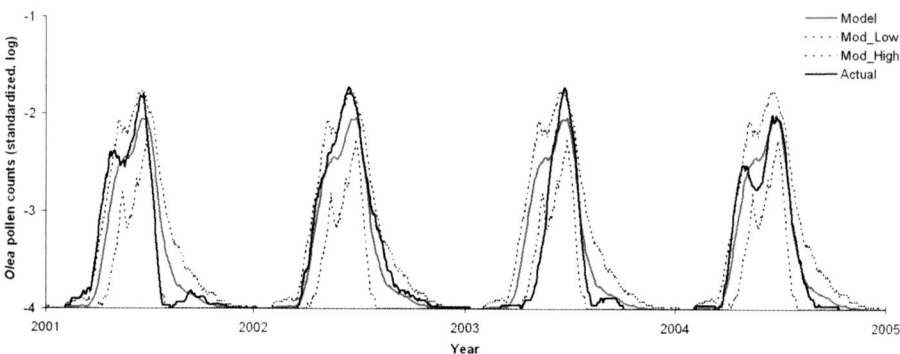

Fig. 4.4 Application of the Box-Jenkins method to forecast olive pollen levels in Thessaloniki for the years 2001–2005 based on a model parameterized for the years 1995–2000. The data and predictions are resolved on a weekly basis per m³ of air. The units in the range have been standardised and logarithmically transformed. We plot $X_t = \log(P_t + 1)$, where P_t is the pollen level, against time t

This is especially true in the case of pollen data, where variability is strongly skewed. Covariates such as temperature and humidity can also be built into this forecasting approach (Moseholm et al. 1987).

Figure 4.4 shows an example of this approach applied to a pollen series (olive) from Thessaloniki, for the years 2001–2005. It is clear that this model can be used to fit pollen data and it should be possible to make predictions of future pollen concentrations, at least for certain groups.

However, the basic Box-Jenkins approach can also be quite limited. Figure 4.5 illustrates some major difficulties of forecasting in the context of pollen dynamics. Noting that the scale is logarithmic, where the model fails to predict, the divergence can be very large (e.g. end of 2001, first cycle). Seasons can be irregular, so in 2003 (third cycle), the season begins later and is shorter than expected. Variations in the seasonality of pollen concentrations are primarily driven by ecological factors that may in turn be driven by climate or ecological interactions leading to shifts of the peak (Ocana-Peinado et al. 2008). The statistical behaviour of the counts does not conform to the usual patterns. For example, though data are extremely right skewed, a log-transformation does not remove it (Aznarte et al. 2007). The background "noise" generates variability on all scales. It cannot be removed by smoothing. For example, in Fig. 4.4, it is clear that the cleft in the peak moves about on a scale of weeks. This multi-scale behaviour (Halley and Inchausti 2004) has major consequences for the design of forecast procedures. In Fig. 4.4, the data have been normalised by year, that is, each year's data have been scaled so as to lead to the same total for each year. Figure 4.5 shows weekly data, without this pre-processing, for the family Poaceae, in Thessaloniki, over the years 1996–2004. Yearly counts vary considerably, so this introduces an extra twist that illustrates the direction we need to take in pollen season forecasting.

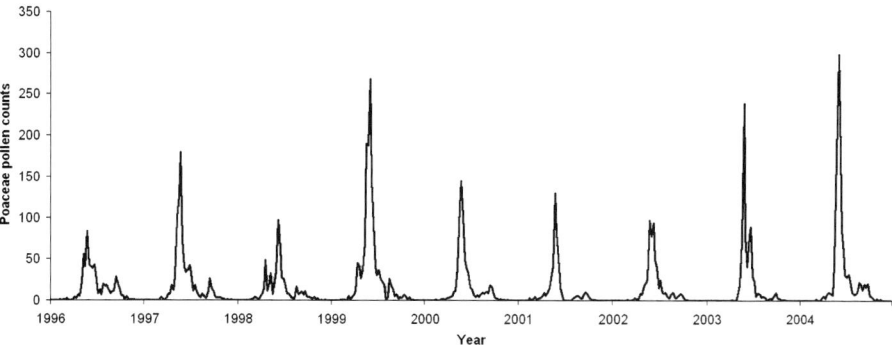

Fig. 4.5 Weekly counts of pollen of the Poaceae family in Thessaloniki, for the years 1996–2004 inclusive. The data are resolved on a weekly basis per m³ of air. Because of the large between-year variability, no model curve is drawn. The time axis begins on January 1, 1996 and ends December 31, 2004. There is no transformation of pollen levels P_t

There is clearly a year to year variation of pollen counts. Thus, we have a more suitable model for pollen prediction:

$$X(t) = M + Tr(t) + A(y_t) \cdot S(m_t) + \varepsilon_t \tag{4.3}$$

Here, the seasonal component is the product of a random annual component A for the year y_t, and the cyclic seasonal component that depends only on the month, m, as before. This modified Box-Jenkins model addresses some of the multiple-scale variability that needs to be included in any attempt to forecast pollen levels.

Other time-series based approaches include neural networks (Arizmendi et al. 1993) or neuro-fuzzy approaches (Aznarte et al. 2007) and functional regression (Ocana-Peinado et al. 2008). To assess the relative success of forecasting, various comparisons have been made, but as yet have been mainly confined to specific places, taxa and timescales. Thus, there is a need for considerable work on the multiple-scale nature of pollen variability, which can be addressed within a suitably modified Box-Jenkins framework or using other time-series based methods. It is too early to say, which of the methods is the best. Investigations into appropriate measures of deviation of models from data are needed to quantify the relative merit of different models.

4.3.1.3 Computational Intelligence

Environmental data are very complex to model due to underlying interrelations between numerous variables of different type. However, as standard statistical techniques may possibly fail to adequately model complex, non-linear phenomena and chemical procedures, the application of Computational Intelligence (CI) methods for forecasting of a wide range of air pollutants and their concentrations at various time scales, perform usually well in atmospheric sciences.

Computational Intelligence methods, such as Neural Networks, Classifications and Regression Trees, Self Organising Maps, Support Vector Machines, etc. are advanced tools for knowledge discovery and forecasting parameters of interest. CI methods can be used for multiple tasks, such as classification, numerical prediction, clustering etc., while the main advantage of these methods is the accuracy combined with computational efficiency. CI techniques such as Artificial Neural Networks (ANNs), Classification and Regression Trees (CART) and Support Vector Machines have already been applied for analysing and forecasting air pollution parameters (Slini et al. 2006; Karatzas and Kaltsatos 2007). The performance of CI methods is similar or in some cases better compared with that of deterministic models, when applied to the atmospheric environment (Kukkonen et al. 2003), thus CI methods are appropriate tools to be applied for the development of operational forecasting, among others.

The application of CI methods for analysing and modelling pollen concentration data has increased in the recent years, since it was identified that methods, such as Artificial Neural Networks and Neuro-Fuzzy models, clearly outperform traditional linear methods in forecasting tasks (Sánchez-Mesa et al. 2002; Ranzi et al. 2003; Aznarte et al. 2007). Most of these applications have taken into account daily average pollen concentrations and meteorological parameters, aiming at forecasting pollen concentration of certain species 1 to 5 days ahead. CI methods have also been applied in order to investigate the relationships between pollen and air pollution with very promising results (Voukantsis et al. 2010), while papers published concerning the use of CI methods for analysing and forecasting pollen data are appearing more and more frequently (Degaudenzi and Arizmendi 1998; Aznarte et al. 2007; Voukantsis et al. 2011).

The rest of this chapter presents a short description of some of the most popular CI methods applied in atmospheric sciences are included, based on Tzima et al. (2007).

Decision trees usually assume that the function f(x) to be learned, is constant in intervals defined by splits on the individual attribute axes. Internal nodes of the tree implement split decisions based on impurity measures (defined in terms of the class distribution of records before and after splitting), while leaf nodes define "neighborhoods" of records, each of which is assigned a specific class attribute value (class label).

In neural networks the target function f(x) is implemented as a composition of other functions $g_i(x)$:

$$f(x) = K\left[\sum_i w_i g_i(x)\right], \tag{4.4}$$

where K is some predefined transfer function, such as a member of the sigmoid family (typical for multi-layer perceptron networks) or a radial basis function (as in RBF Networks). Given a specific task to solve, and a class of functions F, the set of observations is used in order to find the optimal target function that minimizes a predefined cost function. For CI applications, where the solution is dependent on the training data, the cost must necessarily be a function of the observations, such as the mean-squared error between the network's output, f(x), and the target class value y over all the example pairs.

Neural networks can be effectively applied to classification problems, even in the presence of large datasets. However, the resulting model's robustness depends heavily on the appropriate choice of the model (network size and topology), the cost function and the learning algorithm. Inappropriate implementations, combined with bad choice of a training data set, typically impair the classifier's generalisation ability or lead to model overfitting.

The self-organising map (SOM) also referred to as Kohonen Network, is a subtype of artificial neural networks. SOM is based on competitive learning, which runs in an unsupervised manner, aiming at selecting the so called winning neuron that best matches a vector of the input space. In this way, "a continuous input space of activation patterns is mapped onto a discrete output space of neurons by a process of competition of the neurons in the network" (Haykin 1999). This makes SOM one of best methods for modelling a knowledge domain with the aim to reveal topological interrelations and hidden knowledge, via the visualisation of the network's neurons.

SOM is capable of learning from complex, multi-dimensional data without specification of the output, thus making it very appropriate to be applied in pollen and atmospheric quality data. The resulting nonlinear classification consists of clusters that can be interpreted via visual inspection. The method's unsupervised learning algorithm involves a self-organising process to identify the weight factors in the network, reflecting the main features of the input data as a whole. In that process, the input data is mapped onto a lower dimensional (usually two-dimensional) map of output nodes with little or no knowledge of the data structure being required (a characteristic of the method that renders it appropriate for data compression). The output nodes (neurons) represent groups of entities with similar properties, revealing possible clusters in the input data. It should be noted that, although the method is unsupervised in learning, the number of the output nodes and configuration of the output map (number of nodes included, etc.), need to be specified before the learning process.

Rule-based algorithms apply "if...then..." rules of the form (Condition)->y, where Condition is a conjunction of observable attributes and y is the class label, where the values are put. The collection of rules may contain rules that are:

- mutually exclusive (each record is covered by at most one rule) or not (the rule set is ordered or employs a voting scheme);
- exhaustive (each record is covered by at least one rule) or not (a record may not trigger any rules and be assigned to a default class).

Among others, advantages of rule-based algorithms include the fact that they are easy to interpret and highly expressive. Moreover, they are fast to generate and can classify new instances rapidly, with a performance comparable to that of decision trees.

Bayesian classifiers compute conditional probability distributions of future observables given already observed data. More specifically, the analysis usually begins with a full probability model – a joint probability distribution for all attributes including the class – and then uses Bayes' theorem to compute the posterior probability distribution of the class attribute. The classifier's prediction is the value of the class attribute that maximizes this posterior probability. Naïve Bayes classifiers additionally assume independence among all attributes, given the class, when computing posterior probabilities.

Despite the fact that the independence assumptions made by Naïve Bayes classifiers are often inaccurate, the latter have several interesting properties that may prove useful in practice. They are robust to isolated noise points and irrelevant attributes and can handle missing values by ignoring the instance during probability estimate calculations. Moreover, their independence assumption allows for each distribution to be estimated independently as a one dimensional distribution, thus alleviating problems such as the "curse of dimensionality". Finally, another advantage of all Bayesian classifiers is their conceptual and interpretation simplicity, rendering them appropriate for use by non-domain experts.

Support vector machines (SVMs) were introduced by Vapnik in 1963. The original algorithm defines a method for finding the optimal hyperplane that separates, with the maximum margin, a set of positive examples from a set of negative examples. Thus, it is a linear classifier. A later extension of the algorithm, though, proposes the use of the "kernel trick" to maximum-margin hyperplanes: every dot product in the original algorithm is replaced by a non-linear kernel function, allowing the transformation of the feature set to a high-dimensional space, whose hyperplanes are no longer linear in the original input space.

4.3.2 Process-Based Phenological Models

Phenological models determine the entry dates of phenological phases as function of environmental factors. First efforts date back to the idea of Reaumur (1735), who explained spatial and temporal differences in phenological entry dates as a result of different heat sums during plant development. During the past years a number of review articles about phenological modelling have been published so that we will keep this sub-chapter as concise as possible. In addition, we will evaluate the current phenological modelling scene and touch a number of problems, which have been discussed only marginally in the literature. Special sections will deal with the application of phenological models to seasonal pollen modelling. More extensive reviews about phenological models are to be found in Hänninen (1995), Menzel (1997), Chuine et al. (1998, 2003), Chuine (2000), in Schwartz (2003), Schaber and Badeck (2003), Chuine and Belmonte (2004) and Linkosalo et al. (2008). The phenological models introduced here refer to plants of medium to high latitudes, whereas low latitude phenological modelling requires different approaches (Hudson et al. 2010).

Generally two different kinds of phenological models exist. The purely statistical-empirical approach relates the entry dates usually with mean temperatures over certain time periods preceding the phenological occurrence date. No mechanistic details of the relationship between plant physiology and environment are considered. The second kind of phenological models, also called process based models, is based on experimental studies about possible mechanisms, which are supposed to govern the relationship between plant physiology and the environment. During the vegetation period the buds for the next season are created, but remain in an inactive state called dormancy. During autumn and early winter dormancy is overcome through chilling. After dormancy has been released, the development of leaf, flower and shoot buds begins in the following spring. The timing of these events is the crucial point for the plants. If it occurs too early, frost might damage the plant organs, if it happens too late, the plant suffers from a loss of photosynthetic potential (Linkosalo et al. 2006). Although the high-temperature requirements of bud burst and flowering are well established, there is great uncertainty about the mechanisms enabling bud development. Among the basic factors governing the seasonal plant development are:

1. chilling temperature
2. forcing temperature
3. photoperiod
4. water availability

Phenological models may consider at least one or any combination of these four factors. There follows a short description of phenological models, which use individual factors or combinations of them, as found in literature.

4.3.2.1 Models Considering Thermal Forcing Only

The Thermal Time or Spring Warming model or Growing Degree Day models ignore the chilling requirement and consider only the temperature forcing in spring (Linkosalo et al. 2008). They describe the start of bud development in spring, omitting the dormancy phase altogether. Thermal Time models implicitly assume that environmental conditions required to release dormancy have been met before the starting date of temperature sum accumulation. The start date of temperature summation can be fixed or determined via an inverse procedure separately for each phase and station. The entry date t_2 is a function of the starting date of temperature accumulation t_1, and the temperature sum F between t_1 and t_2 above the temperature threshold T_b: $t_2 = f(t_1, T_b, F)$. The state of forcing F (forcing units usually in degree days) is represented by

$$F = \int_{t_1}^{t_2} R_f(T) dt \qquad (4.5)$$

where t_1 is the starting date of temperature accumulation, t_2 the entry date of the phase and R_f the rates of forcing, which are defined as

$$R_f = \begin{cases} 0 & \text{if } T < T_b \\ TT_b & \text{if } T \geq T_b \end{cases} \qquad (4.6)$$

where T is the daily mean temperature and T_b the temperature threshold.

A number of variants and simplified versions of Thermal Time model have been applied to the pollen period and compared with each other (García-Mozo et al. 2000). For instance the threshold temperature may be omitted and daily maximum temperatures may be added up from the end of the chilling period to the entry date. Most works prefer the daily mean temperature as input for the calculation of temperature accumulation, whereas some use alternatively the maximum daily temperature. Weighting of the temperature before summation has also been considered by some authors (Clot 1998; 2001).

An interesting alternative formulation of the thermal accumulation method is described in Aono and Kazui (2008), who consider the thermal energy accumulated during the developmental period of the plants. The daily DTS (number of days transformed to standard temperature) value is a ratio expressing the amount of growth that occurs in one day at the actual daily mean temperature with respect to that which occurs at the temperature set as a standard. The authors claim their method to produce a generally 0.4 days lower RMSE in comparison to the conventional degree day models.

4.3.2.2 Models Considering Chilling Only

For some species it turned out that there exists a useful relationship between the date, when the required chilling hours have been accumulated and the onset date of bud burst. Orlandi et al. (2004) tested two different chilling models and their ability to predict the onset date of bud burst in olive trees. Similarly, it appears that the chilling requirements exert a greater influence than the heat requirements for the start of black alder pollen release in the Mediterranean region (Jato et al. 2000)

4.3.2.3 Models Considering Thermal Forcing and Chilling

The following models include a description of the dormancy and the thermal forcing factor. Chilling requirement must be met before ontogenetic development can commence. Here follows a short description of each of the thermal forcing and chilling models with their underlying speculative assumptions:

- Sequential models are based on the assumption that chilling units must have been accumulated completely before accumulation of heat units can commence (Linkosalo et al. 2006).
- Parallel models consider chilling and forcing factors too, but assume that both processes may proceed in parallel.
- Alternating models assume a negative exponential relationship between the sum of forcing units required for completion of quiescence and the sum of chilling units received (Chuine et al. 2003)
- The deepening-rest model stipulates that the state of chilling must increase, before it can loosen its block on assimilation of heat units again.

- The Four Phases model assumes three phases of dormancy (pre-rest, true-rest and post-rest) before the phase of quiescence. This is formalised by an increasing temperature threshold for forcing during pre-rest and a decreasing temperature threshold for forcing during post-rest, and buds cannot respond to forcing temperature at all during true rest (Chuine et al. 2003).

4.3.2.4 Models Considering Thermal Forcing, Chilling and Photoperiod

The evidence showing that the dormancy is released solely by the chilling requirement is far from solid (Linkosalo 2000). There is in fact evidence that increasing day length has to do with the onset of ontogenetic development.

4.3.2.5 Models for Herbaceous Species Considering Temperature, Photoperiod and Soil Water Availability

Among the commonly recognised environmental factors governing the beginning of flowering of grasses, such as temperature and photoperiod, water availability plays a dominant role, especially in Mediterranean areas (Clary et al. 2004). Although various authors have developed models for predicting daily grass-pollen concentrations (Moseholm et al. 1987; Emberlin et al. 1999; Sánchez-Mesa et al. 2002), few papers have addressed the development of models to forecast the main pollen-season dates, i.e. start date and peak date (Clot 1998; Chuine and Belmonte 2004; Laaidi 2001a; Stach et al. 2008; García-Mozo et al. 2009). The main difficulty in developing forecasting models for this taxon is that grass pollen counts are an amalgam of pollen from many species, and pollen release dynamics prompt a large number of peaks (Férnández-González et al. 1999; Emberlin et al. 1999). García-Mozo et al. (2009) developed process-based models to predict the start- and the peak-date of the grass pollen season. The models take into account the effects of temperature, photoperiod and water availability on the timing of grass flowering in Spain. Apart from predicting the pollen-season start and peak dates, process-based models provide information on (i) the Poaceae response to weather-related factors, (ii) the period during which these factors affect grass growth, and (iii) the relationship between photoperiod, temperature and water availability for the flowering of grasses.

4.3.2.6 Generalised Phenological Models

There are two even more generalised models, which can be summarised in a separate group (Linkosalo et al. 2008). They are based on the idea that a model with a flexible structure will conform to the essential features of phenological control when fitted to a dataset.

- The Unified model was developed by Chuine (2000), where various weighting functions regulate the relationship between temperature and the development of

chilling and forcing. Weighting is summarised by two generalised functions with all together nine parameters, which must be determined by a numerical optimisation procedure.
- The Promotor-Inhibitor model by Schaber and Badeck (2003) is based on the idea that a hypothesised balance or ratio between promotory and inhibitory agents determines the physiological state of development of the plant and its reaction to external driving forces.

4.3.2.7 Thermal Time Models Incorporating Real Time Phenological and Pollen Data

A number of authors experimented successfully with the idea to hinge the Thermal Time model forecast on the observation of entry dates of previously flowering species (Driessen and Moelands 1985; Driessen et al. 1989; Frenguelli and Bricchi 1998; Norris-Hill 1998). This approach requires real time phenological observations or pollen counts of the preceding species, which might not be readily available everywhere.

4.3.2.8 Optimisation of Model Parameters

From the schematic representation of for example the Thermal Time model $t_2 = f(t_1, T_b, F)$ as explained above, it becomes clear that the three parameters t_1, T_b and F have to be determined such that the phenological model yields best results with lowest error values.

In many cases it may be sufficient to work with a-priori fixed parameter values. From experimental evidence, for instance it appears that for a great number of species in the temperate zone 5°C represents an optimum threshold T_b (Frenguelli and Bricchi 1998; Jato et al. 2000). For colder regions 0°C has been suggested (Gerad-Peeters 1998; Clot 1998) and in warmer climates some authors have proposed 12.5°C (Alcalá and Barranco 1992; Galán et al. 2001a).

It is difficult to generalise model parameter values across a number of species or over large areas, as it turned out that the threshold temperature of certain species and phases depends on environmental factors, like the bio-climatic region and altitude. In the case of olive, the optimum threshold temperature was 10°C in Malaga (5 m.a.s.) but 12.5 in Córdoba (123 m.a.s.) within the same bio-climatic belt (thermo-Mediterranean). Different plants in the same locality have different temperature requirements: i.e. in Córdoba province, it has been defined at 11°C for oaks (early spring flowering), and 12.5°C for olive (late spring flowering, Galán et al. 2001a, 2005; García-Mozo et al. 2002. Ribeiro et al. (2006) suggest around 9°C for the olive in Portugal, but Orlandi et al. (2005) use temperatures between 7 and 15°C for the olive, in Spain and Italy, respectively.

For the start date of temperature accumulation t_1, a number of suggestions can be found in the literature. In Europe, 1 January has been suggested for early flowering species (Frenguelli and Bricchi 1998; Ribeiro et al. 2006; Orlandi et al. 2005), whereas 1 March for late flowering ones (Alba and Díaz de la Guardia 1998; Clot 1998).

Today's numerical techniques make it comparatively easy, to determine the optimum values of model parameters. A number of authors rely on LUT (Look Up Table) methods to optimise model parameter values. The phenological model is applied through a range of parameter values with a set increment. With the help of a cost function, which can be the squared error, standard deviation or root mean square error (RMSE), the optimum values are determined (for instance Van Vliet et al. 2002; Crepinsek et al. 2006; Migliavacca et al. 2008). If the dimensionality of the phenological model is not too high (not more than two or three model parameters), the LUT results can be plotted and viewed (Fig. 4.6). In case of phenological models, it turned out that the RSME values form a valley with any number of very differing optimum parameter values along the valley floor with similarly low RSME values. One may conclude that there is no unique solution to the problem and any one from an infinite number of parameter value sets is equally well describing the phenological behaviour. The objective selection of the most adequate set of parameter values constitutes a problem, which has not been solved yet. One way to arbitrarily select the most appropriate set is to calculate a mean from a certain fraction of the best parameter values.

Apart from the graphical visualisation of the minimisation problem, another advantage of the LUT method consists in its robustness. But the computational effort increases quickly with each additional dimension respectively model parameter to be optimised.

The other group of works prefer numerical methods to find the optimum parameter values of the phenological model. Kramer (1994, 1995) applies various numerical procedures (FITNONLINEAR from the GENSTAT package, NAG subroutine E04FCF, Downhill Simplex from Press et al. (1992) or a Newton approximation), whereas Linkosalo et al. (2009) rely on the direct search algorithm of Hooke and Jeeves (http://www.netlib.org./opt/hooke.c). The Simulated Annealing Algorithm provided in Press et al. (1992) has gained some popularity among phenological modellers (for instance Chuine et al. 1998 or Schaber and Badeck 2003). In comparison with the LUT methods, the numerical algorithms are computationally much more efficient, especially, if the phenological model needs considerably more than two or three parameters to be optimised. On the other hand, the procedure may converge or not to the global minimum. The Simulated Annealing Algorithm tries to overcome that problem by introducing a random fluctuation, which helps the procedure to step over local minima. In some cases, it might be an advantage to apply more than one method, especially, if the results of a numerical procedure are doubtful.

To assess the model quality, Galán et al. (2005) provide Root Mean Square Errors (RMSE) for a temperature sum model for olive in Spain. The RMSE range between 6.2 and 7.8. The mean absolute difference between the modelled and the actual date was 4.8 days using independent data. In a similar but older study (Galán et al. 2001a), this number was 4.7 for olive (though not tested on independent data).

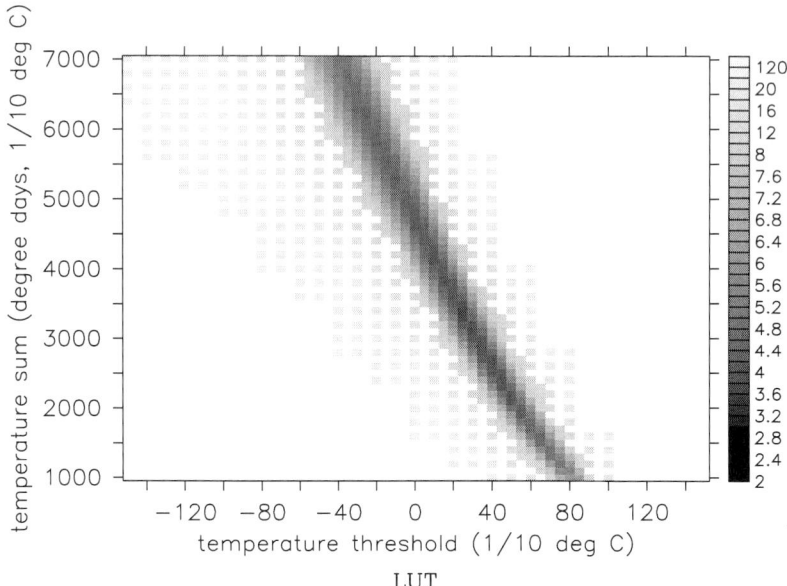

Fig. 4.6 Sample plot of a LUT (Look Up Table): the phenological phase is lilac beginning of flowering at the Austrian station of Kremsmuenster (1951–2004), the minimisation function is the RMSE (Root Mean Square Error) depicted in various shades of grey at starting date (t_j) yearday 71 (2 March). A three step approximation procedure selects only the relevant area and leaves the rest white

4.3.2.9 Application of Process Based Phenological Models to Pollen Season Modelling

The year-to-year variability of the beginning of the flowering season is strongly linked with the year-to-year variability of the atmospheric conditions prior flowering inception. The modelling and forecasting of the start of the flowering season as function of the atmospheric conditions is therefore very useful for the pollen forecast procedure. Once flowering has started, the subsequent temporal development of the pollen concentration follows a certain pattern, which can also be modelled (Linkosalo et al. 2010).

The application of phenological models to predict the start, peak and end of the pollen season of various anemophilous species appears well established and generally yields reliable results (Thibaudon and Lachasse 2005). Nevertheless, the application of phenological models to the pollen season modelling is not straightforward and therefore the underlying ideas should be critically reviewed. The following two assumptions are tacitly expected to be true:

- Assumption 1. If the start of the local flowering season has been observed, local pollen shedding has also started.

- Assumption 2. If pollen of a certain species has been recorded by the local pollen sampler, local pollen shedding has started.

If Assumption 1 holds, one could apply a phenological model to predict the entry of the flowering phase and thus have an indication of the beginning of the local pollen season. This has actually rarely been done because of a number of reasons. Generally, the species range of pollen data is much wider than that of phenological networks. Phenological and pollen networks have been created independently, are not coordinated and are usually run by different organisations, so that pollen and phenological species overlap only to a minor extent in most networks. Only in a few cases the density of phenological data are sufficient to calibrate a phenological model to support the forecast of the beginning of the pollen season.

Assumption 2 implies that pollen is not transported over larger distances and that the local pollen record faithfully reflects the local pollen shedding after flowering of the local plants has commenced. But Estrella et al. (2006) did show that in the case of birch, in Germany, Assumption 2 needs not be true. They found major temporal discrepancies between local phenology and local pollen concentration. Pollen can be transported over large distances from areas with currently flowering plants to the recording site, where plants have either not yet commenced flowering or have stopped flowering already. Therefore, the locally observed entry date of the phenological phase of flowering, the locally recorded beginning of the pollen shedding and the beginning of the pollen season may not be identical.

Usually, Assumption 2 is applied and a phenological model is fitted for a date relevant for the pollen season, like start and end, according to a selection criterion, or peak of the pollen season. In order to convert pollen concentration values to one or more of such dates, a range of definitions has been suggested in literature. Although transport processes influence the local pollen concentration, pollen season start dates (regardless of how they are defined) can be modelled as function of temperature sums.

4.3.3 Special Problems in Pollen and Phenological Modelling

4.3.3.1 Application to Large Areas

When designing a pollen forecast procedure, one has to choose an area large enough to accommodate for a possible long range transport of pollen. This, in turn, requires the modelling of the start and progress of the pollen season over a larger area, beyond national boundaries and with a spatial resolution, much higher than that of existing networks. In complex terrain, for instance, it makes much sense to calculate the phenological entry dates on a grid with a high spatial resolution, either few km or even <1 km, because deviations in elevation between the DEM and real topography can cause a substantial shift of entry dates modelled on a grid (entry dates may vary between 20 and 40 days/1,000 m elevation, Scheifinger et al. 2002). Typically, the density of the phenological, pollen or meteorological networks is not high enough to enable such a procedure for the required spatial resolution.

The spatial robustness of phenological models has been tested with a set of European phenological and pollen data (Chuine 2000; Chuine and Belmonte 2004). In some cases, models fitted with data from one station set predicted pollen season entry dates well at a set of neighbouring stations or even at stations more than 900 km apart. If the model gave reliable results in the area, where it was fitted, the probability was high that it also worked well at distant stations. The geographical range of the applicability of the model parameters appeared to be to some extent also species dependent.

García-Mozo et al. (2008) grouped Spanish pollen stations according to phenological model parameter values. Phenological models fitted with local pollen and meteorological data yielded the best results (75–95% explained variance); phenological models fitted with regionally deduced parameter values resulted in lower explained variances at individual stations (55–85%), whereas phenological models fitted with parameter values deduced on the basis of all stations gave the lowest explained variance (51%). One might conclude from that study that in order to achieve the best fit, only locally deduced model parameters should be applied or a method needs to be found to interpolate the model parameter values to each individual climate station or grid point.

White et al. (1997) developed a phenological model for the onset of greenness of deciduous broad leafed forests and grasslands of the temperate zone based on the Thermal Time model approach including radiation. They found that the temperature sums combined with radiation sums at onset of greenness are a function of average annual temperature and radiation. Applying this relationship they were able to model the onset of greenness over the contiguous US with a high degree of accuracy (mean absolute errors ranged from 5.3 to 7.1 days).

4.3.3.2 Real Time Modelling

Numerical pollen forecast procedures require real-time operational phenological models and the spatial interpolation of phenological entry dates (Helbig et al. 2004; Sofiev et al. 2006). The temporal development of flowering in space with the subsequent pollen emission is the essential input for atmospheric transport models. Phenological real-time observation is still in its infancy and cannot be used for the purpose of pollen forecast. Therefore, phenological models have to simulate the developmental stages of the plants in real time. If the entry of the flowering phase has been calculated in an area, another model must assess the quantity of pollen emitted into the atmosphere, which is then input for the dispersion model. All process-based phenological models are basically suitable for such a real-time application. As real-time phenological observation systems (remote sensing via satellites and digital cameras) and real-time pollen measuring devices are being developed, the question of assimilation of such data into the operation models becomes relevant in the near future.

In some cases, the pollen forecast procedure is designed so that an interpolation of observed or modelled phenological entry dates is not necessary. If the prognostic atmospheric model provides a temperature field with the required spatial resolution, the phenological model can directly be applied to the temperature data on the same grid (Puppi and Zanotti 1992; Kawashima and Takahashi 1995; Hidalgo et al. 2002). Possible model temperature biases have to be taken into account.

4.3.3.3 High Resolution Spatial Representation of Phenological Entry Dates

In other cases it might be desirable to interpolate phenological entry dates observed at network stations or modelled on a comparatively coarse grid to a Digital Elevation Model (DEM) with a higher spatial resolution (for a more detailed overview see Jeanneret and Rutishauser 2010). For pollen modelling purposes, for instance, the spatial distribution of phenological entry dates must be available on a DEM, which resolves the main topographical features of the area. A great mismatch between the spatial resolution of the phenological information and the real topography inevitably leads to an equivalent mismatch between the modelled entry dates on the DEM and on the real topography.

A number of methods have been developed to produce phenological maps. Beginning with Ihne's (1885) work, a good historical overview and a description of the theoretical background of spatial interpolation of phenological observations is given in Puppi and Zanotti (1989). If the terrain is largely flat or the grid is rather coarse, one might just interpolate the entry dates straightforward with Inverse Distance Weighting (IDW, Ahas et al. 2002; Scheifinger et al. 2002). If the area is large (from a few hundred up to a few thousand kilometer in diameter) and it turns out that the relationship between phenological entry dates and space is strict (for instance >70%), a multiple regression model can be applied, where the phenological entry dates are modelled as a function of station longitude, latitude and elevation (Rötzer and Chmielewski 2001). Small scale topographical features are considered by height reduced methods, like reduced detrended Kriging (Badeck et al. 2004) or via a radiation model, which is applied to a high resolution DEM to subsequently provide the spatial weights for the phenological entry dates (Chytry and Tichy 1998). A more complex approach is presented by Puppi and Zanotti (1989), where the phenological entry dates are calculated as a function of a number of independent environmental variables, like altitude, slope, incident solar radiation, tree layer cover, urbanisation or geomorphologic features (sides and bottom of the valley) via a set of regression equations. Geostatistical software can facilitate interpolation and visualisation of interpolated fields (García-Mozo et al. 2006).

4.4 Discussion and Summary

This section is thought to present a few ideas, which pop up when one begins to reflect about the compiled snap shot of the current state of the art about monitoring and modelling of pollen counts, pollen season and phenology.

Each of the highly diverse scientific disciplines contributing to aeropalynology has its own history, where tools and technologies have been developed and data sets collected. These factors could be called science intrinsic factors. Similarly, external factors have evolved related, for instance, with the public health issue of pollinosis. Driven by external factors, the accumulated expertise of each discipline is being summarised by a small scientific community within the field of aeropalynology, which has just begun to exploit their heritage in a most fruitful manner. Pollen transport models, which are being developed by a number of European weather services, can be cited as an example for such a combined interdisciplinary effort, where biologists, meteorologists and physicians work together.

Plant physiology, atmospheric dispersion and the human immune system are rather complex research objects, so that progress in many aspects of aeropalynology is counterbalanced by an increasing number of unresolved questions. Thus, this field appears both challenging and fascinating.

4.4.1 Monitoring

The chapter about monitoring reviews the current situation of three data sources, which are directly related with aeropalynology: phenological observations, pollen counts and remote sensing of the vegetation activity. Up to now, all three data sets more or less co-exist, without much exchange or fusion. Any assimilation of two or all three of them into one is in its infancy, if existing at all. A number of assimilation techniques have been developed in earth sciences, from which suitable ones could be chosen and adapted. Phenological observations, pollen counts and remote sensing information on the state of the vegetation could then be assimilated into pollen emission models and finally into the numerical pollen forecast. Pollen modelling would benefit a great deal from such a data fusion.

Over the last decades, consistent monitoring efforts of various national networks have created a wealth of pollen concentration time series. These constitute a nearly untouched treasure, which is still to be exploited to investigate questions concerning pollen emission, transport and deposition.

New monitoring methods emerge, which allow measuring the allergen content in pollen. This adds a new dimension to the problem of pollen related allergies. Results from research on the allergen content in pollen, like the HIALINE project (http://www.hialine.com/en/klinikum-rechts-der-isar-der-technischen-universitaet-muenchen.php) are expected to make the operational pollen forecasts more specific, which in turn helps the sufferers to improve their avoidance strategies.

4.4.2 Modelling

Although process-based phenological models have been around for a couple of decades, a number of problems remain to be solved, when applying them to statistical and numerical pollen forecast models:

- Model quality
  - Model quality is restricted by the noise inherent in the data. Phenological entry dates are observed subjectively and pollen counts are influenced by plant distribution and atmospheric factors. Models can not be more accurate than the observations they are based on.
  - The Thermal Time Model appears to exhaust the noisy information contained in commonly available observational data sets. Attempts have been made to explicitly incorporate plant physiological concepts into the process-based phenological models, but model quality could not substantially be improved beyond that of the simple Thermal Time models (Chuine 2000; Schaber and Badeck 2003; Linkosalo et al. 2008).
- Operational statistical or numerical pollen forecast ideally requires the model results over a large area on a grid. Up to now, only a few studies have proposed methods to model phenological entry dates over a larger area and a practical solution is still absent.
- As already mentioned in the monitoring section, the assimilation of phenological observations, pollen counts and remotely sensed information about vegetation will emerge as one of the central topics in the field of numerical pollen forecast.

Regression models, where the pollen count is modelled as function of a number of environmental variables, are well established and widely used to reliably improve the operational day to day pollen forecast. More elaborate statistical techniques, like computational intelligence methods, have still to become established for the operational pollen forecast. Some statistical packages offer such methods, like neural networks in the SPSS package, which can be run on personal computers.

The question, which of the models, regression or process-based, is superior, cannot yet be answered. Reviewing the wide range of models for forecasting the start of the pollen season, no superior model can be identified. Laaidi et al. (2003) employed a temperature sum model and a regression model using a number of predictors (air temperature, rainfall, relative humidity, sunshine duration and soil temperature) to forecast the start of the pollen season of *Ambrosia*. The regression model performed better during calibration, but the temperature sum model showed better results when tested on independent data. Chuine et al. (1999) conclude from a model comparison study that there is no single model that accurately predicts the dates of flowering of every species. Depending on the species, different models may perform best. Even among a single species there is not one model performing best, because relationships between the species and the environment may differ according to the climatic region. This emphasises the importance of careful evaluation and testing of various models for predicting the start of the pollen season.

References

Ahas, R., Aasa, A., Menzel, A., Fedotova, V. G., & Scheifinger, H. (2002). Changes in European spring phenology. *International Journal of Climatology, 22*, 1727–1738.

Aizenberg, V., Reponen, T., Grinshpun, S. A., & Willeke, K. (2000). Performance if Air-O-Cell, Burkard, and Button samplers for total enumeration of airborne spores. *American Industrial Hygiene Association Journal, 61*, 855–864.

Alba, F., & Díaz de la Guardia, C. (1998). The effect of air temperature on the starting dates of the *Ulmus*, *Platanus* and *Olea* pollen season in the SE Iberian Peninsula. *Aerobiologia, 14*, 191–194.

Alcalá, A. R., & Barranco, D. (1992). Prediction of flowering time in olive for the Córdoba Olive collection. *HortScience, 27*(11), 1205–1207.

Alcazar, P., Carinanos, P., de Castro, C., Guerra, F., Moreno, C., Dominguez Vilchez, E., & Galán, C. (2004). Airborne plane tree (*Platanus hispanica*) pollen distribution in the city of Cordoba, south-western Spain and possible implications on pollen allergy. *Journal of Investigational Allergology and Clinical Immunology, 14*(3), 238–243.

Andersen, A. A. (1958). New sampler for the collection, sizing and enumeration of viable airborne particles. *Journal of Bacteriology, 76*, 471–484.

Antepara, I., Fernandez, J. C., Gamboa, P., Jauregui, I., & Miguel, F. (1995). Pollen allergy in the Bilbao area (European Atlantic seaboard climate): Pollination forecasting methods. *Clinical and Experimental Allergy, 25*, 133–140.

Aono, Y., & Kazui, K. (2008). Phenological data series of cherry tree flowering in Kyoto, Japan, and its application to reconstruction of springtime temperatures since the 9th century. *International Journal of Climatology, 28*, 905–914. doi:10.1002/joc.1594.

Arelco, A. R. C. (2004). *CIP 10 personal dust sampler users' manual*. [Online] Available from: http://www.arelco.fr/docs/produits/CIP10_M_GB.pdf. Accessed 5 Oct 2010.

Arizmendi, C. M., Sanchez, J. R., Ramos, N. E., & Ramos, G. I. (1993). Time-series prediction with neural nets – applications to airborne pollen forecasting. *International Journal of Biometeorology, 37*(3), 139–144.

Autio, J., & Hicks, S. (2004). Annual variations in pollen deposition and meteorological conditions on the fell Aakenusrunturi in northern Finland: Potential for using fossil pollen as a climate proxy. *Grana, 43*, 31–47.

Aznarte, J. L., Nieto Lugilde, D., Benítez, J. M., Alba Sánchez, F., & de Linares Fernández, C. (2007). Forecasting airborne pollen concentration time series with neural and neuro-fuzzy models. *Expert Systems with Applications, 32*, 1218–1225.

Badeck, F. W., Bondeau, A., Böttcher, K., Doktor, D., Lucht, W., Schaber, J., & Sitch, S. (2004). Responses of spring phenology to climate change. *New Phytologist, 162*(2), 295–309.

BAF. (1995). *Airborne pollens and spores: A guide to trapping and counting*. Aylesford: The British Aerobiology Federation. 1995. ISBN 0-9525617-0-0.

Bertin Technologies (2007). Continuous cyclonic air sampler for outdoor air monitoring. http://www.coriolis-airsampler.com/resources/fichiers/coriolis/CoriolisDelta.pdf. Accessed 5 October/2010.

Bonton, P., Boucher, A., Thonnat, M., Tomczak, R., Hidalgo, P. J., Belmonte, J., & Galán, C. (2001). Colour image in 2D and 3D microscopy for the automation of pollen rate measurement. *Image Analysis and Stereology, 20*, 527–532.

Boucher, A., Hidalgo, P. J., Thonnat, M., Belmonte, J., Galán, C., Bonton, P., & Tomczak, R. (2002). Development of a semi-automatic system for pollen recognition. *Aerobiologia, 18*, 195–201.

Box, G. E. P., Jenkins, G. M., & Reinsel, G. C. (1994). *Time-series analysis: Forecasting and control*. Upper Saddle River, NJ: Prentice-Hall International.

Bringfelt, B., Engstrom, I., & Nilsson, S. (1982). An evaluation of some models to predict airborne pollen concentration from meteorological conditions in Stockholm, Sweden. *Grana, 21*, 59–64.

Bryant, W. M., Jr. (1989). *Yearbook of science and future* (pp. 92–111). Chicago: Encyclopaedia Britannica, Inc.

Buters, J. T., Weichenmeier, I., Ochs, S., Pusch, G., Kreyling, W., & Boere, A. J. (2010). The allergen Bet v 1 in fractions of ambient air deviates from birch pollen counts. *Allergy, 65*, 850–858.

Carinanos, P., Emberlin, J., Galán, C., & Dominguez-Vilches, E. (2000). Comparison of two pollen counting methods of slides from a Hirst type volumetric trap. *Aerobiologia, 16*, 339–346.

Chen, C., Hendriks, E. A., Duin, R. P., Reiber, J. H. C., Hiemstra, P. S., de Weger, L. A., & Stoel, B. C. (2006). Feasibility study for an automatic recognition system for three relevant allergenic pollen, grass, birch and mugwort. *Aerobiologia, 22*, 275–284.

Chmielewski, F.-M. (2008). The international phenological gardens. In J. Nekovar, E., Koch, E., Kubin, P., Nejedlik, T., Sparks, & F. Wielgolaski (Eds.), *The History and current status of plant phenology in Europe*. COST Action 725 (182 p.).Brussels: COST Office.

Chuine, I. (2000). A unified model for budburst of trees. *Journal of Theoretical Biology, 207*, 337–347.

Chuine, I., & Belmonte, J. (2004). Improving prophylaxis for pollen allergies: Predicting the time course of the pollen load of the atmosphere of major allergenic plants in France and Spain. *Grana, 43*, 65–80.

Chuine, I., Cour, P., & Rousseau, D. D. (1998). Fitting models predicting dates of flowering of temperature zone trees using simulated annealing. *Plan, Cell and Environment, 21*, 455–466.

Chuine, I., Cour, P., & Rousseau, D. D. (1999). Selecting models to predict the timing of flowering of temperate trees: Implications for tree phenology modelling. *Plant Cell and Environment, 22*, 1–13.

Chuine, I., Belmonte, J., & Mignot, A. (2000). A modelling analysis of the genetic variation of phenology between tree populations. *Journal of Ecology, 80*, 561–570.

Chuine, I., Kramer, K., & Hänninen, H. (2003). Plant development models. In M. D. Schwartz (Ed.), *Phenology: An integrative environmental science* (p. 564). Dordrecht/Boston/London: Kluwer Academic Publishers.

Chytry, M., & Tichy, L. (1998). Phenological mapping in a topographically complex landscape by combining field survey with an irradiation model. *Applied Vegetation Science, 1*, 225–232.

Clary, J., Savé, R., Biel, C., & de Herralde, F. (2004). Water relations in competitive interactions of Mediterranean grasses and shrubs. *Annals of Applied Biology, 144*(2), 149–155.

Clot, B. (1998). Forecast of the Poaceae pollination in Zurich and Basel (Switzerland). *Aerobiologia, 14*, 267–268.

Clot, B. (2001). Airborne birch pollen in Neuchâtel (Switzerland): Onset, peak and daily patterns. *Aerobiologia, 17*, 25–29.

Comtois, P., Alcázar, P., & Neron, D. (1999). Pollen count statistics and its relevance to precision. *Aerobiologia, 15*, 19–28.

Cour, P. (1974). Nouvelles techniques de détection des flux et retombées polliniques: étude de la sédimentation des pollens et des spores à la surface du sol. *Pollen et spores XVI, 1*, 103–141.

Cox, C. S., & Wathes, C. M. (1995). *Bioaerosol handbook* (621 p.). Boca Raton, FL: Lewis Publisher.

Crepinsek, Z., Kaifez-Bogataj, L., & Bergant, K. (2006). Modelling of weather variability effect on phytophenology. *Ecological Modelling, 194*, 256–265.

Cristofolini, F., & Gottardini, E. (2000). Concentration of airborne pollen of *Vitis vinifera* L. and yield forecast: A case study at S. Michele all'Adige, Trento, Italy. *Aerobiologia, 16*, 125–129.

Damialis, A., Gioulekas, D., Lazopoulou, C., Balafoutis, C., & Vokou, D. (2005). Transport of airborne pollen into the city of Thessaloniki: The effects of wind direction, speed and persistence. *International Journal of Biometeorology, 49*, 139–145.

Damialis, A., Halley, J. M., Gioulekas, D., & Vokou, D. (2007). Long-term trends in atmospheric pollen levels in the city of Thessaloniki, Greece. *Atmospheric Environment, 41*, 7011–7021.

Davies, R. R., & Smith, L. P. (1973). Forecasting the start and severity of the hay fever season. *Clinical Allergy, 3*, 263–267.

Degaudenzi, M. E., & Arizmendi, C. M. (1998). Wavelet based fractal analysis of airborne pollen. *Physical Review E, 59*(2), 6569–6573.

De Linares, C., Nieto-Lugilde, D., Alba, F., Díaz de la Guardia, C., Galán, C., & Trigo, M. M. (2007). Detection of airborne allergen (Ole e 1) in relation to Olea europaea pollen in S Spain. *Clinical & Experimental Allergy, 37*(1), 125–132.

De Linares, C., Días de la Guardia, C., Nieto-Lugilde, D., & Alba, F. (2010). Airborne study of grass allergen (Lol p 1) in different-sized particles. *International Archives of Allergy and Immunology, 152*(1), 49–57.

DeLurgio, S. A. (1998). *Forecasting principles and applications.* New York: McGraw-Hill.

Demokritou, P., Kavouras, I. G., Ferguson, S. T., & Koutrakis, P. (2002). Development of a high volume cascade impactor for toxicological and chemical characterisation studies. *Aerosol Science and Technology, 36*, 925–933.

Dominguez, E., Galán, C., Villamandos, F., & Infante, F. (1991). Handling and evaluation of the data from the aerobiological sampling. *Monografías REA/EAN, 1*, 131–141.

Driessen, M., & Moelands, M. (1985). Estimation of the commencement of the grass pollen season and its prediction by means of the phenological method. *Acta Botanica Neerlandica, 34*, 131. Abstract.

Driessen, M., Van Herpen, A., Moelands, M., & Spiekma, M. (1989). Prediction of the start of the grass pollen season for the western part of the Netherlands. *Grana, 28*, 37–44.

Emberlin, J., & Baboonian, C. (1995). The development of a new method for sampling airborne particles for immunological analysis. Paper presented at XVI European Congress of Allergy and Clinical Immunology, ECACI 95 (pp. 39–43). Madrid: Monduzzi, Bologna.

Emberlin, J., Savage, M., & Woodman, R. (1993a). Annual variations in *Betula* pollen seasons in London 1961–1990. *Grana, 32*, 359–363.

Emberlin, J., Savage, M., & Jones, S. (1993b). Annual variations in grass pollen seasons in London 1961–1990: Trends and forecast models. *Clinical and Experimental Allergy, 23*, 911–918.

Emberlin, J., Jones, S., Bailey, J., Caulton, E., Corden, J., Dubbets, S., Evans, J., McDonagh, N., Millington, W., Mullins, J., Russel, R., & Spencer, T. (1994). Variation in the start of the grass pollen season at selected sites in the United Kingdom. 1987–1992. *Grana, 33*, 94–99.

Emberlin, J., Mullins, J., Corden, J., Jones, S., Millington, W., Brooke, M., & Savage, M. (1999). Regional variations in grass pollen seasons in the UK, long-term trends and forecast models. *Clinical and Experimental Allergy, 29*, 347–356.

Emberlin, J., Smith, M., Close, R., & Adams-Groom, B. (2007). Changes in the pollen seasons of the early flowering trees *Alnus* spp. and *Corylus* spp. in Worcester, United Kingdom, 1996–2005. *International Journal of Biometeorology, 51*, 181–191.

Erbas, B., Chang, J. H., Dharmage, S., Ong, E. K., Hyndman, R., Newbigin, E., & Abramson, M. (2007). Do levels of airborne grass pollen influence asthma hospital admissions? *Clinical and Experimental Allergy, 37*(11), 1641–1647.

Estrella, N., Menzel, A., Kramer, U., & Behrendt, H. (2006). Integration of flowering dates in phenology and pollen counts in aerobiology: Analysis of their spatial and temporal coherence in Germany (1992–1999). *International Journal of Biometeorology, 51*, 49–59.

Férnández-González, D., Valencia-Barrera, R. M., Vega, A., Díaz de la Guardia, C., Trigo, M. M., Cariñanos, P., Guardia, A., Pertiñez, C., & Rodríguez-Rajo, F. J. (1999). Analysis of grass pollen concentrations in the atmosphere of several Spanish sites. *Pollen, 10*, 123–132.

Fiorina, A., Scordamaglia, A., Guerra, L., & Passalacqua, G. (2003). Aerobiologic diagnosis of Brassicaceae-induced asthma. *Allergy, 58*, 829–830.

Fotiou, C., Damialis, A., Krigas, N., Halley, J. M., & Vokou, D. (2011). *Parietaria judaica* flowering phenology, pollen production, viability and atmospheric circulation, and expansive ability in the urban environment: Impacts of environmental factors. *International Journal of Biometeorology, 55*, 35–50.

Frei, T., & Gassner, E. (2008). Climate change and its impact on birch pollen quantities and the start of the pollen season an example form Switzerland for the period 1969–2006. *International Journal of Biometeorology, 52*, 667–674.

Frenguelli, G. (1998). The contribution of aerobiology to agriculture. *Aerobiologia, 14*, 95–100.

Frenguelli, G., & Bricchi, E. (1998). The use of the pheno-climatic model for forecasting the pollination of some arboreal taxa. *Aerobiologia, 14*, 39–44.

Frenguelli, G., Bricchi, E., Romano, B., Mincigrucci, G., & Spieksma, F. Th. M. (1989). A predictive study on the beginning of the pollen season for Gramineae and *Olea europaea* L. *Aerobiologia, 5*, 64–70.

Galán, C., Emberlin, J., Dominguez-Vilches, E., Bryant, R. H., & Villamandos, F. (1995). A comparative analysis of daily variations in the Gramineae pollen counts at Cordoba, Spain and London, UK. *Grana, 34*, 189–198.

Galán, C., Fuillerat, M. J., Comtois, P., & Dominguez-Vilches, E. (1998). A predictive study of Cupressaceae pollen season onset, severity, maximum value and maximum value date. *Aerobiologia, 14*, 195–199.

Galán, C., Alcazar, P., Cariñanos, P., García-Mozo, H., & Dominguez-Vilches, E. (2000). Meteorological factors affecting daily Urticaceae pollen counts in southwest Spain. *International Journal of Biometeorology, 43*, 191–195.

Galán, C., García-Mozo, H., Cariñanos, P., Alcázar, P., & Domínguez-Vilches, E. (2001a). The role of temperature in the onset of the *Olea europaea* L. pollen season in southwestern Spain. *International Journal of Biometeorology, 45*(1), 8–12.

Galán, C., Carinanos, P., García-Mozo, H., Alcazar, P., & Dominguez-Vilches, E. (2001b). Model for forecasting *Olea europaea* L. airborne pollen in South-West Andalusia, Spain. *International Journal of Biometeorology, 45*, 59–93.

Galán, C., Vazquez, L., García-Mozo, H., & Dominguez-Vilches, E. (2004). Forecasting olive (*Olea europea*) crop yield based on pollen emission. *Field Crops Research, 86*, 43–51.

Galán, C., García-Mozo, H., Vazquez, L., Ruiz-Valenzuela, L., Díaz de la Guardia, C., & Trigo-Perez, M. (2005). Heat requirement for the onset of the *Olea europaea* L. pollen season in several places of Andalusia region and the effect of the expected future climate change. *International Journal of Biometeorology, 49*(3), 184–188.

Galán, C., Cariñanos, P., Alcázar, P., & Dominguez-Vilches, E. (2007). *Spanish Aerobiology Network (REA) management and quality manual*. Cordoba: Servicio de Publicaciones Universidad de Córdoba. ISBN 978-84-690-6353-8.

Galán, C., García-Mozo, H., Vázquez, L., Ruiz, L., Díaz de la Guardia, C., & Domínguez-Vilches, E. (2008). Modelling olive (*Olea europaea* L.) crop yield in Andalusia Region, Spain. *Agronomy Journal, 100*(1), 98–104.

García-Mozo, H., Galán, C., Gomez-Casero, M. T., & Dominguez-Vilches, E. (2000). A comparative study of different temperature accumulation methods for predicting the start of the *Quercus* pollen season in Cordoba (South West Spain). *Grana, 39*, 194–199.

García-Mozo, H., Hidalgo, P. J., Galán, C., & Gomez-Casero, M. T. (2001). Catkin frost damage in Mediterranean cork-oak (*Quercus suber* L.). *Israel Journal of Plant Sciences, 49*, 41–47.

García-Mozo, H., Galán, C., Aira, M. J., Belmonte, J., Díaz de la Guardia, C., Fernández, D., Gutierrez, A. M., Rodriguez, F. J., Trigo, M. M., & Dominguez-Vilches, E. (2002). Modelling start of oak pollen season in different climatic zones in Spain. *Agricultural and Forest Meteorology, 110*, 247–257.

García-Mozo, H., Galán, C., Jato, V., Belmonte, J., Díaz de la Guardia, C., Fernández, D., Gutiérrez, M., Aira, M. J., Roure, J. M., Ruiz, L., Trigo, M. M., & Domínguez-Vilches, E. (2006). *Quercus* pollen season dynamics in the Iberian Peninsula: Response to meteorological parameters and possible consequences of climate change. *Annals of Agricultural and Environmental Medicine, 13*, 209–224.

García-Mozo, H., Perez-Badía, R., & Galán, C. (2007a). Aerobiological and meteorological factors influence on olive (*Olea europaea* L.) crop yield in Castilla-La Mancha (Central Spain). *Aerobiologia, 24*(1), 13–18.

García-Mozo, H., Gómez-Casero, M. T., Dominguez-Vilches, E., & Galán, C. (2007b). Influence of pollen emission and weather-related factors on variations in holm-oak (*Quercus ilex* subsp. *ballota*) acorn production. *Environmental and Experimental Botany, 61*, 35–40.

García-Mozo, H., Chuine, I., Aira, M. J., Belmonte, J., Bermejo, D., Díaz de la Guardia, C., Elvira, B., Gutiérrez, M., Rodríguez-Rajo, J., Ruiz, L., Trigo, M. M., Tormo, R., Valencia, R., & Galán, C. (2008). Regional phenological models for forecasting the start and peak of the *Quercus* pollen season in Spain. *Agricultural and Forest Meteorology, 148*, 372–380.

García-Mozo, H., Galán, C., Belmonte, J., Bermejo, D., Candau, P., Díaz de la Guardia, C., Elvira, B., Gutiérrez, M., Jato, V., Silva, I., Trigo, M. M., Valencia, R., & Chuine, I. (2009). Predicting the start and peak dates of the Poaceae pollen season in Spain using process-based models. *Agriculture and Forest Meteorology, 149*, 256–262.

Gerad-Peeters, A. (1998). Cumulative temperatures for prediction of the beginning of ash (*Fraxinus excelsior* L.) pollen season. *Aerobiologia, 14*, 375–381.

Goldberg, C., Buch, H., Moseholm, L., & Weeke, E. V. (1988). Airborne pollen records in Denmark, 1977–1986. *Grana, 27*, 209–217.

Govindaraju, D. R. (1988). The relationship between dispersal ability and levels of gene flow in plants. *Oikos, 52*, 31–35.

Graham, J. A. H., Pavlicek, P. K., Sercombe, J. K., Xavier, M. L., & Tovey, E. R. (2000). The nasal air sampler: A device for sampling inhaled aeroallergens. *Annals of Allergy, Asthma, & Immunology, 84*, 599–604.

Grivas, G., & Chaloulakou, A. (2006). Artificial neural network models for prediction of Pm10 hourly concentrations, in the Greater Area of Athens, Greece. *Atmospheric Environment, 40*(7), 1216–1230.

Guyot, G., Gugon, D., & Riom, J. (1989). Factors affecting the spectral response of forest canopies: A review. *Geocarta International, 3*, 43–60.

Halley, J. M., & Inchausti, P. (2004). The increasing importance of 1/f-noises as models of ecological variability. *Fluctuation and Noise Letters, 4*, R1–R26.

Hänninen, H. (1995). Effects of climatic change on trees from cool and temperate regions: An ecophyisological approach to modelling of budburst phenology. *Canadian Journal of Botany, 73*, 183–199.

Haykin, S. (1999). *Neural networks, a comprehensive foundation.* Upper Saddle River, NJ: Prentice Hall.

Helbig, N., Vogel, B., Vogel, H., & Fiedler, F. (2004). Numerical modelling of pollen dispersion on the regional scale. *Aerobiologia, 3*, 3–19.

Hidalgo, P. J., Mangin, A., Galán, C., Hembise, O., Vazquez, L. M., & Sanchez, O. (2002). An automated system for surveying and forecasting *Olea* pollen dispersion. *Aerobiologia, 18*, 23–31.

Hirst, J. M. (1952). An automatic volumetric spore trap. *The Annals of Applied Biology, 39*(2), 257–265.

Høgda, K. A., Karlsen, S. R., Solheim, I., Tømmervik, H., & Ramfjord, H. (2003). The start dates of birch pollen seasons in Fennoscandia studied by NOAA AVHRR NDVI data. Paper presented at geoscience and remote sensing symposium, 2002. IGARSS'02. *IEEE International, 6*, 3299–3301.

Holben, B. N. (1986). Characteristics of maximum-value composite images for temporal AVHRR data. *International Journal of Remote Sensing, 7*, 1435–1445.

Hudson, I. L., Kim, S. W., & Keatley, M. R. (2010). Modelling the flowering of four Eucalypt species using new mixture transition distribution models. In I. L. Hudson & M. R. Keatley (Eds.), *Phenological research: Methods for environmental and climate change analysis.* Dordrecht/Heidelberg/London/New York: Springer. 521 p.

Huynen, M., Menne, B., Behrendt, H., Bertollini, R., Bonini, S., Brandao, R., Brown-Fährlander, C., Clot, B., D'Ambrosio, C., De Nuntiis, P., Ebi, K. L., Emberlin, J., Orbanne, E. E., Galán, C., Jäger, S., Kovats, S., Mandrioli, P., Martens, P., Menzel, A., Nyenzi, B., Rantio Lehtimäki, A., Ring, J., Rybnicek, O., Traidl-Hoffmann, T., Van Vliet, A., Voigt, T., Weiland, S., & Wickman, M. (2003). Phenology and human health: Allergic disorders. Report of a WHO meeting, Rome, Italy.

Iglesias, I., Rodriguez-Rajo, F. J., & Mendez, J. (2007). Behavior of *Platanus hispanica* pollen, an important spring aeroallergen in northwestern Spain. *Journal of Investigational Allergology and Clinical Immunology, 17*(3), 145–156.

Ihne, E. (1885). Karte der Aufblühzeit von *Syringa vulgaris* in Europa. *Botanisches Centralblatt, 21*, 85–88, 116–121, 150–155.

Jato, V., Frenguelli, G., Rodríguez, F. J., & Aira, M. J. (2000). Temperature requirements of *Alnus* pollen in Spain. *Grana, 39*, 240–245.

Jeanneret, F., & Rutishauser, T. (2010). Phenology for topoclimatological surveys and large-scale mapping. In I. L. Hudson & M. R. Keatley (Eds.), *Phenological research: Methods for environmental and climate change analysis*. Dordrecht/Heidelberg/London/New York: Springer. 521 p.

Kapyla, M., & Penttinen, A. (1981). An evaluation of the microscopical counting methods of the tape in Hirst-Burkard pollen and spore trap. *Grana, 20*, 131–141.

Karatzas, K., & Kaltsatos, S. (2007). Air pollution modelling with the aid of computational intelligence methods in Thessaloniki, Greece. *Simulation Modelling Practice and Theory, 15*(10), 1310–1319.

Karlsen, S. R., Elvebakk, A., Høgda, K. A., & Johansen, B. (2006). Satellite based mapping of the growing season and bioclimatic zones in Fennoscandia. *Global Ecology and Biogeography, 15*, 416–430.

Karlsen, S. R., Solheim, I., Beck, P. S. A., Høgda, K. A., Wielgolaski, F. E., & Tømmervik, H. (2007). Variability of the start of the growing season in Fennoscandia, 1982–2002. *International Journal of Biometeorology, 51*, 513–524. doi:10.1007/s00484-007-0091-x.

Karlsen, S. R., Ramfjord, H., Høgda, K. A., Johansen, B., Danks, F. S., & Brobakk, T. E. (2009a). A satellite-based map of onset of birch (*Betula*) flowering in Norway. *Aerobiologia, 25*, 15–25.

Karlsen, S. R., Høgda, K. A., Ramfjord, H., Brobakk, T. E., & Johansen, B. (2009b). Use of satellite data in near real-time monitoring of the birch flowering in Norway. Paper presented at the 12th Nordic Symposium on Aerobiology, Copenhagen, Denmark, August 28–30.

Kawashima, S., & Takahashi, Y. (1995). Modelling and simulation of mesoscale dispersion processes for airborne cedar pollen. *Grana, 34*, 142–150.

Kawashima, S., Clot, B., Fujita, T., Takahashi, Y., & Nakamura, K. (2007). An algorithm and a device for counting airborne pollen automatically using laser optics. *Atmospheric Environment, 41*, 7987–7993.

Kinnear, P. R., & Gray, C. D. (1999). *SPSS for Windows Made Simple*. Padstow: T.J. International.

Koch, E. (2010). Global framework for data collection – data bases, data availability, future networks, online databases. In I. L. Hudson & M. R. Keatley (Eds.), *Phenological research: Methods for environmental and climate change analysis*. Dordrecht: Springer. 522 p.

Koch, E., Dittmann, E., Lipa, W., Menzel, A., & van Vliet, A. (2005). COST 725 Establishing a European phenological data platform for climatological purposes. *Annalen der Meteorologie 41(2)*, 554–558. DWD.

Koch, E., Demarée, G., Lipa, W., Zach, S., & Zimmermann, K. (2008). History of international phenology networks. In J. Nekovar, E. Koch, E. Kubin, P. Nejedlik, T. Sparks, & F.-E. Wielgolaski (Eds.), *The history and current status of plant phenology in Europe – COST Action 725: Establishing a European Data Platform for Climatological Purposes*. Sastamala: Vammalan Kirjapaino Oy, COST Office. ISBN 978-951-40-2091-9.

Kramer, K. (1994). A modelling analysis of the effects of climatic warming on the probability of spring frost damage to tree species in The Netherlands and Germany. *Plant, Cell and Environment, 17*, 367–377.

Kramer, K. (1995). Phenotypic plasticity of the phenology of seven European tree species in relation to climatic warming. *Plant, Cell and Environment, 18*, 93–104.

Kukkonen, J., Partanen, L., Karppinen, A., Ruuskanen, J., Junninen, H., Kolehmainen, M., Niska, H., Dorling, S., Chatterton, T., Foxall, R., & Cawley, G. (2003). Extensive evaluation of neural network models for the prediction of NO_2 and PM10 concentrations, compared with a deterministic modelling system and measurements in central Helsinki. *Atmospheric Environment, 37*(32), 4539–4550.

Laaidi, M. (2001a). Forecasting the start of the pollen season of Poaceae: Evaluation of some methods based on meteorological factors. *International Journal of Biometeorology, 45*(1), 1–7.

Laaidi, M. (2001b). Regional variations in the pollen season of *Betula* in Burgundy: Two models for predicting the start of the pollination. *Aerobiologia, 17*, 247–254.

Laaidi, M., Thibaudon, M., & Besancenot, J.-P. (2003). Two statistical approaches to forecasting the start and duration of the pollen season of *Ambrosia* in the area of Lyon (France). *International Journal of Biometeorology, 48*, 65–73.

Landsmeer, S. H., Hendriks, E. A., de Weger, L. A., Reiber, J. H., & Stoel, B. C. (2009). Detection of pollen grains in multifocal optical microscopy images of air samples. *Microscopy Research and Technique, 72*, 424–430.

Levetin, E., Rogers, C. A., & Hall, S. A. (2000). Comparison of pollen sampling with a Burkard Spore Trap and a Tauber Trap in a warm temperate climate. *Grana, 39*, 294–302.

Lillesand, T. M., & Kiefer, R. (1994). *Remote sensing and image interpretation* (3rd ed.). New York: Wiley.

Linkosalo, T. (1999). Regularities and patterns in the spring phenology of some boreal trees. *Silva Fennica, 33*, 237–245.

Linkosalo, T. (2000). Mutual regularity of spring phenology of some boreal tree species: Predicting with other species and phenological models. *Canadian Journal of Forest Research, 30*, 667–673.

Linkosalo, T., Häkkinen, R., & Hänninen, H. (2006). Models of the spring phenology of boreal and temperate trees: Is there something missing? *Tree Physiology, 26*, 1165–1172.

Linkosalo, T., Lappalainen, H. K., & Hari, P. (2008). A comparison of phenological models of leaf bud burst and flowering of boreal trees using independent observations. *Tree Phyisology, 28*, 1873–1882.

Linkosalo, T., Häkkinen, R., Terhivuo, J., Tuomenvirta, H., & Hari, P. (2009). The time series of flowering and leaf bud burst of boreal trees (1846–2005) support the direct temperature observations of climatic warming. *Agricultural and Forest Meteorology, 149*(3–4), 453–461.

Linkosalo, T., Ranta, H., Oksanen, A., Siljamo, P., Luomajoki, A., Kukkonen, J., & Sofiev, M. (2010). A double-threshold temperature sum model for predicting the flowering duration and relative intensity of *Betula pendula* and *B. pubescens*. *Agricultural and Forest Meteorology, 150*, 1579–1584.

Litschauer, R. (2003). Untersuchungen zum Reproduktionspotential im Bergwald. *FBVA, 130*, 79–85.

Makinen, Y. (1981). Random sampling in the study of microscopic slides. *Reports from the Aerobiological Laboratory, University of Turku, 5*, 27–43.

Makra, L., Juhasz, M., Borsos, E., & Beczi, R. (2004). Meteorological variables connected with airborne ragweed pollen in southern Hungary. *International Journal of Biometeorology, 49*, 37–47.

Mandrioli, P. (1990). The Italian Aeroallergen Network. Sampling and counting method. *Aerobiologia, 6*, 5–7.

Mandrioli, P., & Ariatti, A. (2001). Aerobiology: Future course of action. *Aerobiologia, 17*, 1–10.

Marletto, V., Puppi Branzi, G., & Sirotti, M. (1992). Forecasting flowering dates of lawn species with air temperatures: Application boundaries of the linear approach. *Aerobiologia, 8*, 75–83.

Mendez, J., Comptois, P., & Iglesias, I. (2005). *Betula* pollen: One of the most important aeroallergens in Ourense, Spain. Aerobiological studies from 1993 to 2000. *Aerobiologia, 21*, 115–123.

Menzel, A. (1997). Phänologie von Waldbäumen unter sich ändernden Klimabedingungen – Auswertung der Beobachtungen in den Internationalen Phänologischen Gärten und Möglichkeiten der Modellierung von Phänodaten. Forstliche Forschungsberichte 164/1997 (147 p.). Forstwissenschaftliche Fakultät der Universität München und der Bayerischen Landesanstalt für Wald und Forstwirtschaft.

Migliavacca, M., Cremonese, E., Colombo, R., Busetto, L., Galvagno, M., Ganis, L., Meroni, M., Pari, E., Rossini, M., Siniscalco, C., & Morra di Cella, U. (2008). European larch phenology in the Alps: Can we grasp the role of ecological factors by combining field observations and inverse modelling? *International Journal of Biometeorology, 52*(7), 587–605.

Moriondo, M., Orlandini, S., De Nuntiis, P., & Mandrioli, P. (2001). Effect of agrometeorological parameters on the phenology of pollen emission and production of olive trees (*Olea europea* L.). *Aerobiologia, 7*, 225–232.

Moseholm, L., Weeke, E. R., & Petersen, B. N. (1987). Forecast of pollen concentrations of Poaceae (grasses) in the air by time series analysis. *Pollen et Spores, 19*(2–3), 305–322.

Mullins, J., & Emberlin, J. (1997). Sampling pollen. *Journal of Aerosol Science, 28*, 365–370.

Myneni, R. B., Keeling, C. D., Tucker, C. J., Asrar, G., & Nemani, R. R. (1997). Increased plant growth in the northern high latitudes from 1981 to 1991. *Nature, 386*, 698–702.

Nekovar, J., Koch, E., Kubin, E., Nejedlik, P., Sparks, T., & Wielgolaski, F. (Eds., 2008). *COST Action 725, The History and current status of plant phenology in Europe* (182 p.). Brussels: COST Office.

Norris-Hill, J. (1995). The modelling of daily Poaceae pollen concentrations. *Grana, 34*, 182–188.

Norris-Hill, J. (1998). A method to forecast the start of the *Betula*, *Platanus* and *Quercus* pollen seasons in North London. *Aerobiology, 14*, 165–170.

Ocana-Peinado, F., Valderrama, M. J., & Aguilera, A. M. (2008). A dynamic regression model for air pollen concentration. *Stochastic Environmental Research and Risk Assessment, 22*, 59–63.

Ong, E. K., Taylor, P. E., & Knox, R. B. (1997). Forecasting the onset of the grass pollen season in Melbourne (Australia). *Aerobiologia, 13*, 43–48.

Orlandi, F., Garcia-Mozo, H., Vazquez Ezquerra, L., Romano, B., Dominguez, E., Galán, C., & Fornaciari, M. (2004). Phenological olive chilling requirements in Umbria (Italy) and Andalusia (Spain). *Plant Biosystems, 138*(2), 111–116.

Orlandi, F., Vazquez, L. M., Ruga, L., Bonofiglio, T., Fornaciari, M., Garcia-Mozo, H., Dominguez, E., Romano, B., & Galán, C. (2005). Bioclimatic requirements for olive flowering in two Mediterranean regions located at the same latitude (Andalucía, Spain, and Sicily, Italy). *Annals of Agricultural and Environmental Medicine, 12*, 47–52.

Orlandi, F., Lanari, D., Romano, B., & Fornaciari, M. (2006). New model to predict the timing of olive (*Olea europaea*) flowering: A case study in central Italy. *New Zealand Journal of Crop and Horticultural Science, 34*, 93–99.

Pallant, J. (2001). *SPSS Survival Manual*. Buckingham: Open University Press.

Pasken, R., & Pietrowicz, J. A. (2005). Using dispersion and mesoscale meteorological models to forecast pollen concentrations. *Atmospheric Environment, 39*, 7689–7701.

Press, W. H., Teukolsky, S. A., Vetterling, W. T., & Flannery, B. P. (1992). *Numerical recipes: The art of scientific computing*. New York: Cambridge University Press. 963 p.

Puppi, G., & Zanotti, A. L. (1989). Methods in phenological mapping. *Aerobiologia, 5*, 44–54.

Puppi, G., & Zanotti, A. L. (1992). Estimate and mapping of the activity of airborne pollen sources. *Aerobiologia, 8*(1), 69–74.

Rantio-Lehtimaki, A., Viander, M., & Koivikko, A. (1994). Airborne birch antigens in different particle sizes. *Clinical and Experimental Allergy, 24*, 23–28.

Ranzi, A., Lauriola, P., Marletto, V., & Zinoni, F. (2003). Forecasting airborne pollen concentrations: Development of local models. *Aerobiologia, 19*, 39–45.

Reaumur, R. A. F. de. (1735). *Observations du thermomètre, faites a Paris pendant l'année 1735, comparée avec celles qui ont été faites sous la ligne, a lísle de France, a Alger et quelque unes de nos isles de l'Amérique*. Memoire de l'Académie des Sciences de Paris.

Ribeiro, H., Cunha, M., & Abreu, I. (2006). Comparison of classical models for evaluating the heat requirements of olive (*Olea europeae* L.) in Portugal. *Journal of Integrative Plant Biology, 48*(6), 664–671. doi:10.1111/j.1744-7909.2006.00269.x.

Rodriguez Rajo, F. J., Dopazo, A., & Jato, V. (2004). Environmental factors affecting the start of the pollen season and concentrations of airborne *Alnus* pollen in two localities of Galicia (NW Spain). *Annals of Agricultural and Environmental Medicine, 11*, 35–44.

Rodriguez Rajo, F. J., Medez, J., & Jato, V. (2005). Factors affecting pollination ecology of *Quercus anemophilous* species in north-west Spain. *Botanical Journal of the Linnean Society, 149*, 283–297.

Rodriguez-Rajo, F. J., Valencia-Barrera, R. M., Vega-Maray, A. M., Suarez, F. J., Fernandez-Gonzalez, D., & Jato, V. (2006). Prediction of airborne *Alnus* pollen concentration by using ARIMA models. *Annals of Agricultural and Environmental Medicine, 13*, 25–32.

Rogers, C. (2006). Knowledge gaps and hot topics in aerobiology. Paper presented at the 8th International Congress in Aerobiology, Neuchatel Switzerland.

Ronneberger, O. (2007). 3D invariants for automated pollen recognition. Thesis. Freiburg im Breisgau: University of Freiburg.

Ronneberger, O., Schultz, E., & Burkhardt, H. (2002). Automated pollen recognition using 3D volume images from fluorescence microscopy. *Aerobiologia, 18*, 107–115.

Rosenzweig, C., Casassa, G., Karoly, D. J., Imeson, A., Liu, C., Menzel, A., Rawlins, S., Root, T. L., Seguin, B., & Tryjanowski, P. (2007). Assessment of observed changes and responses in natural and managed systems. In M. L. Parry, O. F. Canziani, J. P. Palutikof, P. J. van der Linden, & C. E. Hanson (Eds.), *Climate Change 2007: Impacts, adaptation and vulnerability* (Contribution of Working Group II to the Fourth Assessment Report of the Intergovernmental Panel on Climate Change, pp. 79–131). Cambridge: Cambridge University Press.

Rötzer, Th, & Chmielewski, F.-M. (2001). Phenological maps of Europe. *Climate Research, 18*, 249–257.

Sánchez-Mesa, J. A., Galán, C., Martínez-Heras, J. A., & Hervás-Martínez, C. (2002). The use of a neural network to forecast daily grass pollen concentration in a Mediterranean region: The southern part of the Iberian Peninsula. *Clinical and Experimental Allergy, 32*, 1606–1612.

Sánchez-Mesa, J. A., Smith, M., Emberlin, J., Allitt, U., Caulton, E., & Galán, C. (2003). Characteristics of grass pollen seasons in areas of southern Spain and the United Kingdom. *Aerobiologia, 19*, 243–250.

Schaber, J., & Badeck, F.-W. (2003). Physiology-based phenology models for forest tree species in Germany. *International Journal for Biometeorology, 47*, 193–201.

Scheifinger, H., Menzel, A., Koch, E., Peter, Ch, & Ahas, R. (2002). Atmospheric mechanisms governing the spatial and temporal variability of phenological observations in central Europe. *International Journal of Climatology, 22*, 1739–1755.

Schueler, S., Schlunzen, K. H., & Scholz, F. (2005). Viability and sunlight sensitivity of oak pollen and its implications for pollen-mediated gene flow. *Trees, 19*, 154–161.

Schwartz, M. D. (Ed.). (2003). *Phenology: An integrative environmental science*. Dordrecht/Boston/London: Kluwer Academic Publishers. 564 pp.

Siljamo, P., Sofiev, M., Ranta, H., Linkosalo, T., Kubin, E., Ahas, R., Genikhovich, E., Jatczak, K., Jato, V., Nekovar, J., Minin, A., Severova, E., & Shalaboda, V. (2008). Representativeness of point-wise phenological *Betula* data collected in different parts of Europe. *Global Ecology and Biogeography, 17*(4), 489–502.

SKC (2010). Button Aerosol Sampler. http://www.skcinc.com/prod/225-360.asp. Accessed 5 October/2010.

Slini, T., Kaprara, A., Karatzas, K., & Moussiopoulos, N. (2006). PM10 forecasting for Thessaloniki, Greece. *Environmental Modelling & Software, 21*, 559–565.

Smith, M., & Emberlin, J. (2005). Constructing a 7-day ahead forecast model for grass pollen at north London, United Kingdom. *Clinical and Experimental Allergy, 35*, 1400–1406.

Smith, M., & Emberlin, J. (2006). A 30-day-ahead forecast model for grass pollen in north London, United Kingdom. *International Journal of Biometeorology, 50*, 233–242.

Sodeau, J., O'Connor, D., Hellebust, S., & Healy, D. (2010). Real-time spectrocopic measurments of PBAP in field and laboratory envirnments. Paper presented at the 9th International Congress on Aerobiology, Buenos Aires.

Sofiev, M., Siljamo, P., Ranta, H., & Rantio-Lehtimäki, A. (2006). Towards numerical forecasting of long-range air transport of birch pollen: Theoretical considerations and a feasibility study. *International Journal of Biometeorology, 50*, 392–402.

Spieksma, F. T., & Nikkels, A. H. (1998). Airborne grass pollen in Leiden, The Netherlands: Annual variations and trends in quantities and season starts over 26 years. *Aerobiologia, 14*, 347–358.

Spieksma, F. T., & Nikkels, A. H. (1999). Similarity in seasonal appearance between atmospheric birch-pollen grains and allergen in paucimicronic, size-fractionated ambient aerosol. *Allergy, 54*(3), 235–241.

Spieksma, F. T., Kramps, J. A., Van der Linden, A. C., Nikkels, B. H., Plomp, A., Koerten, H. K., & Dijkman, J. H. (1990). Evidence of grass-pollen allergenic activity in the smaller micronic atmosphere aerosol fraction. *Clinical and Experimental Allergy, 20*, 273–280.

Stach, A., Smith, M., Prieto Baena, J. C., & Emberlin, J. (2008). Long-term and short-term forecast models for Poaceae (grass) pollen in Poznań, Poland, constructed using regression analysis. *Environmental and Experimental Botany, 62*, 232–332.

Stark, P. C., Ryan, L. M., McDonald, J. L., & Burge, H. A. (1997). Using meteorologic data to predict daily ragweed pollen levels. *Aerobiologia, 13*, 177–184.

Stephen, E., Raftery, A. E., & Dowding, P. (1990). Forecasting spore concentrations – a time-series approach. *International Journal of Biometeorology, 34*, 87–89.

Sterling, M., Rogers, C., & Levetin, E. (1999). An evaluation of two methods used for microscopic analysis of airborne fungal spore concentrations from the Burkard Spore Trap. *Aerobiologia, 15*, 9–18.

Stöckli, R., & Vidale, P. L. (2004). European plant phenology and climate as seen in a 20-year AVHRR land-surface parameter data set. *International Journal of Remote Sensing, 25*, 3303–3330.

Stöckli, R., Rutishauser, T., Dragoni, D., O'Keefe, J., Thornton, P. E., Jolly, M., Lu, L., & Denning, A. S. (2008). Remote sensing data assimilation for a prognostic phenology model. *Journal of Geophysical Research, 113*, G04021. doi:10.1029/2008JG000781.

Stokstad, E. (2002). A little pollen goes a long way. *Science, 296*, 2314.

Suzuki, M., Tonouchi, M., & Murayama, K. (2008). Automatic measurements of Japanese cedar/cypress pollen concentration and the numerical forecasting at Tokyo metropolitan area.(Paper presented at the International Congress of Biometeorology, September 2008.

Tabachnick, B. G., & Fidell, L. S. (2001). *Using Multivariate Statistics*. New York: Harper Collins.

Tedeschini, E., Rodriguez-Rajo, F. J., Caramiello, R., Jato, V., & Frenguelli, G. (2006). The influence of climate changes in *Platanus* spp. pollination in Spain and Italy. *Grana, 45*, 222–229.

Thibaudon, M., & Lachasse, C. (2005). Phénologie: Intérêt et méthodes en aérobiologie – Interest of phenology in relation to aerobiology. *Revue française d'allergologie et d'immunologie clinique, 45*, 194–199.

Tormo, R., Munoz, A., & Silva, I. (1996). Sampling in aerobiology. Differences between traverses along the length of the slide in Hirst spore traps. *Aerobiologia, 12*, 161–166.

Toro, F. J., Recio, M., Trigo, M. M., & Cabezudo, B. (1998). Predictive models in aerobiology: Data transformation. *Aerobiologia, 14*, 179–184.

Torrigiani Malaspina, T., Cecchi, L., Morabito, M., Onorari, M., Domeneghetti, M. P., & Orlandini, S. (2007). Influence of meteorological conditions on male flower phenology of *Cupressus sempervirens* and correlation with pollen production in Florence. *Trees, 21*(5), 507–514.

Tucker, C. J., Pinzon, J. E., Brown, M. E., Slayback, D., Pak, E. W., Mahoney, R., Vermote, E., & El Saleous, N. (2005). An extended AVHRR 8-km NDVI data set compatible with MODIS and SPOT vegetation NDVI data. *International Journal of Remote Sensing, 26*(20), 4485–5598.

Tzima, F., Karatzas, K., Mitkas, P., & Karathanasis, S. (2007). Using data-mining techniques for PM10 forecasting in the metropolitan area of Thessaloniki, Greece, IEEE International Conference on Neural Networks – Conference Proceedings, art. no. 4371394 (pp. 2752–2757). *Proceedings of the 20th International Joint Conference on Neural Networks* (http://www.ijcnn2007.org. Organized by the IEEE Computational Intelligence Society and by the International Neural Network Society, Orlando, Florida, August 2007).

van Vliet, A. J. H., Overeem, A., de Groot, R. S., Jacobs, A. F. G., & Spieksma, F. T. M. (2002). The influence of temperature and climate change on the timing of pollen release in The Netherlands. *International Journal of Climatology, 22*, 1757–1767.

Vogel, H., Pauling, A., & Vogel, B. (2008). Numerical simulation of birch pollen dispersion with an operational weather forecast system. *International Journal of Biometeorology, 52*, 805–814.

Voukantsis, D., Niska, H., Karatzas, K., Riga, M., Damialis, A., & Vokou, D. (2010). Forecasting daily pollen concentrations using data-driven modelling methods in Thessaloniki, Greece. *Atmospheric Environment, 44*, 5101–5111. doi:10.1016/j.atmosenv.2010.09.006.

Voukantsis, D., Karatzas, D., Jaeger, S., & Berger, U. (2011). Personalized information services for quality of life: The case of airborne pollen induced symptoms. 12th International Conference, EANN 2011, and 7th IFIP WG 12.5 International Conference, AIAI 2011, Corfu, Greece, 15–18 September 2011. *Proceedings*, Part I (L. S. Iliadis, & C. Jayne (Eds.)) (pp. 509–515), ISBN 978-3-642-23956-4. New York: Springer.

Walker, D. A., Epstein, H. E., Jia, G. J., Balser, A., Copass, C., Edwards, E. J., Gould, W. A., Hollingsworth, J., Knudson, J., Maier, H. A., Moody, A., & Raynolds, M. K. (2003). Phytomass, LAI, and NDVI in northern Alaska: Relationship to summer warmth, soil pH, plant functional types, and extrapolation to the circumpolar Arctic. *Journal of Geophysical Research, 108*, 8169.

Ward, S. (2008). The earth observation handbook – Climate change special edition 2008 (esa sP-1315). ESA Communication Production Office, ESTEC, Postbus 299, 2200 AG Noordwijk, The Netherlands, 278 p.

White, M. A., Thornton, P. E., & Running, S. W. (1997). A continental phenology model for monitoring vegetation responses to interannual climatic variability. *Global Biogeochemical Cycles, 11*(2), 217–234.

Winkler, H., Ostrowski, R., & Wilhelm, M. (2001). *Pollenbestimmungsbuch der Stiftung Deutscher Polleninformationsdienst*. Takt Verlag: Paderborn.

Zhang, Z., Zhe, J., Chandra, S., & Hu, J. (2005). An electronic pollen detection method using Coulter counting principle. *Atmospheric Environment, 39*, 5446–5453.

Ziello, C., Böck, A., Jochner, S., Estrella, N., Buters, J., Weichenmeier, I, Behrendt, H., & Menzel, A. (2010). Bio-monitoring of the Zugspitze area: Linking phenological, meteorological and palynological data along an altitudinal gradient. Paper presented at Erste Wissenschaftliche Tagung Umweltforschungsstation Schneefernerhaus, Iffeldorf, Mai 2010.

Chapter 5
Airborne Pollen Transport

Mikhail Sofiev, Jordina Belmonte, Regula Gehrig, Rebeca Izquierdo, Matt Smith, Åslög Dahl, and Pilvi Siljamo

Abstract This chapter reviews the present knowledge and previous developments concerning the pollen transport in the atmosphere. Numerous studies are classified according to the spatial scales of the applications, key processes considered, and the methodology involved. Space-wise, local, regional and long-range scales are distinguished. An attempt of systematization is made towards the key processes responsible for the observed patterns: initial dispersion of pollen grains in the nearest vicinity of the sources at micro-scale, transport with the wind, mixing inside the atmospheric boundary layer and dry and wet removal at the regional scale, and the

---

M. Sofiev (✉)
Air Quality Research, Finnish Meteorological Institute, P.O. Box 503,
Erik Palménin Aukio 1, 00560 Helsinki, Finland
e-mail: mikhail.sofiev@fmi.fi

J. Belmonte
Botany Unit and Institut de Ciència i Tecnologies Ambientals,
Universitat Autònoma de Barcelona, Barcelona, Spain

R. Gehrig
Bio- & Umweltmeteorologie, MeteoSwiss, Zurich, Switzerland

R. Izquierdo
Centre de Recerca Ecològica I Aplicacions Forestals,
Universitat Autònoma de Barcelona,
Barcelona, Spain

M. Smith
National Pollen and Aerobiology Research Unit,
University of Worcester, Worcester, UK

Å. Dahl
Department of Biological and Environmental Sciences,
University of Gothenburg, Gothenburg, Sweden

P. Siljamo
Meteorological Research, Finnish Meteorological Institute, P.O. Box 503,
Erik Palménin Aukio 1, 00560 Helsinki, Finland

long-range dispersion with synoptic-scale wind, exchange between the boundary layer and free troposphere, roles of dry and wet removal, interactions with chemicals and solar radiation at the large scales.

Atmospheric dispersion modelling can pursue two goals: estimation of concentrations from known source (forward problem), and the source apportionment (inverse problem). Historically, the inverse applications were made first, mainly using the simple trajectory models. The sophisticated integrated systems capable of simulating all main processes of pollen lifecycle have been emerging only during last decade using experience of the atmospheric chemical composition modelling.

Several studies suggest the allergen existence in the atmosphere separately from the pollen grains – as observed in different parts of the world. However, there is no general understanding of the underlying processes, and the phenomenon itself is still debated. Another new area with strongly insufficient knowledge is the interactions of airborne allergens and chemical pollutants.

Keywords Airborne pollen • Atmospheric pollen transport • Dispersion modelling

5.1 Introduction

The atmospheric pathway is the fastest and the simplest way for biological agents to spread over terrestrial ecosystems. Many organisms can significantly increase the efficiency of their movements by taking advantage of air currents (Isard et al. 2005). Biota that is present in the atmosphere ranges in size from very small (viruses, bacteria, pollen and spores) to quite large (seeds, aphids, butterflies and moths, songbirds, and waterfowl) (Gage et al. 1999; Westbrook and Isard 1999). The link between these biological systems and the atmosphere is the key to understanding the population dynamics of and diseases spread by, aerobiota. Within this chapter, the emphasis is placed on identifying biologically and medically relevant temporal and spatial scales of atmospheric motions and meteorological parameters, which help control the abundance and distribution of airborne biota, such as pollen and other aeroallergens.

Biologically-relevant dispersion of bioaerosols affects the structure of ecosystems, since pollen is responsible for gene flow (Ellstrand 1992; Ennos 1994; Burczyck et al. 2004; Belmonte et al. 2008), and it contributes in determining the spatial distribution of plant species (Ellstrand 1992; Smouse et al. 2001; Sharma and Khanduri 2007; Schmidt-Lebuhn et al. 2007; Belmonte et al. 2008). Therefore, understanding the pollen gene dispersal is instrumental for the interpretation of the biogeographic range of plants and plant conservation issues. A review by Di-Giovanni and Kevan (1991) on the factors affecting the pollen dispersion in natural habitats can be recommended for the corresponding processes.

Apart from gene flow, the transport of bioaerosols causes concern because of its potential to distribute pathogens and allergens, which can affect human health, agriculture, and farming (Belmonte et al. 2000; Griffin et al. 2001a, b; Taylor 2002; Brown and Hovmoller 2002; Shinn et al. 2003; Garrison et al. 2003; Wu et al. 2004; Kellogg et al. 2004; Griffin 2007; Paz and Broza 2007; Polymenakou et al. 2008).

5 Airborne Pollen Transport

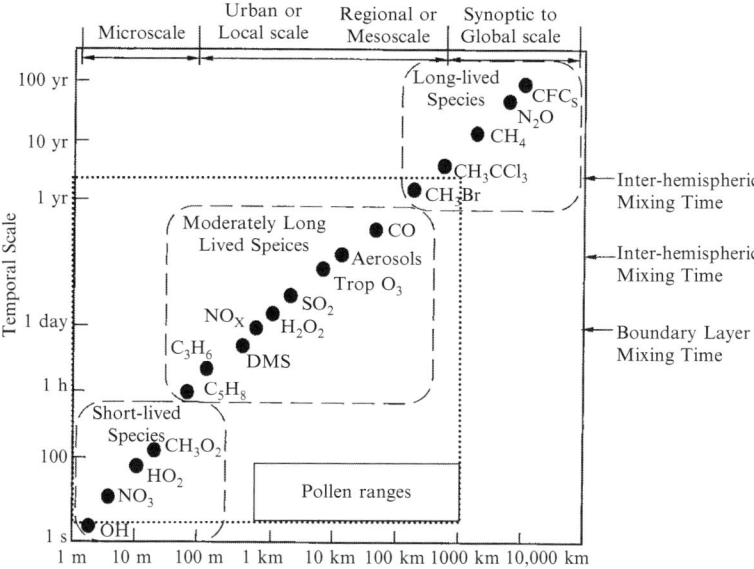

Fig. 5.1 Spatial and temporal scales of variability of the atmospheric constituents (Modified from Seinfeld and Pandis 2006)

The pollen records from aerobiological monitoring sites have traditionally been interpreted as if the grains always originate from the local environment. Consequently, pollen forecasting tools have been designed by taking into account only local meteorological variables and phenological observations in the neighbourhood. This view is currently changing to acknowledge much broader bioaerosol movements, based on increasing evidence of pollen and spore transport at much greater distances, including continental (Belmonte et al. 2000,2008; Sofiev et al. 2006a; Siljamo et al. 2007, 2008b, c; Skjøth et al. 2009) and intercontinental scales (Prospero et al. 2005; Kellogg and Griffin 2006; Rousseau et al. 2008).

5.1.1 Basic Terms

5.1.1.1 Spatial and Temporal Scales

Pollen-related processes in the atmosphere take place at a wide range of scales. For the needs of the current chapter the following terms will be used following the classical definitions after Seinfeld and Pandis (2006) – Fig. 5.1:

- Processes at the micro-scale are connected with pollen release and take place within a few metres from the plants
- Local-scale processes include initial dispersion of grains that happen within the nearest kilometre from the source

- Regional and meso-scales are considered as synonyms and cover the processes responsible for the dispersion and removal of the bulk of pollen grains – at distances of up to a hundred kilometres.
- The hierarchy of the scales related to long-range transport consists of synoptic, continental, and global scales, which include processes of up to 1,000–2,000 km, up to 5,000 km, and over 5,000 km, respectively.

Connections between the above spatial and corresponding temporal scales are shown in Fig. 5.1.

Distinction between the scales is not always unequivocal and specific borders can vary depending on the specific application and criteria used. For the purposes of this review, we will also consider the cases as "local" or "regional" if the conditions at both source and the receptor points can be (roughly) described by a single or few observation stations. In the long-range transport case, such a description always requires many stations distributed over the area and atmospheric modelling as a way to evaluate the transport conditions.

5.1.1.2 Pollen Life Time in the Atmosphere

Atmospheric lifetime is the key parameter for each tracer, which has a direct connection to its spatial scale of distribution and temporal scale of its variations (Seinfeld and Pandis 2006). Several example species are marked in Fig. 5.1 following their characteristic lifetimes. With pollen and allergens, the situation is more complicated. Indeed, the pollen atmospheric lifetime of a few days (due to substantial gravitational sedimentation) defines it as a local-scale pollutant with some minor connection to regional scales (Sofiev et al. 2006a). However, as discussed below, in many cases the released amount is so large that the medical impact can be substantial at much larger distances – up to continental scales. This ambiguity is reflected in Fig. 5.1 where the pollen-related processes are delineated by a separate rectangular.

5.1.1.3 Aerobiological Phases of Pollen and Other Biogenic Aerosols in the Atmosphere

A schematic model for passively transported biological substances includes presentation of the biological material, its release in the atmosphere, dispersion, transformation, deposition, and impact (Fig. 5.2, extended from Isard and Gage 2001). Sometimes, after deposition, the grains can be subjected to refloating processes, thus being resuspended back into the air. In this chapter we will consider only release, transport, transformation, and deposition phases as the most important and studied so far.

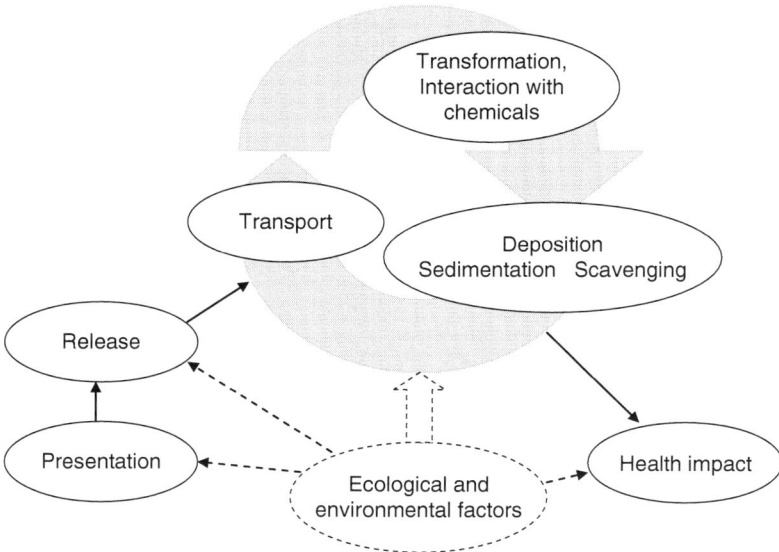

Fig. 5.2 Phases of aerobiological processes related to transport of chemically inactive biogenic aerosol (Extended from Isard and Gage 2001)

5.1.1.4 Types of Biogenic Aerosols

The most-studied biogenic aerosol is evidently the pollen grain. It transfers the male gametophyte to the female reproductive organs, which is termed as pollination process. Once pollen grains are deposited on the reproductive female organs the recognition process starts via protein exchange to allow the germination and liberation of the male gametes for fecundation. Several proteins inside the grain are considered to be allergens since, in some occasions, the human immune system can react on their presence by triggering allergenic reactions. Direct studies of these proteins in connection to atmospheric transport are very limited, therefore the main attention in this chapter will be given to pollen grains. The allergen-related studies will be included wherever possible. Other types of biogenic particulates, such as spores, seeds, etc., are beyond the scope of the review.

5.1.1.5 Main Features of Pollen and Allergen as Atmospheric Tracers

A pollen grain, from the point of view of the atmospheric transport, is a very large but comparatively light aerosol. Aerodynamic features of most of the pollen grains (except for the largest ones with diameter over 100 μm) allow the classical model considerations when the particle is considered to be embedded into all atmospheric flows including the small-scale turbulent eddies (Sofiev et al. 2006a). The life span of the particles in the atmosphere strongly depends on their deposition intensity, i.e.

the sum of dry and wet deposition. For coarse particles, the gravitational settling is the most important deposition pathway (Seinfeld and Pandis 2006), which makes the sedimentation velocity the primary parameter deciding the atmospheric life time of pollen. For birch, it is about 1.2 cm sec^{-1} (Sofiev et al. 2006a), which corresponds to the lifetime ranging from a few tens of hours up to a couple of days depending on the vertical transport and mixing.

Pollen is not soluble in water but can easily be scavenged via impaction (in case of sub-cloud scavenging). Processes occurring inside the clouds are very poorly studied but one can expect that the grains can be embedded into the forming droplets and scavenged together with them.

A pollen allergen is usually a sub-micron aerosol with high water solubility (Vrtala et al. 1993; Taylor et al. 2004). It can therefore be scavenged with precipitation but has no substantial dry deposition velocity (see Chap. 19 of Seinfeld and Pandis 2006). As a result, one can expect the allergen to stay in air much longer than pollen – days and, possibly, weeks in the case of no precipitation.

Among the transformation processes occurring during the atmospheric transport, the most frequent one is a loss and gain of water depending on the air humidity, an ability of the pollen grain called harmomegathy (Wodehouse 1935). Sometimes this situation provokes the pollen grain rupture and allergen release. Secondly, it is a loss of viability due to temperature and UV radiation. Thirdly, a chemical damage of the grains by strong oxidants can take place in polluted environments. Finally, the release of the pollen content due to the grain rupture or (pseudo-) germination has been confirmed by several studies (Pacini et al. 2006; Motta et al. 2006; Taylor et al. 2002, 2004, 2006).

5.2 Meteorological Drivers of Pollen Dispersion in the Atmosphere

The pollen lifecycle in the atmosphere starts from the release from the anthers – arguably the smallest-scale relevant atmospheric process. In many anemophilous trees, the anther's burst is a result of dehydration due to high temperature, solar exposure, low humidity and moderate wind (see Helbig et al. 2004; Linskens and Cresti 2000 and references therein). In other plants, floral parts, such as the filaments in grasses actively push the anthers into an exposed position, and anther's dehiscence may be a result of passive desiccation or active reabsorption. Reabsorption enables the anther opening at any time of day, whereas evaporation may occur only during dry hours. The Asteraceae, which includes the genera of *Artemisia* and *Ambrosia*, is characterized by the secondary pollen presentation, which in this case means that the pollen mass is pushed to the top of the flower where it is exposed to atmospheric stress (von Wahl and Puls 1989; Kazlauskas et al. 2006). In this case, the pollen detachment from the capitulum can take place later than the anther opening. In particular, it could take place at any time during the day when the wind and turbulence are strong enough to pick up the pollen from the flower. In the Urticaceae, pollen emission is explosive and is caused by the spring-like release of the anthers caused by dehydration. A more detailed account of pollen release can be found in the Chap. 3 of this book and dedicated reviews, such as (Pacini and Hesse 2004).

Once the pollen grains are released into the atmosphere, the mechanical force of air flow, either induced by mean wind or by turbulent eddies, becomes the only process that keeps the grains in the air. The initial mixing and uplift driven by the turbulent motions largely determines the fraction of the released grains that will come to larger-scale dispersion (e.g. Gregory 1961). At the meso-scale, the mean wind becomes the main transport force while the turbulent mixing keeps the grains aloft and further redistributes them along the vertical – inside and beyond the atmospheric boundary layer. At regional and large scales, both horizontal and vertical mean-wind components are responsible for the transport on-par with turbulent mixing. At all scales, the processes of dry deposition, by impact or gravitational sedimentation, as well as wet deposition by scavenging with precipitation, are responsible for the removal of grains from the atmosphere. However, their importance varies being the highest at large scales.

5.3 Micro- and Local Scales

Micro- and local-scale transport of pollen grains includes the initial dispersion of pollen grains from the anthers and the transport over the first kilometre(s) from the source. An outlook of the main factors affecting the dispersion can be found in (Di-Giovanni and Kevan 1991). The knowledge about micro- to local scale pollen transport is based on field experiments with pollen release from a well-defined source. The dispersion of the pollen plume at these scales can be modelled with various types of dispersion models (Lagrangian, Gaussian plume model, large eddy simulation, or quasi-mechanistic models – e.g. Aylor et al. 2006; Arritt et al. 2007; Chamecki et al. 2009; Jarosz et al. 2003, 2004; Klein et al. 2003; Kuparinen 2006; Kuparinen et al. 2007; Schueler and Schlünzen 2006).

Experiments studying the dispersion and deposition of pollen were already made in the 1960s and 1970s by Raynor, Ogden and Hayes. For ragweed pollen, Raynor et al. (1970) used a set of point and area sources surrounded by four to five rings of samplers located concentrically from 1 to 69 m from the centre of the plot. The samples were taken at four heights from 0.5 to 4.6 m. It was found that about half of the ragweed pollen grains reaching the edge of the source are still airborne some 55–65 m further away. Extrapolation towards greater distances indicates that about 1% of the pollen grains remain airborne at 1 km. Raynor et al. (1970) state that since pollen clearly disperses similarly to other (small) particles, calculation of pollen dispersion by use of existing diffusion models should be practical. Further studies for Timothy grass and maize pollen were published by Raynor et al. (1972a, b).

Particular interest to the pollen transport was caused by the gene flow problem at local scales. Gleaves (1973) has developed a local-scale empirical gene-flow model and performed a series of sensitivity studies of the exchange efficiency of the genetic material depending on the mutual position of the plants, transport conditions, etc. A motivation for the study was that, despite "pollen can be blown hundreds of miles, the bulk of cross-pollination occurs over very short distances". Therefore, the analysis has been performed for the transport range of a few tens of metres. Govindaraju (1988) has statistically demonstrated that the pollination mechanism directly influences the gene flow and showed that the dependence is the strongest for the wind-pollinating plants, being much weaker for animal- and self-pollinating species. Quantitative estimates for the tree species can be found in Govindaraju (1989).

Several recent field experiments have been designed to estimate the three-dimensional plume shape of airborne pollen grains, as well as the amount deposited. The outcome was also used for evaluation of the dispersion models. With numerous samplers placed around the source, instrument towers, and occasionally even remote-piloted planes and aircrafts, pollen concentrations up to several tens of meters above the ground were measured. The experiments for maize were made by Jarosz et al. (2003, 2004), Aylor et al. (2003, 2006), Boehm et al. (2008), Klein et al. (2003), Arritt et al. (2007); for ragweed pollen by Chamecki et al. (2009); and for oilseed rape pollen by McCartney and Lacey (1991).

The introduction of genetically modified organisms (GMO) in agriculture favoured the research of pollen transport processes on local scales for establishing a basis for assessment of the gene flow from these plants to other crops or to natural vegetation. An overview of these efforts is given by the European Environment Agency (2002). This report considers the significance of pollen-mediated gene flow from six major crops: oilseed rape, sugar beet, potatoes, maize, wheat and barley.

The EEA (2002) report confirmed that the majority of airborne pollen is deposited at very short distance from the pollen source, although out-crossing of maize pollen has been recorded at up to 800 m from the source and, in extreme cases, there is the evidence of wind transfer of oilseed rape pollen up to at least 1.5 km (Timmons et al. 1995). Studies of maize pollen dispersal from small plots (e.g. Raynor et al. 1972b; Jarosz et al. 2003) have suggested that there is relatively little impact of pollen beyond a few hundred meters from the source. However, the impact of pollen may extend to much greater distances because the dispersal distribution has a long extending tail (Aylor et al. 2003). The tail is expected to be much more evident for large source areas (Aylor et al. 2006). Brunet et al. (2004) used measurements taken by aircraft and found the viable pollen throughout the entire depth of the atmospheric boundary layer. Their results imply that updrafts due to turbulent eddies in the boundary layer overcome the terminal fall velocity of maize pollen grains and transport pollen to considerable heights, so that it could travel long distances before settling (Arritt et al. 2007).

One of the key parameters originating from the local-scale studies is the estimation of the pollen source strength, as seen at various scales, and its dependence on the plant type and atmospheric conditions.

A detailed micro-scale experiment by Laursen et al. (2007) included 15-min measurements of the key meteorological parameters accompanied by the pollen concentration monitoring in the nearest vicinity of the plants. An empirical multi-linear regression fit was obtained for concentrations of *Artemisia* grains as a function of temperature, humidity, and wind speed. As was shown, wind and temperature are synergetic, i.e. higher wind and temperature both promote pollen release. Interestingly, relative humidity was not included into the equation (although measured during the experiment).

Aylor et al. (2006) estimated the maize pollen release rates 10–700 grains per square meter per second depending on weather factors and the day-to-day and diurnal variation of pollen production. In many cases pollen is not released steadily but reacts on wind gusts that "shake" the anthers. Grasses are known to disperse pollen at a certain time of the day, which varies between the species and environmental conditions. Some grass species exhibit a bimodal diurnal pattern (Davidson 1941; Subba Reddi et al. 1988).

Ambrosia pollen is initially shed from the staminate flowers in large pollen clumps containing hundreds of grains. Turbulent wind stress breaks these pollen clumps quickly into smaller ones, so that at regional scales *Ambrosia* pollen is released as a largely homogeneous plume. Modelling with large eddy simulation by Chamecki et al. (2009) provided an estimate of the total pollen emission of a ragweed field and its deposition partition inside the field and downwind. About 60% of pollen is deposited inside the field, 14% was deposited downwind inside the simulation domain, which is about 1,000 m long, while 26% of the total pollen remained airborne and left the domain.

Bricchi et al. (2000) found that the highest *Platanus* pollen deposition is recorded close to the plants and decreases quite quickly at a distance greater than 400 m. More than 88% of the pollen is deposited within a range of 2,750 m from the source.

A measuring experiment of local *Betula* pollen dispersion was presented by Michel et al. (2010), aiming at 3-D representation of the plume originating from an isolated birch stand.

A study by Spieksma et al. (2000) about the influence of nearby stands of *Artemisia* on pollen concentrations at street-level versus roof-top-level shows that the street-level concentrations were in average 4.8 times higher than the ones at 25 m on the roof.

5.4 Regional Scale

Regional-scale pollen dispersion over distances of about a hundred kilometres (or slightly more) poses different challenges in comparison with local scales. First of all, only a fraction of emitted pollen reaches these scales – so-called "regional component" (Faegri et al. 1989) or "escape fraction" (Gregory 1961) of the emission. Secondly, the observations at the receptor location are not fully representative for the pollen release area. They can, however, provide rough estimates for the transport conditions between the source and receptor points.

This connection of regional-scale meteorology and climatology with the pollen dispersion has been known for decades and is used in practically all studies. For example, the relation between measurements of *Castanea sativa* pollen concentration and the meteorological variables registered in different locations of Po Valley (Italy) were studied by Tampieri et al. (1977). Peeters and Zoller (1988) and Gehrig and Peeters (2000) connected the *Betula* and *Castanea* pollen distribution in Switzerland with such factors as altitude, prevailing meteorological patterns during spring time, etc.

The situations of strong regional pollen dispersion and the conditions favouring the regional-scale transport of *Juniperus ashei* pollen were described by Van de Water et al. (2003) and Van de Water and Levetin (2001). The authors qualitatively estimated the conditions favouring or precluding the released pollen to be transported regionally. A trajectory model HYSPLIT and five observation stations were used to obtain a complex pattern of the regional pollen distribution and to evaluate the allergenic threat in the region of Tulsa city. Conditions found to be favouring the regional-scale transport were: sunshine level of greater than 65%, temperature over 5 °C, relative humidity less than 50% and wind speed exceeding 1.8 m/s. The considered region was about 1,000 by 1,000 km with a characteristic transport distance being about 300–400 km in case of favourable conditions.

Many studies rely on single-point observations at a receptor site, thus concentrating on evaluation of the origin of the observed pollen. Strong indication of the regional-scale transport in such cases is high concentrations of pollen during night time when the local release is low.

Examples of these studies are: the multi-species study of Damialis et al. (2005) in Thessaloniki, the study of the regional impact of Copenhagen birch sources (Skjøth et al. 2008b), impact assessment of the southern England birch population onto London (Skjøth et al. 2009), evaluation of the role of regional and long-range transport on the Lithuanian pollen seasons by Veriankaité et al. (2010); the airborne monitoring during a cruise across the East Mediterranean Sea (Waisel et al. 2008). Most of the above studies provided qualitative connections between the regional-scale meteorological patterns and pollen distribution.

Some information helping to detect the pollen transport can be obtained using the local phenological observations together with the pollen counts. For example, the differences between flowering dates recorded by phenology network and pollen counts of *Betula*, Poaceae and *Artemisia* observed in Germany were correlated with the regional and large-scale transport by Estrella et al. (2006). Similar comparisons were used by Veriankaité et al. (2010), Siljamo et al. (2008b) and others. However, such comparison has to be performed with care because of very large uncertainties of the phenological observations (Siljamo et al. 2008a).

In contrast with the above studies stressing the episodic features of the regional transport, several assessments have concentrated on its long-term characteristics and impact. The comparison of multi-annual data sets of airborne and deposited pollen in northern Finland by Ranta et al. (2008) were used for qualitative description of the main transport directions in the region. The other long-term studies included the redistribution of genetically modified creeping bentgrass pollen (Van de Water et al. 2007); and paleopalynological studies (Romero et al. 2003; Hooghiemstra et al. 2006).

5.5 Long-Range Transport (LRT)

The large-scale dispersion of atmospheric constituents is controlled by synoptic-, continental-, or hemispheric- scale meteorological phenomena. In particular, it means that even combined observations at both source and receptor points cannot describe the transport conditions. In fact, in many cases the mere connection between the sources and receptors is difficult to establish due to complicated large-scale dispersion patterns. Under such constraints, the indication of the LRT can be either foreign pollen grains, which cannot be produced locally, or "wrong" time of appearance of the grains, which is significantly outside the local flowering period. Application of dispersion models in either forward or inverse mode is practically inevitable for analysis of such cases.

Studies of exotic grains observed in various parts of the globe started at least half a century ago, mainly in application to polar altitudes where the diversity of plants is comparatively low and exotic pollen is easier to find (Nichols 1967; Ritchie and Lichti-Federovich 1967; Janssen 1973; Ritchie 1974). Probably, these were the first unequivocal proofs of existence of such long-range dispersion of pollen material. Similar studies continued and extended towards temperate zones in later years (Hicks 1985; Porsbjerg et al. 2003; Hicks et al. 2001; Campbell et al. 1999; Rousseau et al. 2003, 2004, 2006). In some cases, specific meteorological conditions could be identified as a driving force for the exotic pollen appearance (Campbell et al. 1999). However, usually the number of the exotic grains found is small and the practical importance of the trans-continental transport for the human allergy and gene flow has proven to be low.

More significant amounts of pollen are episodically transported from Northern Africa, which has been identified as a source area in a number of studies. For instance, Van Campo and Quet (1982) identified several pollen types that had been transported from North Africa to Montpellier (France) together with mineral desert dust. Further south, *Cannabis sativa* (marihuana) pollen originating in Morocco was detected in Malaga, Southern Spain (Cabezudo et al. 1997) and *Cannabis*, *Cupressus*, *Pinus*, *Platanus* and *Sambucus* pollen were observed in Cordoba (South Spain) exclusively during African-dust events (Cariñanos et al. 2004). In addition, the source areas of several LRT episodes in Tenerife (Canary Islands) that originated from Mediterranean region were traced to the Saharan sector and the Sahel (Izquierdo et al. 2011).

A strong association of biological particles with desert dust was also suggested by Kellogg and Griffin (2006). Dust clouds generated by storm activity over arid land can result in mineral particles combined with viruses, bacteria (Hua et al. 2007; Hervàs et al. 2009), fungal spores (Griffin 2004, 2007; Griffin and Kellogg 2004; Griffin et al. 2001a, 2003, 2007, 2006; Kellog and Griffin 2006; Wu et al. 2004; Garrison et al. 2006; Schlesinger et al. 2006; Lee et al. 2007) and pollen being raised to altitudes over 2 km and then transported for thousands of kilometres, i.e. at a planetary scale. For example, viable microorganisms and fungal spores from Africa were sampled at Barbados after being transported by African dust plumes (Prospero et al. 2005). Intercontinental mineral dust transport has been the subject of much attention over decades (Guerzoni and Chester 1996; Goudie and Middleton 2001; Prospero et al. 1996; Zhang et al. 1997), but further research is needed on the biological component associated to the dust.

Finally, the study of aerobiological long-range transport has stimulated a new research line about the viability of pollen (Bohrerova et al. 2009), bacteria (Hervàs et al. 2009) and fungal spores (Gorbushina et al. 2007) after the long distance dispersion.

In Europe, numerous long range transport episodes have been identified in Fennoscandia (Ranta et al. 2006; Oikonen et al. 2005). Franzen and Hjelmroos (1988) observed pollen transport from Germany, Holland and England to the Swedish coast, and Franzen (1989) and Franzen et al. (1994) documented the arrival of pollen grains to Fennoscandia from the Mediterranean. A strong episode of the ragweed pollen transport from southern Europe to Finland was recorded in 2005 and traced back to the source areas in Hungary. A particularly specific spring season in Europe took place in 2006, when a strong plume of birch pollen was transported from Russia over the whole Europe and reached Iceland. The pollen cloud was mixed with the dense smoke from wild-land fires, which allowed its easy identification and follow-up by chemical observations and air quality models (e.g. Saarikoski et al. 2007).

As examples of transport episodes from East to West Europe, there are episodes of *Betula* pollen coming from Poland and Germany to Denmark (Skjøth et al. 2007) and from Russia to Finland (Siljamo et al. 2008b). In the opposite direction, *Betula* pollen arrived in Lithuania from Latvia, Sweden, Denmark, Belarus, Ukraine, Moldava, Germany and Poland (Veriankaité et al. 2010). Also, episodes of the pollen of the tree species *Fagus sylvatica* (beech) reaching Catalonia were traced back to central Europe (Belmonte et al. 2008). Complicated large-scale patterns have been found for Moscow by Siljamo et al. (2008b), who also reported strong bi-directional exchange of birch pollen between Russia and Finland.

The LRT episodes are not specifics of the tree pollen only. Thus, Smith et al. (2005) has registered Poaceae pollen in the UK originating from the continental Europe. The pollen spectrum of Lithuania was also affected by long range transport of several pollen types coming from other European regions (Sauliene et al. 2007).

As an outcome of a series of studies, *Ambrosia* distribution pathways over Europe were found to be highly irregular and episodic. Thus, the pollen registered in Poland, the Balkans and Italy was shown to originate near the Pannonian Plain (Cecchi et al. 2006, 2007; Stach et al. 2007; Kasprzyk 2008; Smith et al. 2008; Šikoparija et al. 2009) and Ukraine (Kasprzyk et al. 2010). In contrast, other studies recorded the transport in the opposite direction from France, Italy and Croatia to Hungary (Makra and Palfi 2007), as well as from France to Switzerland (Clot et al. 2002). There is also evidence of *Ambrosia* pollen transport to Catalonia (northeastern Spain) from France (Belmonte et al. 2000). Furthermore, an increasing risk of LRT *Ambrosia* episodes was detected in Scandinavia due to the rapid spread of ragweed in North-Northeast Europe (Dahl et al. 1999).

In North America, Raynor and Hayes (1983) applied the trajectory model to cases of pollen wet deposition in Albany (New York) after long-distance transport from the source areas located at South-Southwest. The arrival of *Juniperus ashei* pollen released in southern Oklahoma and Texas to Tulsa has been reported by Rogers and Levetin (1998) and Van de Water et al. (2003).

In South America, extra-regional pollen of *Celtis* coming from the northeast and of *Nothofagus* from the southwest has been found to contribute to Mar de Plata City (Argentina) pollen records (Gassmann and Pérez 2006).

In the north-western India, bioaerosols collected during dust storms sporadically contained pollen from Himalayan species (Yadav et al. 2007). The presence of pollen grains from trees forming forests at much lower latitudes has been evidenced in the Arctic environment (Bourgeois 2000; Savelieva et al. 2002; Rousseau et al. 2003, 2004, 2005, 2006, 2008). Extra-regional pollen transport has also been found in Antarctic (Wynn-Williams 1991), Arctic (Campbell et al. 1999), Australia (Salas 1983; Hart et al. 2007), and New Zealand (Moar 1969).

5.6 Release of Pollen Allergens from Grains

Probably the most important process that takes place during the pollen transport is the release of allergen from the grains. In a dry atmosphere, pollen is very stable and can keep its content over years (Stanley and Linskens 1974). However, under specific conditions the allergen release can take place within minutes: (i) a high relative air humidity; (ii) thunderstorms and heavy rain; (iii) high concentrations of air pollutants. According to Behrendt et al. (1997) under humid conditions, allergens are released from pollen grains in the process that resembles pollination. The existence of free allergen in air and its importance for the pollination season were confirmed by a series of field observations, such as Busse et al. (1972) for *Ambrosia*, Stewart and Holt (1985) for grass, and Rantio-Lehtimaki and Matikainen (2002), Rantio-Lehtimaki (2002), Matikainen and Rantio-Lehtimaki (1999), and Rantio-Lehtimaki et al. (1994) for birch.

Pollen allergens are generally glycoproteins. The majority of them are found in a limited number of protein families. Their biological functions are presumably related to the recognition, attachment, growth and development of the pollen tube on and within the pistil, i.e. to hydrolysis of proteins, polysaccharides, and lipids, binding of metal ions and lipids, and to the cytoskeleton (Radauer et al. 2008). Allergen activity can be detected, depending on the specific role of the protein in question, both before and after germination (Alché et al. 2002; Buters et al. 2010). These proteins were found in organelles, such as mitochondria, polysaccharide particles, starch granules, and endoplasmatic reticulum (Behrendt and Becker 2001; Rodríguez-García et al. 1995). Allergens are sometimes stored in the ectexine of the outer pollen wall. Rapid elution and water solubility is considered an important prerequisite for a protein to behave as a major allergen (Grote et al. 2001; Gupta et al. 1995).

The allergen release from pollen is responsible for bio-availability of the allergen (Behrendt and Becker 2001) but it also releases into the atmosphere the new particles with entirely different transport features. Indeed, using the formulations of Sofiev et al. (2006a) and applying them to the allergen size distribution observed by Taylor et al. (2004) or Miguel et al. (2006), one obtains the sedimentation velocity for allergen to be 0.01–0.1 mm sec^{-1}, compared to 12 mm sec^{-1} for pollen. Allergens can stay in air much longer than pollens and thus be transported over much longer distances. In contrast to pollen grains, which due to their size do not penetrate into the lower human airways, the allergen aerosols are respirable and can provoke stronger immune response than pollen itself (Motta et al. 2006).

The allergen liberation is well reproduced in laboratories and can occur within a few tens of minutes in favourable conditions (Behrendt et al. 1997; Behrendt and

Becker 2001; Taylor et al. 2002, 2004, 2007). The release can take place through the pollen walls or through the pollen tube wall. The liberation is pH- and temperature dependent (Behrendt et al. 1997). It is yet unclear how frequent this phenomenon is present in the real atmosphere because the field measurement of allergen are extremely complicated and often give semi-qualitative results based on the indirect indicators (Matikainen and Rantio-Lehtimaki 1999; Buters et al. 2010).

Schäppi et al. (1997) found that atmospheric concentrations of birch pollen allergen are correlated with birch pollen counts, but also that concentration of fine particles associated to Bet v 1 dramatically increase with light rainfall. They suggested that deposited pollen was stimulated to germinate, and that these particles were liberated as pollen tubes dried out when the rain was over. In Derby, UK, the pollen levels affected the number of emergency visits for asthma during days with light rainfall, but not during dry days (Lewis et al. 2000).

During thunderstorms, outbreaks of allergenic reactions have been registered and a high fraction of the patients were sensitive to both grass pollen and fungal spores (Pulimood et al. 2007). It was suggested that pollen grains, spores and other bioaerosols are carried into a cloud base, where they burst. Cold outflows transport the debris downwards where it can cause the observed asthma exacerbations (Marks et al. 2001).

In contrast to results from other studies, Buters et al. (2010) in Munich did not find any Bet v 1 in the fraction 0.12–2.5 µm. A hypothesis was suggested that either no allergen was present in the particles of this size range or the allergen was absorbed to diesel soot particles and became invisible for the technique used. Diesel soot particles bound to the grass allergen Lol p 1 under in vitro conditions were previously found by Knox et al. (1997).

An indirect way to detect the allergen in the air is to monitor the long-range transport plumes. Due to sharply different features of the particles, pollen grains are deposited faster than allergen and, after some travel time, the only aerosols present in air will be the allergen. The other option is to use size-segregated observations, thus separating pollens and allergens in different filters. That way, however, is difficult because of: (i) non-ideal filtering of coarse particles during sampling, which results in contamination of the fine-particle filter with pollen grains, (ii) very low allergen concentrations in ambient air – a few nanograms m^{-3} which are difficult to measure, (iii) addiction of allergen to the black carbon aerosols originating from diesel exhaust, which can make the detection of the allergens complicated or impossible.

5.7 Inter-Action Between Pollen and Chemical Pollutants

At all spatial and temporal scales, pollen, spores and, if present in a free form in the air, allergens are subjected to chemical and physical interactions with other atmospheric constituents. Chemical pollution can stress both the pollinating plants and the pollen grain, which cause qualitative and quantitative changes of the pollen content. The impact starts already during the pollen formation. For example, Aina et al. (2010) found an increased amount of the allergenic proteins in the grass *Poa annua* if the plant is grown in soil contaminated with cadmium. Similarly, the pollen vitality of *Parietaria judaica* was found maximal in soils enriched with heavy metals (Fotiou et al. 2010).

Processes taking place in the air during pollen transport are very poorly studied and one can only guess the type and intensity of the involved reactions. In an attempt to classify these processes, one can consider the physical transformations (rupture of grains, phase transformations of allergen particles, coagulation with other aerosols, etc.); chemical transformations (oxidation and nitration), and biological transformations of the particles (loss of viability, germination). Some of these processes can lead to substantial changes in the atmospheric features of the particles affecting their life time and dispersion features, and some are of importance from medical and biological points of view. In this sub chapter, we will consider only changes that are induced by chemical and physical transformations and result in alterations of the atmospheric features of the particles. The other aspects of the interactions are considered in other chapters of this book.

During transport, the water content of the grains is adapted to ambient humidity via water exchange through the pollen walls (Traidl-Hoffmann et al. 2003). In a very humid environment, a process resembling germination (so-called pseudogermination) can be triggered resulting in an abrupt release of the pollen content into the air (Motta et al. 2006; Traidl-Hoffmann et al. 2003; Grote et al. 2000, 2001, 2003). Together with water, pollen also absorbs dissolved heavy metals, nitrate and sulphur (WHO 2003).

The other process revealed in laboratory conditions refers to damage of pollen by aggressive chemicals, first of all ozone and nitrogen oxides. The related chemical processes are oxidation and nitration. According to Motta et al. (2006), treatment of pollen grains by ozone leads to a substantial increase of the fraction of damaged pollen already for the ozone concentrations of 100 ppb, which can easily be observed in the real atmosphere. For NO_2, the levels of a few ppm needed for the non-negligible impact were much too high to represent any real-life case, which questions the role of nitrogen oxides in the pollen rupture. Apart from NO_2 and O_3, oxidation agents, such as hydroxyl radicals, can damage pollen or oxidise allergen but, to our knowledge, no experiments were performed with such species.

Chemical pollution can cause morphological changes of the pollen grains. Among the effects are the collapse and thinning of the exine (Shahali et al. 2009), which increases a bioavailability of the content. Stronger leaching was one of the explanations of observed decreased allergen detection in grass pollen exposed to car exhausts (Peltre et al. 1991), as well as to O_3, SO_2 and NO_2 (Rogerieux et al. 2007).

The reaction between the pollen allergens and air pollutants is nitration (Franze et al. 2005), which may augment the allergenic potential of the allergen.

Interactions of pollen with fine aerosols of anthropogenic origin are confirmed by observations (Behrendt and Becker 2001), who showed that small carbon particles stick to the surface of pollen grains. This process provokes pseudogermination and pollen rupture. However, comparatively fast depositing pollens can serve as a scavenging agent for small particles. The latter process, however, can be non-negligible only for the episodes with extremely high pollen concentrations in the air.

Interaction of allergen and diesel exhaust probably rests on the coagulation mechanism. Knox et al. (1997) showed that grass allergen molecules tend to bind to micrometre-size aggregates formed by black carbon particles. Since the resulting aggregate is in the respirable range and is much less soluble than the allergen itself, these particles tend to be transported even further than allergen and are capable of penetrating deep into human airways.

Apart from allergens, pollen grains also contain an array of non-allergenic but pro-inflammatory pollen-associated lipid mediators (PALMs) and enzymes that are suggested to be involved in the pathogenesis of allergic diseases (Gunawan et al. 2008; Gilles et al. 2010). Pollen collected in streets with a heavy traffic released significantly more PALMs than pollen collected in rural surroundings (Risse et al. 2000). A similar effect was found by Behrendt et al. (1999) for grains treated with volatile organic compounds.

5.8 Modelling the Pollen Dispersion

Modelling studies of pollen dispersion in the atmosphere are not numerous and concentrate on just two main directions. The most widely used model-based approach is an inverse-modelling analysis of the observational results and determination of the source regions affecting the specific monitoring site. This direction is dominated by simple backwards trajectory analysis.

The second application area – forward pollen dispersion – aims at simulation of the actual pollen life cycle: production, release into the air, transport, deposition, and sometimes, also the health impact. The outcome of such studies is a set of concentration and deposition maps of the particular pollen type. This task normally requires comprehensive systems, detailed input information of various kinds, and substantial computational resources. Compromising approaches can be applied in this area too but they are quite rare.

5.8.1 Inverse Studies and Analysis of Observational Results

The inverse modelling studies usually pursue one of the following objectives: (i) to find out the reasons of the observed peculiar behaviour of pollen concentrations, (ii) to outline the source areas most frequently and significantly affecting certain region. The first task normally covers a short period of time, up to a few days that immediately precede the considered episode, while the second task requires the analysis of a period of a few years to accumulate sufficient information for the statistical analysis.

The most common and simple method for back-tracing the origin of the observed pollen is the backward trajectories computed using a Lagrangian trajectory model (the Lagrangian trajectory approach). Within this method, a few trajectories directed backwards in time are started from the location of the pollen monitor and evaluated a few hours or days back in time. The resulting trajectories roughly show the direction from where the air masses arrived at the observation place during the observation time period. Since the method is qualitative, extra measures can be taken to increase the value of the outcome: clusters or ensembles of trajectories can be started from different points (e.g. Smith et al. 2005; Stach et al. 2007), trajectories can also be started at different heights, a random shift of start time can be applied, etc.

The backward trajectory methodology has been used for studying the origin of the observed pollen by Gassmann and Pérez (2006) for *Celtis* and *Nothofagus* pollen in Argentina, by Stach et al. (2007) and Smith et al. (2008) for *Ambrosia* in Poland, by Skjøth et al. (2007) for birch in Poland, by Šaulienė and Veriankaitė (2006) and Veriankaité et al. (2010) for birch in Lithuania, by Cecchi et al. (2007) for *Ambrosia* in Central Italy, by Mahura et al. (2007) and Skjøth et al. (2007, 2008b) for birch in Denmark, by Skjøth et al. (2009) for birch in London, etc. The bulk of the works were based on inverse trajectories computed by the NOAA HYSPLIT model (Draxler and Hess 2010), but also by own systems, such as THOR (Skjøth et al. 2002) in Denmark. A manual construction of trajectories via summing-up the wind vectors was used by Giner et al. (1999) for evaluation of the *Artemisia* source areas affecting the station at the eastern coast of Spain.

Apart from the above regional-scale assessments, trajectories were also used for back-tracking the transoceanic transport. In particular, many studies quoted in Sect. 5.5 (Long-Range Transport) have used trajectories to understand the origin of the exotic pollen (e.g. Rousseau et al. 2003, 2004, 2006).

A next step in the complexity of the methodology is consideration of a large number of Lagrangian particles – from a few thousands and up to a few millions. Each particle is transported with wind, thus drawing a trajectory, and is randomly relocated at every model time step, thus reflecting the impact of atmospheric diffusion. This approach is called as the "Lagrangian particle random-walk" method. In comparison with few trajectories, the large number of particles allows reproducing the actual shape of the clouds and makes it possible to include linear transformation processes, dry, and wet deposition. Most of the presently used Lagrangian models follow that paradigm: FLEXPART (Stohl et al. 2005), SILAM-L (Sofiev et al. 2006b), DERMA (Sorensen 1998), SNAP (Saltbones et al. 2001).

The statistical analysis of a great number of back trajectories from receptor sites has turned out to be an efficient tool to identify sources and sinks of atmospheric trace substances or to reconstruct their average spatial distribution (Ashbaugh 1983; Seibert et al. 1994). A trajectory statistical method was successfully applied to identify source areas for beech pollen recorded in Catalonia (Belmonte et al. 2008) and for pollen of various species recorded on Tenerife (Izquierdo et al. 2011).

Lagrangian particle models, being more realistic than Lagrangian trajectory ones, still inherit some weaknesses of the Lagrangian approach – first of all, limited spatial representativeness of a single Lagrangian particle. As a result, even for regional studies an astronomical number of particles have to be evaluated (Veriankaité et al. 2010; Siljamo, et al. 2008b; Sofiev et al. 2006a). It was also shown that the trajectory models show difficulties in mountain regions, where airflow patterns are known to be complex (Pérez-Landa et al. 2007a, b) and hard to explore with few trajectories. It has therefore been suggested that more sophisticated (Eulerian) approaches could be applied for such environments (Smith et al. 2008; Šikoparija et al. 2009).

Compared with the back-trajectories approaches, adjoint dispersion modelling offers a more rigorous instrument to identify the sources (Marchuk 1982). The idea of the method is to explicitly compute the sensitivity of the observed values to emission fluxes, chemical transformations and meteorological processes, which can affect the particular observation. Adjoint methods (both explicitly identified in the studies and implicitly included into the numerical systems) are extensively used in observational analysis at a local scale (Kuparinen 2006; Kuparinen et al. 2007; Rannik et al. 2003), in the source apportionment of greenhouse gas emissions (Bergamaschi et al. 2005), analysis of regional air quality (e.g. Saarikoski et al. 2007; Prank et al. 2008, 2010), and for the regional-scale source apportionment of the observed pollen (Veriankaité et al. 2010; Sofiev et al. 2006a, b; Siljamo et al. 2008b).

An example of comparison of the three methodologies is a study of Veriankaité et al. (2010), where the Lagrangian trajectory model HYSPLIT was used in combination with the SILAM model, which was run in two setups: as a back-tracking Lagrangian particle model and as an Eulerian model in adjoint mode. Despite qualitatively similar conclusions of all three analyses, a general recommendation was favouring the most comprehensive and least subjective approach of Eulerian adjoint modelling. It was noted that the results of trajectory analysis suffered from limited representativeness of a single trajectory and depended on subjective selection of the trajectory setup. The problems can become acute in meteorologically complicated situations, such as passage of a frontal line, strong cyclonic activity, etc.

As a post-processing of the model output for quantitative source identification, a number of analytical inversion techniques have been developed and applied in air pollution research (Wotawa et al. 2003; Seibert and Frank 2004; Stohl et al. 2009). Such methods have not yet been applied to the problem of identifying pollen sources. The same is true for variational methods of source apportionment (Sofiev and Atlaskin 2007).

5.8.2 Forward Simulations and Forecasting of the Pollen Seasons

Forward model simulations aim at predicting the pollen emission from plants, dispersion in the atmosphere and computing on its basis the concentrations of the grains at any distance from the sources. Similar to the inverse modelling, both Lagrangian and Eulerian approaches have been developed for these tasks.

A series of works addressed the local-scale dispersion of pollen from genetically modified plants: see review of Kuparinen (2006), applications for maize by Arritt et al. (2007), Jarosz et al. (2004), and Aylor et al. (2006), theoretical considerations of Novotny and Perdang (2003) and Fraile et al. (2006), meso-scale Lagrangian model of Kuparinen et al. (2007), etc.

In the 1980s, the significance of atmospheric conditions for the regional pollen dispersion was considered and the trajectory model applied for prediction of the transport direction (Davis and Main 1984, 1986). Another work was connected with the transport of marihuana pollen from Northern Africa to Europe (Cabezudo et al. 1997). The transport of *Junipeerus Ashei* pollen towards the city of Tulsa, USA, was studied by Van de Water et al. (2003) and Van de Water and Levetin (2001), who used a separately estimated release of pollen in combination with trajectory and weather pattern analysis. The result was a qualitative forecast of the threat to the downwind regions, covering a territory of about 1,000 km^2. The outcome was encouraging: during the two seasons of 1998–1999, only a single occurrence of "high" or "very high" pollen concentrations in the city of Tulsa was not directly linked to "moderate" or "severe" forecast threat predictions. Other applications include the study of Pasken and Pietrowicz (2005) for oak, and the study by Kawashima and Takahashi (1999) for Japanese cedar (both emission and dispersion were predicted).

The results of trajectory model applications in 1990s showed a principal possibility to predict the pollen distribution by means of dispersion modelling. However, actual forecasting for large areas, such as Europe, could be approached only by integrated modelling systems, which incorporate descriptions of all the parts of the pollen cycle: production, release, transport, transformation, and deposition (Helbig et al. 2004; Sofiev et al. 2006a, 2009; Siljamo et al. 2004a, b, 2006, 2007, 2008a, b, c; Vogel et al. 2008). The corresponding biological, mathematical and physical sub-models are then driven only by historical, actual and forecasted meteorological information.

The input information for the pollen dispersion models consists of: (i) map of distribution of the source plants (see maps in Sofiev et al. 2006a; Skjøth et al. 2008a, 2010), (ii) information for the phenological model covering the whole region of computations and predicting both start and end of the flowering season (Helbig et al. 2004; Linkosalo et al. 2010; Siljamo et al. 2007, 2008c; Efstathiou et al. 2011), (iii) information on the pollen season severity obtained from observations or predicted from historical meteorological and pollen data (Ranta and Satri 2007), (iv) meteorological data produced by numerical weather prediction models. One has to warn against the climatology-based approaches, when the model is driven by long-term averaged parameters, such as the flowering date expressed as Julian date of the year: such approaches are bound to fail in every specific situation (Sofiev et al. 2006a).

Observational information on pollen distribution, to our knowledge, is not used in any of the presently existing systems. The same is true for the remote-sensing observations, despite showing that they provide valuable information about the phenological stages of some plants (Høgda et al. 2002).

The pollen emission and dispersion computations are usually split into two stages: prior to the season and during the season. Before the season, the only important process is the accumulation of the heat sum and other relevant parameters, which control the pollen development. Upon fulfilling some condition (e.g., reaching the prescribed heat sum threshold), the actual pollination starts. During that stage, the model follows the propagation of the flowering season but also responds to the short-term meteorological stress, such as cold weather, diurnal cycle of temperature and humidity, precipitation, etc. These variations can be parameterised as a function of daytime (e.g. Mahura et al. 2009) or predicted dynamically from the basic meteorological parameters (Linkosalo et al. 2010; Helbig et al. 2004; Siljamo et al. 2008b).

5.9 Discussion

In this section we will briefly touch on a few aspects of the pollen and allergen dispersion in the atmosphere, which bring together the above sections and illustrate interconnections between them. The topics are by no means exhaustive and the reader is referred to the cited literature sources for a more comprehensive analysis.

5.9.1 Quantification of Pollen Loads at Different Scales

As shown above, the pollen dispersion in the atmosphere is relevant at a very wide range of scales (Fig. 5.1) – from micro-scale up to continental. There are very few pollutants, which distribution can be relevant for so large variety of spatial and temporal scales. The main reason for this is a large variety of processes where the pollen impact is important – especially, biological and medical.

For successful fertilization, the density of the pollen cloud must be thousands of grains per cubic metre, though this level varies greatly for different species. At lower concentrations, probability of fertilization becomes too small to be biologically important (Gleaves 1973). Such concentrations usually can be obtained only in the source vicinity, i.e. biologically relevant distances hardly exceed tens or hundreds of metres for herbaceous species and, possibly, a few kilometres for wind-pollinating trees.

With regard to human allergy, concentrations much lower than those near the source are important (see the chapter on the pollen allergy threshold). In some cases, a few tens of grains per cubic metre can be sufficient to trigger the first symptoms. As demonstrated by the above-quoted studies, such concentrations can easily be observed at hundreds and, possibly, thousands of kilometres distance from the sources.

Finally, the inter-continental transported pollen, though existing (see the LRT section), can hardly play any practical role due to extremely low concentrations at the receptor point. One can, however, use exotic pollen as a natural tracer, which can reveal the pathway and frequency of the inter-continental exchange. Pollen, therefore, can serve as a marker of distribution of other species, whose observations may be not possible or very complicated.

A very wide range of relevant scales poses serious challenges to pollen monitoring and modelling. To date, the most widely used methodology of pollen monitoring is the Hirst (1952) trap. This is a low-volume device (about 0.01 m^3 min^{-1}), which is prone to significant uncertainties. The problem exacerbates further due to sharp gradients of pollen concentrations occurring in the vicinity of the sources: moving the trap higher or lower or repositioning it may lead to strong changes.

From the modelling point of view, the major challenge is also the near-source distribution and a fraction of pollen that reaches the regional scale of transport. All regional models currently ignore the complexity of the micro-scale processes replacing them with some variations of the "escape fraction" of Gregory (1961). The micro-scale models, to the contrary, concentrate of reproducing the patterns in the nearest vicinity of the source while the regional tails of the distributions are considered unimportant for the applications.

These deficiencies in both current monitoring and modelling techniques are to be addressed in future research in order to end up with comprehensive descriptions of pollen dispersion at all relevant spatial and temporal scales.

5.9.2 Variety of Modelling Approaches: Any Chance for Harmonization?

A variety of model types used for the pollen-related simulations is largely caused by historical reasons and conditions when the availability and simplicity of the systems were the primary selection criterion. However, the capabilities of the Lagrangian trajectory models are still sufficient for some tasks, first of all, for analysis of some specific measurements obtained at a single station. As long as a simple direction of transport is enough, these models are the reasonable choice. For more sophisticated observation analysis one has to rely on Lagrangian particle or Eulerian systems. Concerning the prediction of the pollen concentrations (the forward modelling tasks), the deficiencies of Lagrangian approaches (first of all, limited representativeness of a single particle) become impossible to overcome, so for these tasks the comprehensive Eulerian models are inevitable.

A specific problem of Lagrangian trajectory systems is selection of the type of trajectories. For about half a century it has been known that, for example, isobaric trajectories should not be used because they can produce entirely wrong results (Danielsen 1961). The isentropic trajectories are somewhat better but can still deviate from the true pathways, which can be drawn only by fully 3D trajectories. In that sense, the Lagrangian particle models, usually based on 3D trajectories, are preferable.

Considering the comprehensive systems for the pollen forecasts, one has to pay special attention to the phenological models as a part of the emission sub-system. There have been numerous phenological models developed for many species (see the Chap. 3, "Pollen season", and Sect. 4.3.2 "Process based phenological models", this issue). However, practically all of these models have been developed and evaluated within a limited territory. An attempt to use these models outside their native region leads to unpredictable results. The experience of SILAM system (presently, the only European-scale pollen forecasting system) shows that none of the existing phenological models can satisfactorily describe the birch phenology over the whole continent. The thermal-sum model used in SILAM had to be reparameterised for each sub-region independently (Sect. 4.3.3.1 "Application to large areas"). Understanding the differences between the regions and the reasons why the same taxa require different accumulated heat in different regions is a challenge for the future.

References

Aina, R., Asero, R., Ghiani, A., Marconi, G., Albertini, E., & Citterio, S. (2010). Exposure to cadmium-contaminated soils increases allergenicity of *Poa annua* L. pollen. *Allergy, 65*, 1313–1321. ALL2364 [pii], 10.1111/j.1398-9995.2010.02364.x [doi].

Alché, J. D., Castro, A. J., & Rodriguez-García, M. I. (2002). Localization of transcripts corresponding to the major allergen from olive pollen (Ole e I) by electron microscopic non-radioactive in situ RT-PCR. *Micron, 33*, 33–37.

Arritt, R. W., Clark, C. A., Goggi, S., Lopez Sanchez, H., Westgate, M. E., & Riese, J. M. (2007). Lagrangian numerical simulations of canopy air flow effects on maize pollen dispersal. *Field Crops Research, 102*, 151–162.

Ashbaugh, L. L. (1983). A statistical trajectory technique for determining air pollution source regions. *Journal of the Air Pollution Control Association, 33*, 1096–1098.

Aylor, D. E., Schultes, N. P., & Shields, E. J. (2003). An aerobiological framework for assessing cross-pollination in maize. *Agricultural and Forest Meteorology, 119*, 111–129.

Aylor, D. E., Boehm, M. T., & Shields, E. J. (2006). Quantifying aerial concentrations of maize pollen in the atmospheric surface layer using remote-piloted airplanes and Lagrangian stochastic modelling. *Journal of Applied Meteorology and Climatology, 45*, 1003–1015.

Behrendt, H., & Becker, W.-F. (2001). Localization, release and bioavailability of pollen allergens: The influence of environmental factors. *Current Opinion in Immunology, 13*, 709–715.

Behrendt, H., Becker, W. M., Fritzsche, C., Sliwa-Tomczok, W., Tomczok, J., Friedrichs, K. H., & Ring, J. (1997). Air pollution and allergy: Experimental studies on modulation of allergen release from pollen by air pollutants. *International Archives of Allergy and Immunology, 113*, 69–74.

Behrendt, H., Tomczok, J., Sliwa-Tomczok, W., Kasche, A., Ebner von Eschenbach, C., Becker, W. M., & Ring, J. (1999). Timothy grass (*Phleum pratense* L.) pollen as allergen carriers and initiators of an allergic response. *International Archives of Allergy and Immunology, 118*, 414–418.

Belmonte, J., Vendrell, M., Roure, J. M., Vidal, J., Botey, J., & Cadahía, A. (2000). Levels of *Ambrosia* pollen in the atmospheric spectra of Catalan aerobiological stations. *Aerobiologia, 16*, 93–99.

Belmonte, J., Alarcón, M., Avila, A., Scialabba, E., & Pino, D. (2008). Long-range transport of beech (*Fagus sylvatica* L.) pollen to Catalonia (north-eastern Spain). *International Journal of Biometeorology, 52*, 675–687. doi:10.1007/s00484-008-0160-9.

Bergamaschi, P., Krol, M., Dentener, F., Vermeulen, A., Meinhardt, F., Graul, R., Ramonet, M., Peters, W., & Dlugokencky, E. J. (2005). Inverse modelling of national and European CH4 emissions using the atmospheric zoom model TM5. *Atmospheric Chemistry and Physics, 5*, 2431–2460.

Boehm, M. T., Aylor, D. E., & Shields, E. J. (2008). Maize pollen dispersal under convective conditions. *Journal of Applied Meteorology and Climatology, 47*, 291–307.

Bohrerova, Z., Bohrer, G., Cho, K. D., Bolch, M. A., & Linden, K. G. (2009). Determining the viability response of pine pollen to atmospheric conditions during long-distance dispersal. *Ecological Applications, 19*(3), 656–667.

Bourgeois, J. C. (2000). Seasonal and interannual pollen variability in snow layers of arctic ice caps. *Review of Palaeobotany and Palynology, 108*, 17–36.

Bricchi, E., Frenguelli, G., & Mincigrucci, G. (2000). Experimental results about *Platanus* pollen deposition. *Aerobiologia, 16*, 347–352.

Brown, J. K. M., & Hovmoller, M. S. (2002). Aerial dispersal of pathogens on the global and continental scales and its impact on plant disease. *Science, 297*, 537–541.

Brunet, Y., Foueillassar, X., Audran, A., Garrigou, D., & Dayau, S. (2004). *Evidence for longrange transport of viable maize pollen*. Reprints, 16th Conference on Biometeorology and Aerobiology. Vancouver, Canada. American Meteorological Society, CD-ROM, P4A.2.

Burczyk, J., DiFazio, S. P., & Adams, W. T. (2004). Gene flow in forest trees: How far do genes really travel? *Forest Genetics, 11*, 1–14.

Busse, W. W., Charles, E. R., & Hoehne, J. H. (1972). Where is the allergic reaction in ragweed asthma? II. Demonstration of ragweed antigen in airborne particles smaller than pollen. *The Journal of Allergy and Clinical Immunology, 50*, 289–293.

Buters, J. T., Weichenmeier, I., Ochs, S., Pusch, G., Kreyling, W., Boere, A. J., Schober, W., & Behrendt, H. (2010). The allergen Bet v 1 in fractions of ambient air deviates from birch pollen counts. *Allergy, 65*(7), 850–858. doi:ALL2286 [pii], 10.1111/j.1398-9995.2009.02286.x.

Cabezudo, B., Recio, M., Sanchez-Laulhe, J. M., Trigo, M. M., Toro, F. J., & Polvorinos, F. (1997). Atmospheric transportation of marihuana pollen from North Africa to the southwest of Europe. *Atmospheric Environment, 31*, 3323–3328.

Campbell, I. D., McDonald, K., Flannigan, M. D., & Kringayark, J. (1999). Long-distance transport of pollen into the Arctic. *Nature, 399*, 29–30.

Cariñanos, P., Galán, C., Alcázar, P., & Domínguez, E. (2004). Analysis of the particles transported with dust-clouds reaching Córdoba, southwestern Spain. *Archives of Environmental Contamination and Toxicology, 46*, 141–146.

Cecchi, L., Morabito, M., Paola Domeneghetti, M., Crisci, A., Onorari, M., & Orlandini, S. (2006). Long distance transport of ragweed pollen as a potential cause of allergy in central Italy. *Annales of Allergy Asthma and Immunology, 96*, 86–91.

Cecchi, L., Torrigiani Malaspina, T., Albertini, R., Zanca, M., Ridolo, E., Usberti, I., Morabito, M., Dall' Aglio, P., & Orlandini, S. (2007). The contribution of long-distance transport to the presence of *Ambrosia* pollen in central northern Italy. *Aerobiologia, 23*, 145–151.

Chamecki, M., Meneveau, C., & Parlange, M. B. (2009). Large eddy simulation of pollen transport in the atmospheric boundary layer. *Journal of Aerosol Science, 40*, 241–255.

Clot, B., Schneiter, D., Tercier, Ph, Gehrig, R., Peeters, A., Thibaudon, M., & Clot, B. (2002). *Ambrosia* pollen in Switzerland: Local production or transport? *Allergie et Immunologie, 34*, 126–128.

Dahl, A., Strandhede, S.-V., & Wihl, J.-A. (1999). Ragweed: An allergy risk in Sweden? *Aerobiologia, 15*, 293–297.

Damialis, A., Gioulekas, D., Lazopoulou, C., Balafoutis, C., & Vokou, D. (2005). Transport of airborne pollen into the city of Thessaloniki: The effects of wind direction, speed and persistence. *International Journal of Biometeorology, 49*, 139–145.

Danielsen, E. F. (1961). Trajectories: Isobaric, isentropic and actual. *Journal of Meteorology, 18*, 479–486.

Davis, J. M., Main, C. E. (1984). A regional analysis of the meteorological aspects of the spread and development of blue mold on tobacco. *Boundary-Layer Meteorology, 28*, 271–304.

Davis, J. M., Main, C. E. (1986). Applying atmospheric trajectory analysis to problems in epidemiology. *Plant Disease, 70*, 490–497.

Davidson, A. (1941). A note on anthesis in some common grasses near Johannesburg, and the relation of anthesis to collection of pollen for medical purposes. *Journal of South African Botany, 7*, 145–152.

Di-Giovanni, F., & Kevan, P. G. (1991). Factors affecting pollen dynamics and its importance to pollen contamination: A review. *Canadian Journal of Forest Research, 21*, 1155–1170.

Draxler, R. R., Hess, G. D. (2010). Description of the HYSPLIT_4 modelling system. NOAA Technical Memorandum ERL ARL-224NOAA Air Resources Laboratory; 27 pp.

http://www.arl.noaa.gov/ready/hysplit4.html, http://www.arl.noaa.gov/documents/reports/rl-224. pdf. Last access 25 July 2012.
Efstathiou, C., Isukapalli, S., & Georgopoulos, P. (2011). A mechanistic modelling system for estimating large-scale emissions and transport of pollen and co-allergens. *Atmospheric Environment, 45*(13), 2260–2276. Available from: http://dx.crossref.org/10.1016%2Fj. atmosenv.2010.12.008. Last access 25 July 2012.
Ellstrand, N. C. (1992). Gene flow by pollen: Implications for plant conservation genetics. *Oikos, 63*, 77–86.
Ennos, R. A. (1994). Estimating the relative rates of pollen and seed migration among plant-populations. *Heredity, 72*, 250–259.
Estrella, N., Menzel, A., Krämer, U., & Behrendt, H. (2006). Integration of flowering dates in phenology and pollen counts in aerobiology: Analysis of their spatial and temporal coherence in Germany (1992-1999). *International Journal of Biometeorology, 54*, 49–59.
European Environment Agency. (2002). Genetically modified organisms (GMOs): The significance of gene flow through pollen transfer. *Environmental Issue Report* No 28, ISBN: 92-9167-411-7, Copenhagen, 75 pp. http://www.eea.europa.eu/publications/environmental_issue_report_ 2002_28. Last access 25 July 2012.
Faegri, K., Iversen, J., & Krzywinski, K. (1989). *Textbook of pollen analysis*. Toronto: Wiley. 328 pp.
Fotiou, C., Damialis, A., Krigas, N., Halley, J. M., & Vokou, D. (2010). *Parietaria judaica* flowering phenology, pollen production, viability and atmospheric circulation, and expansive ability in the urban environment: Impacts of environmental factors. *International Journal of Biometeorology, 55*, 35–50. doi:10.1007/s00484-010-0307-3.
Fraile, R., Calvo, A. I., Castro, A., Fernandez-Gonzalez, D., & Garcia-Ortega, E. (2006). The behavior of the atmosphere in long-range transport. *Aerobiologia, 22*, 35–45.
Franze, T., Weller, M. G., Niessner, R., & Poschl, U. (2005). Protein nitration by polluted air. *Environmental Science and Technology, 39*, 1673–1678.
Franzen, L. (1989). A dustfall episode on the Swedish west-coast, October 1987. *Geografiska Annaler Series A, Physical Geography, 71*, 263–267.
Franzen, L., & Hjelmroos, M. (1988). A coloured snow episode on the Swedish west coast, January 1987. A quantitative study of air borne particles. *Geografiska Annaler Series A, Physical Geography, 70*, 235–243.
Franzen, L., Hjelmroos, M., Kallberg, P., Brorstromlunden, E., Juntto, S., & Savolainen, A. L. (1994). The yellow-snow episode of northern Fennoscandia, March-1991: A case-study of long-distance transport of soil, pollen and stable organic-compounds. *Atmospheric Environment, 28*, 3587–3604.
Gage, S., Isard, S. A., & Colunga, G. M. (1999). Ecological scaling of aerobiological dispersal processes. *Agricultural and Forest Meteorology, 97*, 249–261.
Garrison, V. H., Shinn, E. A., Foreman, W. T., Griffin, D. W., Holmes, C. W., Kellogg, C. A., Majewski, M. S., Richardson, L. L., Ritchie, K. B., & Smith, G. W. (2003). African and Asian dust: From desert soils to coral reefs. *Bioscience, 53*, 469–480.
Garrison, V. H., Foreman, W. T., Genualdi, S., Griffin, D. W., Kellogg, C. A., Majewski, M. S., Mohammed, A., Ramsubhag, A., Shinn, E. A., Simonich, S. L., & Smith, G. W. (2006). Saharan dust – a carrier of persistent organic pollutants, metals and microbes to the Caribbean? *Revista Biologia Tropical* (Int. J. Trop. Biol. ISSN-0034-7744) *54*(3), 9–21.
Gassmann, M. I., & Pérez, C. F. (2006). Trajectories associated to regional and extra-regional pollen transport in the southeast of Buenos Aires province, Mar Del Plata (Argentina). *International Journal of Biometeorology, 50*, 280–291. doi:10.1007/s00484-005-0021-8.
Gehrig, R., & Peeters, A. G. (2000). Pollen distribution at elevations above 1000 m in Switzerland. *Aerobiologia, 16*, 69–74.
Gilles, S., Jacoby, D., Blume, C., Mueller, M. J., Jakob, T., Behrendt, H., Schaekel, K., & Traidl-Hoffmann, C. (2010). Pollen-derived low-molecular weight factors inhibit 6-sulfo LacNAc+dendritic cells' capacity to induce T-helper type 1 responses. *Clinical and Experimental Allergy, 40*, 269–278. doi:CEA3369 [pii], 10.1111/j.1365-2222.2009.03369.x.

Giner, M. M., Garcia, C. J. S., & Selles, G. J. (1999). Aerobiology of *Artemisia* airborne pollen in Murcia (SE Spain) and its relationship with weather variables: Annual and intradiurnal variations for three different species. Wind vectors as a tool in determining pollen origin. *International Journal of Biometeorology, 43*, 51–63.

Gleaves, J. T. (1973). Gene flow mediated by wind-borne pollen. *Heredity, 31*, 355–366.

Gorbushina, A. A., Kort, R., Schuite, A., Lazarus, D., Schnetger, B., Brumsack, H., Broughton, W. J., & Favet, J. (2007). Life in Darwin's dust: Intercontinental transport and survival of microbes in the nineteenth century. *Environmental Microbiology, 9*(12), 2911–2922. doi:10.1111/j.1462-2920.20.2007.01461.x.

Goudie, A. S., & Middleton, N. J. (2001). Saharan dust storms: Nature and consequences. *Earth-Science Reviews, 56*, 179–204.

Govindaraju, D. R. (1988). Relationship between dispersal ability and levels of gene flow in plants. *Oikos, 52*, 31–35.

Govindaraju, D. R. (1989). Estimates of gene flow in forest trees. *Biological Journal of the Linnean Society, 37*, 345–357.

Gregory, P. H. (1961). *The microbiology of the atmosphere*. New York: Interscience. 251 pp.

Griffin, D. W. (2004). Terrestrial microorganism at an altitude of 20,000 m in Earth's atmosphere. *Aerobiologia, 20*, 135–140.

Griffin, D. W. (2007). Atmospheric movement of microorganisms in clouds of desert dust and implications for human health. *Clinical Microbiology Reviews, 20*, 459–477.

Griffin, D. W., & Kellogg, C. A. (2004). Dust storms and their impact on ocean and human health: Dust in Earth's atmosphere. *EcoHealth, 1*, 284–295.

Griffin, D. W., Kellogg, C. A., & Shinn, E. A. (2001a). Dust in the wind: Long range transport of dust in the atmosphere and its implications for global public and ecosystems health. *Global Change and Human Health, 2*, 20–33.

Griffin, D. W., Garrison, V. H., Herman, J. R., & Shinn, E. (2001b). African desert dust in the Caribbean atmosphere: Microbiology and public health. *Aerobiologia, 17*, 203–213.

Griffin, D. W., Kellogg, C. A., Garrison, V. H., Lisle, J. T., Borden, T. C., & Shinn, E. A. (2003). Atmospheric microbiology in the northern Caribbean during African dust events. *Aerobiologia, 19*, 143–157.

Griffin, D. W., Westplhal, D. L., & Gray, M. A. (2006). Airborne microorganisms in the African desert dust corridor over the mid-Atlantic ridge, Ocean Drilling Program, Leg 209. *Aerobiologia, 22*, 211–226. doi:10.1007/s10453-006-9033-z.

Griffin, D. W., Kubilay, N., Koçak, M., Gray, M. A., Borden, T. C., & Shinn, E. A. (2007). Airborne desert dust and aeromicrobiology over the Turkish Mediterranean coastline. *Atmospheric Environment, 41*, 4050–4062.

Grote, M., Vrtala, S., Niederberger, V., Valenta, R., & Reichelt, R. (2000). Expulsion of allergen-containing materials from hydrated rye grass (*Lolium perenne*) pollen revealed by using immunogold field emission scanning and transmission electron microscopy. *The Journal of Allergy and Clinical Immunology, 105*, 1140–1145.

Grote, M., Vrtala, S., Niederberger, V., Wierman, R., Valenta, R., & Reichelt, R. (2001). Release of allergen-bearing cytoplasm from hydrated pollen: A mechanism common to variety of grass (Poaceae) species revealed by electron microscopy. *The Journal of Allergy and Clinical Immunology, 108*, 109–115.

Grote, M., Valenta, R., & Reichelt, R. (2003). Abortive pollen germination: A mechanism of allergen release in birch, alder and hazel revealed by immunogold electron microscopy. *The Journal of Allergy and Clinical Immunology, 111*, 1017–1023.

Guerzoni, S., & Chester, R. (1996). *The impact of desert dust across the Mediterranean*. Dordrecht: Kluwer Academic Publishers. 389 pp.

Gunawan, H., Takai, T., Kamijo, S., Wang, X. L., Ikeda, S., Okumura, K., & Ogawa, H. (2008). Characterization of proteases, proteins, and eicosanoid-like substances in soluble extracts from allergenic pollen grains. *International Archives of Allergy and Immunology, 147*, 276–288. doi:000144035 [pii] 10.1159/000144035 [doi].

Gupta, N., Sriramarao, P., Kori, R., & Rao, P. V. (1995). Immunochemical characterization of rapid and slowly released allergens from the pollen of *Parthenium hysterophorus*. *International Archives of Allergy and Immunology, 107*, 557–565.

Hart, M. A., de Dear, R., & Beggs, P. J. (2007). A synoptic climatology of pollen concentrations during the six warmest months in Sydney, Australia. *International Journal of Biometeorology, 51*, 209–220.

Helbig, N., Vogel, B., Vogel, H., & Fiedler, F. (2004). Numerical modelling of pollen dispersion on the regional scale. *Aerobiologia, 20*(1), 3–19.

Hervàs, A., Camarero, L., Reche, I., & Casamayor, E. O. (2009). Viability and potential for immigration of airborne bacteria from Africa that reach high mountain lakes in Europe. *Environmental Microbiology, 11*(6), 1612–1623.

Hicks, S. (1985). Modern pollen deposition records from Kuusamo, Finland. *Grana, 24*, 167–184.

Hicks, S. H., Tinsley, A., Huusko, C., Jensen, M., Hättestrand, M., Gerasimedes, A., & Kvavadze, E. (2001). Some comments on spatial variation in arboreal pollen deposition: First records from the Pollen Monitoring Programme (PMP). *Review of Palaeobotany and Palynology, 117*, 183–194.

Hirst, J. M. (1952). An automatic volumetric spore trap. *Annals of Applied Biology, 39*(2), 257–265.

Høgda, K. A., Karlsen, S. R., Solheim, I., Tommervik, H., & Ramfjord, H. (2002). *The start dates of birch pollen seasons in Fennoscandia studied by NOAA AVHRR NDVI data*. Proceeding of IGARSS. June 24–28, 2002, Toronto. ISBN 0-7803-7536-X.

Hooghiemstra, H., Lezine, A. M., Leroy, S. A. G., Dupont, L., & Marret, F. (2006). Late quaternary palynology in marine sediments: A synthesis of the understanding of pollen distribution patterns in the NW African setting. *Quaternary International, 148*, 29–44.

Hua, N., Kobayashi, F., Iwasaka, Y., Shi, G., & Naganuma, T. (2007). Detailed identification of desert-originated bacteria carried by Asian dust storms to Japan. *Aerobiologia, 23*, 291–298.

Isard, S. A., & Gage, S. H. (2001). *Flow of life in the atmosphere: An airscape approach to understanding invasive organisms*. East Lansing: Michigan State University Press. 240 pp.

Isard, S. A., Gage, S. H., Comtois, P., & Russo, J. M. (2005). Principles of the atmospheric pathway for invasive species applied to soybean rust. *Bioscience, 55*(10), 851–861.

Izquierdo, R., Belmonte, J., Avila, A., Alarcón, M., Cuevas, E., & Alonso-Pérez, S. (2011). Source areas and long-range transport of pollen from continental land to Tenerife (Canary Islands). *International Journal of Biometeorology, 55*(1), 67–85. doi: 10.1007/s00484-010-0309-1.

Janssen, C. R. (1973). Local and regional pollen deposition. In J. H. B. Birks & R. G. West (Eds.), *Quaternary plant ecology* (pp. 30–43). Oxford: Blackwell Scientific.

Jarosz, N., Loubet, B., Durand, B., McCartney, A., Foueillassar, X., & Huber, L. (2003). Field measurements of airborne concentration and deposition rate of maize pollen. *Agricultural and Forest Meteorology, 119*, 37–51.

Jarosz, N., Loubet, B., & Huber, L. (2004). Modelling airborne concentrations and deposition rate of maize pollen. *Atmospheric Environment, 38*, 5555–5566.

Kasprzyk, I. (2008). Non-native *Ambrosia* pollen in the atmosphere of Rzeszow (SE Poland); Evaluation of the effect of weather conditions on daily concentrations and starting dates of the pollen season. *International Journal of Biometeorology, 52*, 341–351. doi:10.1007/s00484-007-0129.

Kasprzyk, I., Myszkowska, D., Grewling, Ł., Stach, A., Šikoparija, B., Skjøth, C. A., & Smith, M. (2010). The occurrence of *Ambrosia* pollen in Rzeszów, Kraków and Poznań, Poland: Investigation of trends and possible transport of Ambrosia pollen from Ukraine. *International Journal of Biometeorology*. doi:10.1007/s00484-010-0376-3.

Kawashima, S., & Takahashi, Y. (1999). An improved simulation of mesoscale dispersion of airborne cedar pollen using a flowering-time map. *Grana, 38*, 316–324.

Kazlauskas, M., Sauliene, I., & Lankauskas, A. (2006). Airborne *Artemisia* pollen in Siauliai (Lithuania) atmosphere with reference to meteorological factors during 2003-2005. *Acta Biologica Universitatis, 6*, 1–2.

Kellogg, C. A., & Griffin, D. W. (2006). Aerobiology and the global transport of desert dust. *Trends in Ecology and Evolution, 21*, 638–644.

Kellogg, C. A., Griffin, D. W., Garrison, V. H., Peak, K. K., Royall, N., Smith, R. R., & Shinn, E. A. (2004). Characterization of aerosolized bacteria and fungi from desert dust events in Mali, West Africa. *Aerobiologia, 20*, 99–110.

Klein, E. K., Lavigne, C., Foueilassar, X., Gouyon, P. H., & Laredo, C. (2003). Corn pollen dispersal: Quasi-mechanistic models and field experiments. *Ecological Monographs, 73*, 131–150.

Knox, R. B., Suphioglu, C., Taylor, P., Desai, R., Watson, H. C., Peng, J. L., & Bursill, L. A. (1997). Major grass pollen allergen Lol p 1 binds to diesel exhaust particles: Implications for asthma and air pollution. *Clinical and Experimental Allergy, 27*, 246–251.

Kuparinen, A. (2006). Mechanistic models for wind dispersal. *Trends in Plant Science, 11*, 298–301.

Kuparinen, A., Markkanen, T., Riikonen, H., & Vesala, T. (2007). Modelling air-mediated dispersal of spores, pollen and seeds in forested areas. *Ecological Modelling, 208*, 177–188.

Laursen, S. C., Reiners, W. A., Kelly, R. D., & Gerow, K. G. (2007). Pollen dispersal by *Artemisia tridentata* (Asteraceae). *International Journal of Biometeorology, 51*, 465–481.

Lee, A. K. Y., Lau, A. P. S., Cheng, J. Y. W., Fang, M., & Chan, C. K. (2007). Source identification analysis for the airborne bacteria and fungi using a biomarker approach. *Atmospheric Environment, 41*, 2831–2843.

Lewis, S. A., Corden, J. M., Forster, G. E., & Newlands, M. (2000). Combined effects of aerobiological pollutants, chemical pollutants and meteorological conditions on asthma admissions and A & E attendances in Derbyshire UK, 1993-96. *Clinical and Experimental Allergy, 30*, 1724–1732.

Linkosalo, T., Ranta, H., Oksanen, A., Siljamo, P., Luomajoki, A., Kukkonen, J., & Sofiev, M. (2010). A double-threshold temperature sum model for predicting the flowering duration and relative intensity of *Betula pendula* and *B. pubescens*. *Agricultural and Forest Meteorology, 150*(12), 1579–1584.

Linskens, H. F., & Cresti, M. (2000). Pollen allergy as an ecological phenomenon: A review. *Plant Biosystems, 134*(3), 341–352.

Mahura, A., Korsholm, S. U., Baklanov, A. A., & Rasmussen, A. (2007). Elevated birch pollen episodes in Denmark: Contributions from remote sources. *Aerobiologia, 23*, 171–179.

Mahura, A., Baklanov, A., & Korsholm, U. (2009). Parameterization of the birch pollen diurnal cycle. *Aerobiologia, 25*, 203–208.

Makra, L., & Palfi, S. (2007). Intra-regional and long-range ragweed pollen transport over southern Hungary. *Acta Climatologica Et Chorologica, 40–41*, 69–77.

Marchuk, G. I. (1982). *Mathematical modelling in the environmental problems*. Moscow: "Nauka" publisher. 320 pp, (in Russian).

Marks, G. B., Colquhoun, J. R., Girgis, S. T., Koski, M. H., Treloar, A. B., Hansen, P., Downs, S. H., & Car, N. G. (2001). Thunderstorm outflows preceding epidemics of asthma during spring and summer. *Thorax, 56*, 468–471.

Matikainen, E., & Rantio-Lehtimaki, A. (1999). Semiquantitative and qualitative analysis of pre seasonal airborne birch pollen allergens in different particle sizes. *Grana, 37*, 293–297.

McCartney, H. A., & Lacey, M. E. (1991). Wind dispersal of pollen from crops of oilseed rape (*Brassica napus* L.). *Journal of Aerosol Science, 22*, 467–477.

Michel, D., Rotach, M. W., Gehrig, R., & Vogt, R. (2010). Experimental investigation of micrometeorological influences on birch pollen emission. *Arbeitsberichte der MeteoSchweiz, 230*, 37 pp.

Miguel, A. G., Taylor, P. E., House, J., Glovsky, M. M., & Flagan, R. C. (2006). Meteorological influences on respirable fragment release from Chinese Elm pollen. *Aerosol Science and Technology, 40*(9), 690–696.

Moar, N. T. (1969). Possible long-distance transport of pollen to New Zealand. *New Zealand Journal of Botany, 7*, 424–426.

Motta, A. C., Marliere, M., Peltre, G., Sterenberg, P. A., & Lacroix, G. (2006). Traffic-related air pollutants induce the release of allergen-containing cytoplasmic granules from grass pollen. *International Archives of Allergy and Immunology, 139*, 294–298.

Nichols, H. (1967). Pollen diagrams from Sub-Arctic Central Canada. *Science, 155*, 1665–1668.

Novotny, E., & Perdang, J. (2003). Simulations of pollen transport by wind. *SGR (Scientist for Global Responsibility)*, 3 pp.

Oikonen, M., Hicks, S., Heino, S., & Rantio-Lehtimaki, A. (2005). The start of the birch pollen season in Finnish Lapland: Separating non-local from local birch pollen and the implication for allergy sufferers. *Grana, 44*, 181–186.

Pacini, E., & Hesse, M. (2004). Cytophysiology of pollen presentation and dispersal. *Flora, 199*, 273–285.

Pacini, E., Guarnieri, M., & Nepi, M. (2006). Pollen carbohydrates and water content during development, presentation and dispersal: A short review. *Protoplasma, 228*, 73–77.

Pasken, R., & Pietrowicz, J. A. (2005). Using dispersion and mesoscale meteorological models to forecast pollen concentrations. *Atmospheric Environment, 39*, 7689–7701.

Paz, S., & Broza, M. (2007). Wind direction and its linkage with *Vibrio cholerae* dissemination. *Environmental Health Perspectives, 115*(2), 195–200.

Peeters, A. G., & Zoller, H. (1988). Long range transport of *Castanea sativa* pollen. *Grana, 27*, 203–207.

Peltre, G., Derouet, L., & Cerceau-Larrival, M. T. (1991). Model treatments simulating environmental action on allergenic *Dactylis glomerata* pollen. *Grana, 30*, 59–61.

Pérez-Landa, G., Ciais, P., Gangoiti, G., Palau, J. L., Carrara, A., Gioli, B., Miglietta, F., Schumacher, M., Millán, M. M., & Sanz, M. J. (2007a). Mesoscale circulations over complex terrain in the Valencia coastal region, Spain – Part 2: Modeling CO2 transport using idealized surface fluxes. *Atmospheric Chemistry and Physics, 7*, 1851–1868.

Pérez-Landa, G., Ciais, P., Sanz, M. J., Gioli, B., Miglietta, F., Palau, J. L., Gangoiti, G., & Millán, M. M. (2007b). Mesoscale circulations over complex terrain in the Valencia coastal region, Spain – Part 1: Simulation of diurnal circulation regimes. *Atmospheric Chemistry and Physics, 7*, 1835–1849.

Polymenakou, P. N., Mandalakis, M., Stephanou, E. G., & Tselepides, A. (2008). Particle size distribution of airborne microorganisms and pathogens during an intense African dust event in the Eastern Mediterranean. *Environmental Health Perspectives, 116*, 292–296.

Porsbjerg, C., Rasmussen, A., & Backer, A. (2003). Airborne pollen in Nuuk, Greenland, and the importance of meteorological parameters. *Aerobiologia, 19*, 29–37.

Prank, P., Sofiev, M., Kaasik, M., Ruuskanen, T., Kukkonen, J., & Kulmala, M. (2008). The origin and formation mechanics of aerosol during a measurement campaign in Finnish Lapland, evaluated using the regional dispersion model SILAM. In C. Borrego & A. I. Miranda (Eds.), *Air pollution modeling and its application XIX* (NATO science for peace and security series-C: Environmental security, pp. 530–538). Berlin: Springer.

Prank, M., Sofiev, M., Denier van der Gon, H. A. C., Kaasik, M., Ruuskanen, T., & Kukkonen, J. (2010). A refinement of the emission data for Kola Peninsula based on inverse dispersion modelling. *Atmospheric Chemistry and Physics Discussions, 10*, 15963–16006.

Prospero, J. M., Barett, K., Churcha, T., Dentener, F., Duce, R. A., Galloway, J. N., Levy, H., II, Moody, J., & Quinn, P. (1996). Atmospheric deposition of nutrients to the North Atlantic Basin. *Biogeochemistry, 35*, 27–73.

Prospero, J. M., Blades, E., Mathison, G., & Naidu, R. (2005). Interhemispheric transport of viable fungi and bacteria from Africa to the Caribbean with soil dust. *Aerobiologia, 21*, 1–19.

Pulimood, T. B., Corden, J. M., Bryden, C., Sharples, L., & Nasser, S. M. (2007). Epidemic asthma and the role of the fungal mold *Alternaria alternata*. *The Journal of Allergy and Clinical Immunology, 120*, 610–617. doi:S0091-6749(07) 00970-0 [pii], 10.1016/j.jaci.2007.04.045 [doi].

Radauer, C., Bublin, M., Wagner, S., Mari, A., & Breiteneder, H. (2008). Allergens are distributed into few protein families and possess a restricted number of biochemical functions. *The Journal of Allergy and Clinical Immunology, 121*, 847.e7–852.e7. DOI: S0091-6749(08)00163-2.

Rannik, U., Markkanen, T., Raittila, J., Hari, P., & Vesala, T. (2003). Turbulence statistics inside and over forest: Influence on footprint prediction. *Boundary-Layer Meteorology, 109*, 163–189.

Ranta, H., & Satri, P. (2007). Synchronised inter-annual fluctuations of flowering intensity affects the exposure to allergenic tree pollen in north Europe. *Grana, 46*(4), 274–284.

Ranta, H., Kubin, E., Siljamo, P., Sofiev, M., Linkosalo, T., Oksanen, A., & Bondestam, K. (2006). Long distance pollen transport cause problems for determining the timing of birch pollen season in Fennoscandia by using phenological observations. *Grana, 45*(4), 297–304.

Ranta, H., Sokol, C., Hicks, S., Heino, S., & Kubin, E. (2008). How do airborne and deposition pollen samplers reflect the atmospheric dispersal of different pollen types? An example from northern Finland. *Grana, 47*, 285–296.

Rantio-Lehtimaaki, A., Viander, M., & Koivikko, A. (1994). Airborne birch pollen antigens in different particle sizes. *Clinical and Experimental Allergy, 24*, 23–28.

Rantio-Lehtimaki, A. (2002). Siitepolyallergeenit sisalyss. *Allergia & Asthma, 2*, 25–27 (in Finnish).

Rantio-Lehtimaki, A., & Matikainen, E. (2002). Pollen allergen reports help to understand preseason symptoms. *Aerobiologia, 18*, 135–140.

Raynor, G. S., & Hayes, J. V. (1983). Testing of the air resources laboratories trajectory model on cases of pollen wet deposition after long-distance transport from known source regions. *Atmospheric Environment, 17*(2), 213–220.

Raynor, G. S., Ogden, E. C., & Hayes, J. V. (1970). Dispersion and deposition of ragweed pollen from experimental sources. *Journal of Applied Meteorology, 9*, 885–895.

Raynor, G. S., Ogden, E. C., & Hayes, J. V. (1972a). Dispersion and deposition of timothy pollen from experimental sources. *Agricultural Meteorology, 9*, 347–366.

Raynor, G. S., Ogden, E. C., & Hayes, J. V. (1972b). Dispersion and deposition of corn pollen from experimental sources. *Agronomy Journal, 64*, 420–427.

Risse, U., Tomczok, J., Huss-Marp, J., Darsow, U., & Behrendt, H. (2000). Health-relevant interaction between airborne particulate matter and aeroallergens (pollen). *Journal of Aerosol Science, 31*, 27–28.

Ritchie, J. C. (1974). Modern pollen assemblages near arctic tree Line, Mackenzie Delta Region, Northwest-Territories. *Canadian Journal of Botany, 52*, 381–396.

Ritchie, J. C., & Lichti-Federovich, S. (1967). Pollen dispersal phenomena in Arctic-Subarctic Canada. *Review of Palaeobotany and Palynology, 3*, 255–266.

Rodríguez-García, M. I., Fernández, M. C., Alché, J. D., & Olmedilla, A. (1995). Endoplasmic reticulum as a storage site for allergenic proteins in pollen grains of several Oleaceae. *Protoplasma, 187*, 111–116. doi:10.1007/bf01280238.

Rogerieux, F., Godfrin, D., Senechal, H., Motta, A. C., Marliere, M., Peltre, G., & Lacroix, G. (2007). Modifications of *Phleum pratense* grass pollen allergens following artificial exposure to gaseous air pollutants (O_3, NO_2, SO_2). *International Archives of Allergy and Immunology, 143*, 127–134. doi:000099079 [pii] 10.1159/000099079 [doi].

Rogers, C. A., & Levetin, E. (1998). Evidence of long-distance transport of mountain cedar pollen into Tulsa, Oklahoma. *International Journal of Biometeorology, 42*, 65–72.

Romero, O. E., Dupont, L., Wyputta, U., Jahns, S., & Wefer, G. (2003). Temporal variability of fluxes of eolian-transported freshwater diatoms, phytoliths, and pollen grains off Cape Blanc as reflection of land-atmosphere-ocean interactions in northwest Africa. *Journal of Geophysical Research-Oceans, 108*(C5), 3153–3164. doi:10.1029/2000JC000375/2003.

Rousseau, D. D., Duzer, D., Cambon, G. V., Jolly, D., Poulsen, U., Ferrier, J., Schevin, P., & Gros, R. (2003). Long distance transport of pollen to Greenland. *Geophysical Research Letters, 30*, 1765. doi:10.1029/2003GL017539.

Rousseau, D. D., Duzer, D., Etienne, J.-L., Cambon, G., Jolly, D., Ferrier, J., & Schevin, P. (2004). Pollen record of rapidly changing air trajectories to the North Pole. *Journal of Geophysical Research, 109*, D06116. doi:10.1029/2003JD003985.

Rousseau, D. D., Schevin, P., Duzer, D., Cambon, G., Ferrier, J., Jolly, D., & Poulsen, U. (2005). Pollen transport to southern Greenland: New evidences of a late spring long distance transport. *Biogeosciences Discussions, 2*(4), 709–715.

Rousseau, D. D., Schevin, P., Duzer, D., Cambon, G. V., Ferrier, J., Jolly, D., & Poulsen, U. (2006). New evidence of long distance pollen transport to southern Greenland in late spring. *Review of Palaeobotany and Palynology, 141*, 277–286. doi:10.1016/j.revpalbo.2006.05.001.

Rousseau, D. D., Schevin, P., Ferrier, J., Jolly, D., Andreasen, T., Ascanius, S. E., Hendriksen, S. E., & Poulsen, U. (2008). Long-distance pollen transport from North America to Greenland in spring. *Journal of Geophysical Research-Biogeosciences, 113*. doi:10.1029/2007JG000456.

Saarikoski, S., Sillanpää, M., Sofiev, M., Timonen, H., Saarnio, K., Teinilä, K., Karppinen, A., Kukkonen, J., & Hillamo, R. (2007). Chemical composition of aerosols during a major biomass burning episode over northern Europe in spring 2006: Experimental and modelling assessments. *Atmospheric Environment, 41*, 3577–3589.

Salas, M. R. (1983). Long-distance pollen transport over the southern Tasman Sea: Evidence from Macquarie Island. *New Zealand Journal of Botany, 21*, 285–292.

Saltbones, J., Bartnicki, J., & Foss, A. (2001). Handling of fallout processes from nuclear explosions in a severe nuclear accident program (SNAP). *met.no report no. 157/2003*, http://ebookbrowse.com/saltbones-pdf-d25441461. Last access 25 July 2012, 13 pp. (shortened English version).

Šaulienė, I., & Veriankaitė, L. (2006). Application of backward air mass trajectory analysis in evaluating airborne pollen dispersion. *Journal of Environmental Engineering and Landscape Management, 14*(3), 113–120.

Sauliene, I., Veriankaite, L., & Lankauskas, A. (2007). The analysis of the impact of long distance air mass to airborne pollen concentration. Cross-border cooperation in researches of biological diversity. *Acta Biologica Universitatis Daugapiliensis Supplement, 1*, 61–74.

Savelieva, L. A., Dorozhkina, M. V., & Pavlova, E. Y. (2002). Modern annual deposition and aerial pollen transport in the Lena Delta. *Polarforschung, 70*, 115–122.

Schäppi, G. F., Suphioglu, C., Taylor, P. E., & Knox, R. B. (1997). Concentrations of the major birch tree allergen Bet v 1 in pollen and respirable fine particles in the atmosphere. *The Journal of Allergy and Clinical Immunology, 100*, 656–661. doi:S0091-6749(97)70170-2.

Schlesinger, P., Mamane, Y., & Grishkan, I. (2006). Transport of microorganisms to Israel during Saharan dust events. *Aerobiologia, 22*, 259–273. doi:10.1007/s10453-006-9038-7.

Schmidt-Lebuhn, A. N., Seltmann, P., & Kessler, M. (2007). Consequences of the pollination system on genetic structure and patterns of species distribution in the Andean genus *Polylepis* (Rosaceae): A Comparative study. *Plant Systematics and Evolution, 266*, 91–103. doi:10.1007/s00606-007-0543-0.

Schueler, S., & Schlünzen, K. H. (2006). Modelling of oak pollen dispersal on the landscape level with a mesoscale atmospheric model. *Environmental Modelling and Assessment, 11*, 179–194.

Seibert, P., Kromp-Kolb, H., Balterpensger, U., Jost, D.T., Schwikowski, M., Kasper A., Puxbaum, H. (1994). Trajectory analysis of aerosol measurements at high alpine sites. In P. M. Borrel, P. Borrell, T. Cvitas, W. Seiler (Eds.), *Transport and transformation of pollutants in the troposphere* (pp. 689–693). The Hague: Academic.

Seibert, P., & Frank, A. (2004). Source-receptor matrix calculation with a Lagrangian particle dispersion model in backward mode. *Atmospheric Chemistry and Physics, 4*, 51–63. http://www.atmos-chem-phys.net/4/51/2004/. Last access 25 July 2012.

Seinfeld, J. H., & Pandis, S. N. (2006). *Atmospheric chemistry and physics: From air pollution to climate change* (2nd ed.). New York: Wiley Interscience, Wiley. 1203 pp.

Shahali, Y., Pourpak, Z., Moin, M., Zare, A., & Majd, A. (2009). Impacts of air pollution exposure on the allergenic properties of Arizona cypress pollens. *Journal of Physics Conference Series, 151*, 012027. doi: 10.1088/1742-6596/151/1/012027.

Sharma, C. M., & Khanduri, V. P. (2007). Pollen-mediated gene flow in Himalayan long needle pine (*Pinus roxburghii* Sargent). *Aerobiologia, 23*, 153–158. doi:10.1007/s10453-007-9056-0.

Shinn, E. A., Griffin, D. W., & Seba, D. B. (2003). Atmospheric transport of mold spores in clouds of desert dust. *Archives of Environmental Health, 58*, 498–504.

Šikoparija, B., Smith, M., Skjøth, C. A., Radišić, P., Milkovska, S., Šimić, S., & Brandt, J. (2009). The Pannonian plain as a source of *Ambrosia* pollen in the Balkans. *International Journal of Biometeorology, 53*, 263–272.

Siljamo, P., Sofiev, M., & Ranta, H. (2004a). An approach to simulation of long-range atmospheric transport of natural allergens: An example of birch pollen. In C. Borrego & A. -L. Norman, (Eds.), *Air pollution modelling and its applications XVII* (pp. 331–340). New York: Springer (2007). ISBN-10: 0-387-28255-6.

Siljamo, P., Sofiev, M., Ranta, H., Kalnina, L., & Ekebom, A. (2004b). *Long-range atmospheric transport of birch pollen. Problem statement and feasibility studies.* Proceedings of Baltic HIRLAM workshop, St. Petersburg, November 17–20, 2003. (pp. 100–103). HIRLAM publications, SMHI Norrkoping, Sweden.

Siljamo, P., Sofiev, M., Severova, E., Ranta, H., & Polevova, S. (2006). On influence of long-range transport of pollen grains onto pollinating seasons. *Developments in Environmental Science, 6.* DOI: 10.1016/S1474-8177(70)06074-3. C. Borrego & E. Renner (Eds.). *Air pollution modelling and its applications XVIII,* (pp. 708–716) Amsterdam: Elsevier.

Siljamo, P., Sofiev, M., & Ranta, H. (2007). An approach to simulation of long-range atmospheric transport of natural allergens: An example of birch pollen. In C. Borrego & A.-L. Norman (Eds.), *Air pollution modeling and its applications XVII,* (pp. 331–339). New York: Springer.

Siljamo, P., Sofiev, M., Ranta, H., Linkosalo, T., Kubin, E., Ahas, R., Genikhovich, E., Jatczak, K., Jato, V., Nekovar, J., Minin, A., Severova, E., & Shalaboda, V. (2008a). Representativeness of point-wise phenological *Betula* data observed in different parts of Europe. *Global Ecology and Biogeography, 17*(4), 489–502. doi:10.1111/j.1466-8238.2008.00383.x.

Siljamo, P., Sofiev, M., Severova, E., Ranta, H., Kukkonen, J., Polevova, S., Kubin, E., & Minin, A. (2008b). Sources, impact and exchange of early-spring birch pollen in the Moscow region and Finland. *Aerobiologia, 24,* 211–230. doi:10.1007/s10453-008-9100-8.

Siljamo, P., Sofiev, M., Linkosalo, T., Ranta, H., & Kukkonen, J. (2008c). Development and application of biogenic emission term as a basis of long-range transport of allergenic pollen. In C. Borrego & A. I. Miranda (Eds.), *Air pollution modelling and its application XIX. NATO Science for Peace and Security Series C: Environmental Security* (pp. 154–162). Dordrecht: Springer.

Skjøth, C. A., Hertel, O., & Ellermann, T. (2002). Use of the ACDEP trajectory model in the Danish nation-wide background monitoring programme. *Physics and Chemistry of the Earth, 27,* 1469–1477.

Skjøth, C. A., Sommer, J., Stach, A., Smith, M., & Brandt, J. (2007). The long-range transport of birch (*Betula*) pollen from Poland and Germany causes significant pre-season concentrations in Denmark. *Clinical and Experimental Allergy, 37,* 1204–1212.

Skjøth, C. A., Geels, C., Hvidberg, M., Hertel, O., Brandt, J., Frohn, L. M., Hansen, K. M., Hedegard, G. B., Christensen, J. H., & Moseholm, L. (2008a). An inventory of tree species in European essential data input for air pollution modelling. *Ecological Modelling, 217,* 292–304.

Skjøth, C. A., Sommer, J., Brandt, J., Hvidberg, M., Geels, C., Hansen, K. M., Hertel, O., Frohn, L. M., & Christensen, J. H. (2008b). Copenhagen – a significant source to birch (*Betula*) pollen? *International Journal of Biometeorology, 52,* 453–462.

Skjøth, C. A., Smith, M., Brandt, J., & Emberlin, J. (2009). Are the birch trees in Southern England a source of pollen in North London? *International Journal of Biometeorology, 53,* 75–86.

Skjøth, C. A., Smith, M., Sikoparija, B., Stach, A., Myszkowska, D., Kasprzyk, I., Radisic, P., Stjepanovic, B., Hrga, I., Apatini, D., Magyar, D., Páldy, A., & Ianovici, N. (2010). A method for producing airborne pollen source inventories: An example of *Ambrosia* (ragweed) on the Pannonian Plain. *Agricultural and Forest Meteorology, 150,* 1203–1210.

Smith, M., Emberlin, J., & Kress, A. (2005). Examining high magnitude grass pollen episodes at Worcester, United Kingdom, using back-trajectory analysis. *Aerobiologia, 21*(2), 85–94.

Smith, M., Skjøth, C. A., Myszkowska, D. A. U., Puc, M., Stach, A., Balwierz, Z., Chlopek, K., Piotrowska, K., Kasprzyk, I., & Brandt, J. (2008). Long-range transport of *Ambrosia* pollen to Poland. *Agricultural and Forest Meteorology, 148,* 1402–1411.

Smouse, P., Dyer, R. J., Westfall, R. D., & Sork, V. L. (2001). Two-generation analysis of pollen flow across a landscape. I. Male gamete heterogeneity among females. *Evolution, 55,* 260–271. doi:10.1111/j.0014-3820.2001.tb01291.

Sofiev, M., & Atlaskin, E. (2007). An example of application of data assimilation technique and adjoint modelling to an inverse dispersion problem based on the ETEX experiment. In C. Borrego & A. Norman (Eds.), *Air pollution modelling and its application XVII* (pp. 438–448). New York: Springer.

Sofiev, M., Siljamo, P., Ranta, H., & Rantio-Lehtimaki, A. (2006a). Towards numerical forecasting of long-range air transport of birch pollen: Theoretical considerations and a feasibility study. *International Journal of Biometeorology, 50,* 392–402.

Sofiev, M., Siljamo, P., Valkama, I., Ilvonen, M., & Kukkonen, J. (2006b). A dispersion modelling system SILAM and its evaluation against ETEX data. *Atmospheric Environment, 40,* 674–685. doi:10.1016/j.atmosenv.2005.09.069.

Sofiev, M., Bousquet, J., Linkosalo, T., Ranta, H., Rantio-Lehtimaki, A., Siljamo, P., Valovirta, E., & Damialis, A. (2009). Pollen, allergies and adaptation. Chapter 5. In K. Ebi, G. McGregor & I. Burton (Eds.), Biometeorology and adaptation to climate variability and change (pp. 75–107). ISBN 978-4020-8920-6, Dordrecht: Springer.

Sorensen, J. H. (1998). Sensitivity of the DERMA long-range Gaussian dispersion model to meteorological input and diffusion parameters. *Atmospheric Environment, 32*, 4195–4206.

Spieksma, F. T. M., van Noort, P., & Nikkels, H. (2000). Influence of nearby stands of *Artemisia* on street-level versus roof-top-level ratio's of airborne pollen quantities. *Aerobiologia, 16*, 21–24.

Stach, A., Smith, M., Skjøth, C. A., & Brandt, J. (2007). Examining *Ambrosia* pollen episodes at Poznan (Poland) using back-trajectory analysis. *International Journal of Biometeorology, 51*, 275–286.

Stanley, R. G., & Linskens, H. F. (Eds.). (1974). *Pollen: Biology – biochemistry – management*. Berlin: Springer.

Stewart, G. A., & Holt, P. G. (1985). Submicronic airborne allergens. *The Medical Journal of Australia, 143*(9), 426–427.

Stohl, A., Forster, C., Frank, A., Seibert, P., & Wotawa, G. (2005). Technical note: The Lagrangian particle dispersion model FLEXPART version 6.2. *Atmospheric Chemistry and Physics, 5*, 2461–2474.

Stohl, A., Seibert, P., Arduini, J., Eckhardt, S., Fraser, P., Greally, B. R., Lunder, C., Maione, M., Mühle, J., O'Doherty, S., Prinn, R. G., Reimann, S., Saito, T., Schmidbauer, N., Simmonds, P. G., Vollmer, M. K., Weiss, R. F., & Yokouchi, Y. (2009). An analytical inversion method for determining, regional and global emissions of greenhouse gases: Sensitivity studies and application to halocarbons. *Atmospheric Chemistry and Physics, 9*, 1597–1620. www.atmos-chem-phys.net/9/1597/2009/.

Subba Reddi, C., Reddi, N. S., & Atluri Janaki, B. (1988). Circadian patterns of pollen release in some species of Poaceae. *Review of Palaeobotany and Palynology, 54*, 11–42.

Tampieri, F., Mandrioli, P., & Puppi, G. L. (1977). Medium range transport of airborne pollen. *Agricultural Meteorology, 18*, 9–20.

Taylor, D. A. (2002). Dust in the wind. *Environmental Health Perspectives, 110*, A80–A87.

Taylor, P. E., Flagan, R. C., Valenta, R., & Glovsky, M. M. (2002). Release of allergens as respirable aerosols: A link between grass pollen and asthma. *The Journal of Allergy and Clinical Immunology, 109*, 51–56.

Taylor, P. E., Flagan, R. C., Miguel, A. G., Valenta, R., & Glovsky, M. M. (2004). Birch pollen rupture and the release of aerosols of respirable allergens. *Clinical and Experimental Allergy, 34*, 1591–1596.

Taylor, P. E., Card, G., House, J., Dickinson, M. H., & Flagan, R. C. (2006). High-speed pollen release in the white mulberry tree, *Morus alba* L. sexual plant. *Reproduction, 19*, 19–24.

Taylor, P. E., Jacobson, K. W., House, J. M., & Glovsky, M. M. (2007). Links between pollen, atopy and the asthma epidemic. *International Archives of Allergy and Immunology, 144*, 162–170. doi:000103230 [pii] 10.1159/000103230 [doi].

Timmons, A., O'Brien, E., Charters, Y., Dubbels, S., & Wilkinson, M. (1995). Assessing the risks of wind pollination from fields of genetically modified *Brassica napus* ssp. *oleifera. Euphytica, 85*, 417–423.

Traidl-Hoffmann, C., Kasche, A., Menzel, A., Jakob, T., Thiel, M., Ring, J., & Behrendt, H. (2003). Impact of pollen on human health: More than allergen carriers? *International Archives of Allergy and Immunology, 131*, 1–13. doi:10.1159/000070428.

Van Campo, M., & Quet, L. (1982). Pollen and red dust transport from South to North of the Mediterranean area. *Comptes Rendus des Seances de l'Academie des Sciences Serie III Sciences de la Vie, 295*, 61–64.

Van de Water, P. K., & Levetin, E. (2001). The contribution of upwind pollen sources to the characterization of *Juniperus ashei* phenology. *Grana, 40*, 133–141.

Van de Water, P. K., Keever, T., Main, C. E., & Levetin, E. (2003). An assessment of predictive forecasting of *Juniperus ashei* pollen movement in the Southern Great Plains, USA. *International Journal of Biometeorology, 48*, 74–82.

Van de Water, P., Watrud, L. S., Lee, E. H., Burdick, C., & King, G. A. (2007). Long-distance GM pollen movement of creeping bentgrass using modeled wind trajectory analysis. *Ecological Applications, 17*(4), 1244–1256.

Veriankaité, L., Siljamo, P., Sofiev, M., Sauliené, I., & Kukkonen, J. (2010). Modelling analysis of source regions of long range transported birch pollen that influences allergenic seasons in Lithuania. *Aerobiologia, 26*, 47–62.

Vogel, H., Pauling, A., & Vogel, B. (2008). Numerical simulations of birch pollen dispersion with an operational weather forecast system. *International Journal of Biometeorology, 52*, 805–814.

von Wahl, P.-G., & Puls, K. E. (1989). The emission of mugwort polen (*Artemisia vulgaris* L.) and its flight in the air. *Aerobiologia, 5*, 55–63.

Vrtala, S., Grote, M., Duchene, M., van Ree, R., Kraft, D., Scheiner, O., & Valenta, R. (1993). Properties of tree and grass pollen allergens: Reinvestigation of the linkage between solubility and allergenicity. *International Archives of Allergy and Immunology, 102*, 160–169.

Waisel, Y., Ganor, E., Epshtein, V., Stupp, A., & Eshel, A. (2008). Airborne pollen, spores, and dust across the East Mediterranean sea. *Aerobiologia, 24*, 125–131.

Westbrook, J. K., & Isard, S. A. (1999). Atmospheric scales of biotic dispersal. *Agricultural and Forest Meteorology, 97*, 263–274.

WHO. (2003). *Phenology and human health: Allergic disorders*. Copenhagen: WHO Regional Office for Europe. 55 pp.

Wodehouse, R. P. (1935). *Pollen grains. Their structure, identification and significance in science and medicine*. New York: MacGraw-Hill.

Wotawa, G., De Geer, L.-E., Denier, P., Kalinowski, M., Toivonen, H., D'Amours, R., Desiato, F., Issartel, J.-P., Langer, M., Seibert, P., Frank, A., Sloan, C., & Yamazawa, H. (2003). Atmospheric transport modelling in support of CTBT verification – overview and basic concepts. *Atmospheric Environment, 37*, 1565–1573.

Wu, P. C., Tsai, J. C., Li, F. C., Lung, S. C., & Su, H. J. (2004). Increased levels of ambient fungal spores in Taiwan are associated with dust events from China. *Atmospheric Environment, 38*, 4879–4886.

Wynn-Williams, D. D. (1991). Aerobiology and colonization in Antarctica: The BIOTAS programme. *Grana, 30*, 380–393.

Yadav, S., Chauhan, M. S., & Sharma, A. (2007). Characterisation of bio-aerosols during dust storm period in N-NW India. *Atmospheric Environment, 41*, 6063–6073.

Zhang, W. Y., Arimoto, R., & An, Z. S. (1997). Dust emission from Chinese desert sources linked to variations in atmospheric circulation. *Journal of Geophysical Research, 102*(23), 28041–28147.

Chapter 6
Impact of Pollen

Letty A. de Weger, Karl Christian Bergmann, Auli Rantio-Lehtimäki,
Åslög Dahl, Jeroen Buters, Chantal Déchamp, Jordina Belmonte,
Michel Thibaudon, Lorenzo Cecchi, Jean-Pierre Besancenot,
Carmen Galán, and Yoav Waisel

Abstract The impact of pollen on human health is primarily evident in allergic diseases. Sensitized patients can respond to pollen by symptoms of nose, eyes and bronchi. Pollen threshold levels for sensitization are unknown; instead most studies focus on the *prevalence* of sensitization for different pollen species. The pollen thresholds for symptom development vary among the different studies. Factors that influence the threshold level of a pollen species for symptom development are discussed. (i) Differences in response are observed among individual patients, but also among (ii) ethnic populations, (iii) changes in response to pollen concentrations during the pollen season occur, (iv) the amount of allergens carried by the pollen grains can differ in per region, from day to day and from year to year, and finally (v) threshold levels are affected by environmental factors, like weather conditions

L.A. de Weger (✉)
Department of Pulmonology, Leiden University Medical Center,
P.O. Box 9600, 2300 RC Leiden, The Netherlands
e-mail: l.a.de_weger@lumc.nl

K.C. Bergmann
Charite – Universitatsmedizin Berlin, Klinik fur Dermatologie, Venerologie und Allergologie,
Allergie – Centrum – Charite, Chariteplatz 1, 10117 Berlin, Germany

A. Rantio-Lehtimäki
Aerobiology Unit, Turun Yliopisto, FIN-20014, Turku, Finland

Å. Dahl
Department of Plant and Environmental Sciences, University of Gothenburg,
P.O. Box 461, SE405 30 Gothenburg, Sweden

J. Buters
Centre for Allergy and Environment Laboratory for Toxikology and Exposure Research,
Technische Universität München, Biedersteinerstrasse 29, 80802 München, Germany

C. Déchamp
Association Française d`Etude des Ambroisies AFEDA,
25 Rue Ambroise Paré, 69800 Saint-Priest, France

(temperature, pressure and storms), and air pollutants.The diversity of factors that influence the health impact of pollen has hampered the definition of a straight forward relationship between pollen and the severity of symptoms. However, within the public, the policymakers and the pharmaceutical industry there is a need for a definition of threshold pollen levels. A first approach to meet this need could be to define preliminary threshold values for different regions, followed by a validation of these preliminary threshold levels with patient symptom scores that can be collected by using new information and communication technology (ICT).

Finally, the possible role of pollen in non allergic diseases is discussed, especially non-allergic respiratory diseases, cardio- and cerebrovascular diseases, and psychiatric diseases, including suicide and suicide attempt.

Keywords Airborne allergens • Allergenic pollen • Allergy • Pollutants • Threshold levels

6.1 Introduction

Charles Blackley (1820–1900) was the first to demonstrate that pollen has an impact on human health. He performed the first pollen provocation tests on himself by applying several pollen types to the nostrils, the conjunctiva, the tongue and lips, by inhaling it and by inoculating the upper and lower limbs. Furthermore, he

---

J. Belmonte
Facultat de Biociències, Departament de Biologia Animal,
Biologia Vegetal i Ecologia, Unitat Botànica, Universitat Autònoma de Barcelona , Edifici C, E-08193 Bellaterra, Barcelona, Spain

M. Thibaudon
Réseau National de Surveillance Aérobiologique (RNSA),
Chemin des Gardes, 69610 St Genis l'Argentière, France

L. Cecchi
Interdepartmental Centre of Bioclimatology, University of Florence,
Piazzale delle Cascine 18, 50144 Florence, Italy

J.-P. Besancenot
Climat et Santé, Faculté de Médecine, Centre Universitaire d'Épidémiologie de Population,
BP 87900, 21079 Dijon Cedex, France

C. Galán
Faculty of Science, Department of Botany, Ecology and Plant Physiology, University of Córdoba,
Campus de Rabanales, Edificio Celestino Mutis, 14071 Córdoba, Spain

Y. Waisel (Deceased)

demonstrated that the amount of pollen in the atmosphere was correlated with the severity of his own symptoms (Waite 1995). Nowadays, we know that proteins carried by pollen (allergens) can evoke specific responses in the immune system. First of all, individuals can get sensitized to the allergens (sensitization) and subsequently these IgE-sensitized individuals can respond to the allergens with allergic symptoms. These can consist of nasal symptoms, eye symptoms, or bronchial reactions, or a combination of these. The level of allergen exposure can vary for instance, with the location, the weather or the time of year and it is a relevant determinant for both sensitization and symptom development. This paper focuses on the threshold values of allergen required for sensitization or for the development of symptoms, and on the various factors that can influence these threshold values. Furthermore, allergenic pollen not only appeared to interact with the human immune system to elicit an allergic response in sensitized individuals but it has other effects as well. This is described in the last part of this chapter which focuses on the effect of pollen on non-allergic diseases (Traidl-Hoffmann et al. 2003).

6.1.1 Threshold Values for Sensitization

To our knowledge, studies on threshold values for sensitization are not available and instead of threshold values this paper focuses on the prevalence of sensitization in different areas, which is indicative of the sensitization rate. Prevalence of sensitization is determined by both host genetic factors as well as by environmental factors, like temperature, atmospheric pressure, storms or air-pollution. Environmental factors that are associated with asthma, rhinoconjunctivitis and eczema have been recently discussed by Asher et al. (2010). The effect of pollen exposure on the prevalence of sensitization in a population is not clear: two regional scale studies showed a positive correlation between the pollen exposure and the prevalence of sensitization (Frei and Gassner 2008; Charpin et al. 1993), while a world wide study involving children showed a weak inverse relation, suggesting high exposure to pollen in early life may give some protection against the risk of acquiring respiratory allergy (Burr et al. 2003). Sensitization against inhalant allergens is identified either by a skin prick test or by determination of allergen-specific IgE in peripheral blood. However, a positive result of either of these tests indicates sensitization to a particular allergen, but does not fully correlate with the clinical relevance of this sensitization, i.e. symptom development upon exposure. False positive and false negative results can both occur. The characteristics of an allergen that distinguish it from other pollen components that do not result in sensitization, are incompletely understood (Radauer et al. 2008; Traidl-Hoffmann et al. 2009). Furthermore, pollen exposure is not the only environmental factor that determines sensitization. Other factors include air pollution (Riedl and Diaz-Sanchez 2005), and microbial exposures. Environmental pollution, such as diesel soot particles, can change the immune response in making non-reactive individuals react to an allergen (Diaz-Sanchez et al. 1999). Also non-protein substances derived from pollen for example the "pollen-associated lipid mediators" (PALMS) and adenosine influence the human immune system by making it more prone to trigger a Th2-sensitization response (Traidl-Hoffmann et al. 2005; Gilles et al. 2010). Thus there seems to be more than allergen alone in the sensitization process leading to a Th2-dominated allergy to pollen.

6.1.2 Threshold Levels for Symptom Development

The determination of allergen threshold levels for symptom development is hampered by many factors that influence the outcome. Sensitized patients differ in the extent of exposure to allergen required for symptom development. Furthermore, airborne allergen is thought to be present not only in intact pollen, but also in small respirable particles (Taylor et al. 2004; Spieksma et al. 1995a; Grote et al. 2003). However, whereas most of our knowledge on threshold values and symptom development is based on pollen counts in ambient air determined by a pollen sampler in a certain area, data from monitoring allergen in ambient air are still scarce (Buters et al. 2008). This chapter provides an overview of the current knowledge on pollen threshold values on symptom development and the various factors that influence these threshold values (Fig. 6.1).

6.2 Determination of Threshold Levels for Symptom Development

A large variety of studies has focused on the determination of pollen threshold levels that cause symptoms. Therapeutic intervention studies are especially useful. These deal with the determination of pollen (or allergen) thresholds that cause previously defined symptoms (e.g. "at least five sneezes and/or a symptom score of at least 2 on a scale from 0 to 3"). Although natural exposure to pollen under conditions of daily life is the most relevant, it has several disadvantages (see also Sect. 6.2.3, "Ambient Air Studies"). Therefore more controlled settings have been developed like provocation tests and exposure chambers. These study designs have their own advantages and disadvantages that will be discussed in the following section.

6.2.1 Provocation Tests

In an experimental challenge model the allergen is delivered onto the nasal mucosa by dripping, by paper disc, by pump sprays (Akerlund et al. 2005) or by eye drops on the eyes. In these studies the delivered dose is expressed in arbitrary units such as "Allergen Units" (Abelson et al. 1994), "standardized quantity units" (Korsgren

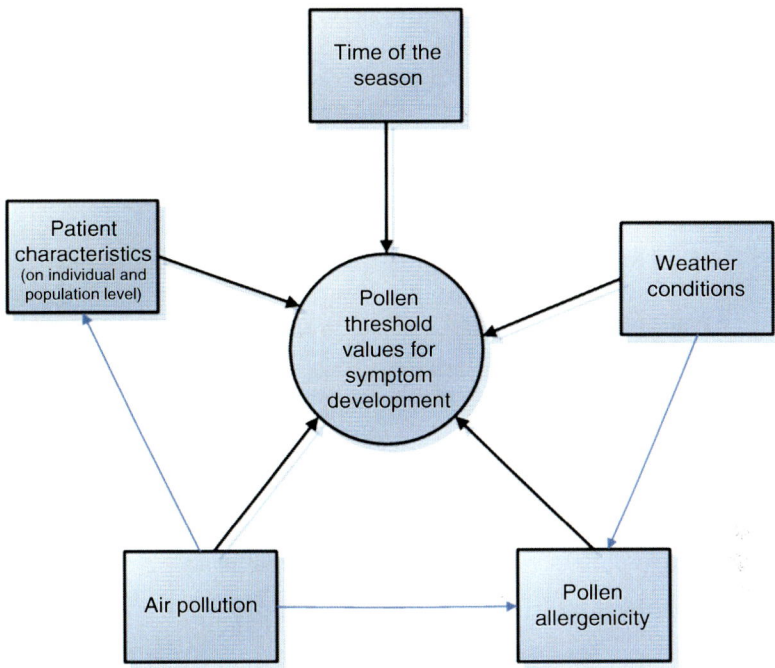

Fig. 6.1 Schematic representation of the main factors that influence the pollen threshold values for symptom development (*black arrows*). Some of these factors can also influence each other (*blue arrows*)

et al. 2007) or "biologic units" (Bonini et al. 1990). These units were defined in most cases for standardization of allergen preparations, and their relation to pollen and/or allergen load is often unclear. Therefore, these studies will not give information on pollen thresholds, but they indicate the different threshold levels to which individual patients will respond. The patients are challenged with standardized increasing concentrations of the allergen until previously defined symptoms develop (Bonini et al. 1990; Abelson et al. 1994; Ahlholm et al. 1998; Ahlstrom-Emanuelsson et al. 2004; Ciprandi et al. 2009). The concentration at which these symptoms develop is the threshold concentration. A drawback of this approach is the repeated challenge of allergen to the nasal mucosa or the conjunctiva, which may prime these tissues and may cause a non-specific increase in the sensitivity and responsiveness (Connell 1968; Canonica and Compalati 2009). Furthermore, whereas these challenge studies will not reveal relevant information on pollen thresholds, they are perfectly suitable for (i) identification of the allergen to which a patient responds; (ii) examining the pathogenesis of allergic rhinitis or conjunctivitis; and (iii) assessing the response of individual patients to therapeutic intervention because of the standardized delivery of the allergen on the surface to be tested. These studies can show clearly that individual patients have different threshold levels for symptom development (Ciprandi et al. 2009).

6.2.2 Pollen Chambers

Special chambers have been developed for controlled human exposure to pollen. These chambers avoid the unpredictable levels of pollen and weather conditions during a study period. Different pollen chambers are developed, either hosting up to 14 persons (Horak et al. 2003), or 160 persons at the same time (Day and Briscoe 1999). The pollen chambers deliver a controlled pollen level over a period and allow assessment of the response throughout the study period. Pollen levels used in these chambers are usually quite high, e.g. 3,500 grains/m^3 for ragweed (Day and Briscoe 1999), 4,000 pollen/m^3 or 1,500 pollen/m^3 for grass pollen (Day and Briscoe 1999; Horak et al. 2003; Krug et al. 2003), to ensure the desired effect of producing a full spectrum of symptom severity in allergic persons without affecting non-allergic participants.

Because of the requirement of high pollen concentrations, the pollen chambers have not yielded good studies on pollen thresholds for symptom development that are relevant in real life. In contrast, the chamber provocations have been very helpful assessing the effect of anti-allergic drugs (Day et al. 2000, 2009).

6.2.3 Ambient Air Studies

Exposure to natural pollen under "real life" conditions is the most relevant situation but also the most difficult to study, since the conditions are not well controlled and the exposure of patients to pollen will be variable due to the fact that they are outdoors during different times at different locations. A study design that tries to control these latter variables are the so-called "park studies" (Georgitis et al. 2000; Akerlund et al. 2005). In these studies, patients are allowed to sit, walk and recreate in a park on a day where at several locations pollen traps monitor the local pollen concentrations. This design has the advantage that parks can be found in almost any city (in contrast to pollen chambers) and patients are in close vicinity to pollen traps during the study period so that exposure can be measured and they can be monitored regularly. The limitations of these park studies are that they are time-consuming to organize and they require effort and time from the patient. Therefore, these studies are often performed on 1 day with high pollen counts to ensure symptom development. For threshold studies the patients should be monitored on subsequent days during the season. Although such daily studies in a park would be valuable for threshold studies, to our knowledge they have never been used for this purpose.

Another method of studying symptoms under natural conditions is to ask patients during their daily life to fill in a record of their symptoms (diary studies). Studies with this kind of approach assess the most relevant situation for the patients, i.e. daily life. However, since there is no general consensus on the method to determine the threshold level under these "real life" conditions, the designs of these studies differ. Furthermore, the "freedom of the patient" leads to variable pollen exposures among individual patients. These aspects are relevant in the interpretation of the results. The local pollen concentration at the patient location is variable and unknown. Often, pollen is monitored by a pollen sampler placed on a central roof top location to give a representative sample of regionally distributed pollen. The results are used as an average estimate of the concentration during the past 24 h in an area defined by a certain vegetation, climate regime and land use. If pollen traps were situated at ground level, the result would be affected by an overrepresentation of pollen produced by sources in the immediate vicinity of the trap. Although the results have been shown to correlate with symptom scores, medication use, and consultations for rhinoconjunctivitis and asthma in several studies, they are not equal to the pollen concentration that individual patients experience during their activities in the street. The difference in the exposure between persons in different parts of a city will vary among pollen types, and is probably more significant for pollen released at low height than for example birch pollen in Scandinavia which is produced in large amounts at tree-top level.

Bi-hourly pollen counts are also possible. These counts reflect the time that the pollen grains arrive at the trap, not the time at which the pollen grains are emitted. Furthermore, these counts will still not give us the pollen levels that the patients encounter. Therefore personal sampling of the environment of the patient is required. Several methods can be used to study the personal exposure to aeroallergens, like battery-operated personal samplers (Fiorina et al. 1997; Mitakakis et al. 2000; Okuda et al. 2005), nasal air samplers (Mitakakis et al. 2000), pollen trapped in nose (Okuda et al. 2005) and hair (Wiltshire 2006). However, to our knowledge none of these methods have been used for threshold studies. The time that the patient is outside and thus exposed to pollen is dependent on the patient's activities schedule and therefore variable. Furthermore, the activity pattern during the exposed time period will be variable and this will result in a different exposure to pollen (Mitakakis et al. 2000); e.g. a jogger will breathe in more pollen-loaded air than a sun-bather. The size of the group of patients differs between studies and also the criterion of "symptom development" at which the threshold pollen concentration should be determined diverges between different studies. For instance, in one study the pollen threshold level is defined as the pollen count at which 90% of all patients experience their first symptoms (Viander and Koivikko 1978), while another study uses the pollen concentration at which, on a scale from 0 to 3, the mean symptoms score of the whole population exceeds 1 (Florido et al. 1999). Standardization of the method to determine pollen threshold levels (size of the group of patients and of criteria for threshold pollen concentrations) will be necessary for a good evaluation of the threshold concentration in different regions and among different populations.

6.2.4 *Clinical Network*

In France the assessment of threshold levels for symptom development is not only based on pollen counts but also on the information obtained from a network of allergists. More than 100 allergists from several areas in France provide weekly information on the number of pollinosis patients, the symptom-type, and the severity of the symptoms (Thibaudon et al. 2008). With these clinical data it was possible to draw three areas in France with different threshold levels for the various pollen types (see also Table 6.1). Furthermore, these clinical data have proven to be a tool to study the change in health impact from pollen during a season and to monitor the change in health impact from year to year.

6.3 Allergenic Pollen Species and Their Threshold Values

In the past decades several studies have been published on threshold levels of different pollen types. Studies on the most relevant pollen types in Europe and the Mediterranean countries are discussed in this section.

In aerobiological studies, pollen of species belonging to the same genus or of genera belonging to the same family usually cannot be distinguished under the light microscope as they share the same characteristics (they are stenopalynous). For this reason the different allergenic pollen or spore expositions refer to airborne pollen that represent a family (e.g. Poaceae), a genus (e.g. *Plantago*), and rarely a single species (e.g. *Olea europaea*).

Furthermore it is important to note that when we try to define the pollen risk threshold, the definition depends on the methodology used for pollen monitoring, i.e. the sampler type or the sampler height position. For instance, most threshold studies in ambient air determine pollen levels using a pollen sampler on a roof top, usually at a different height from personal exposure. This is one of the reasons why threshold values differ among the different studies (Table 6.1). Greater differences can be observed when comparing both daily and intradiurnal pollen content in the air at human height or roof top height, showing higher or lower differences depending on pollen size, meteorological conditions and atmospheric stability during the sampling period (Rantio-Lehtimäki et al. 1991; Galán et al. 1995; Alcázar et al. 1998, 1999a, b). The various pollen types will be discussed in alphabetical order and studies on the threshold values and the prevalence are summarized in Tables 6.1 and 6.2, respectively.

6.3.1 Alnus, Betulaceae Family

Alder is a rather common tree in all temperate climatic regions, mainly restricted to wet soil conditions. Predominant species of the genus are black alder (*A. glutinosa*) and grey alder (*A. incana*). The flowering period occurs from winter to early spring, depending on the climatic characteristic of the region. Patients sensitive to alder pollen may exhibit reactions towards other pollen of the Fagales order, although the cross reactions appear to be strongest among the species belonging to the same family e.g. within the family of Betulaceae, e.g. alder/birch. The major allergen of alder Aln g 1 is closely related to the major allergens of birch (Bet v 1), hazel (Cor a 1), hornbeam (Car b 1) and oak (Que a 1) (Matthiessen et al. 1991). The alder pollen season precedes the season of birch pollen and may act as a primer making patients more sensitive to birch pollen (Emberlin et al. 2007).

6 Impact of Pollen

Table 6.1 Studies on pollen threshold levels published in peer-reviewed journals

Country(city) (Reference)	Criteria	Pollen threshold (grains/m^3, 24 h)
Alnus		
Poland (Rapiejko et al. 2007)	First symptoms	45
	Symptoms present in all patients	85
Poland (Hofman et al. 1996)	?	50
Finland (Rantio-Lehtimäki et al. 1991)	Low	<10
	Moderate	10–100
	Abundant	>100
France[a] (Thibaudon 2003)	Low	6–13
	Moderate	14–227
	High	>227
Ambrosia		
Canada (Comtois and Gagnon 1988)	First symptoms	1–3
Canada (Banken and Comtois 1990)	Majority of allergic patients show symptoms	13
France (Déchamp et al. 1997)	First symptoms	First pollen grains
Russia (Ostroumov 1989)	First symptoms	20–25
Austria (Jäger 1998)	Majority of sensitive patients show symptoms	20
Hungary (Juhász 1998)	60–80% of the sensitive patients show symptoms	50
France (North)[a] (Thibaudon 2003)	Low	>30
	Moderate	>80
France (Mid-France)[a] (Thibaudon 2003)	Low	>1
	Moderate	>2
	High	>10
France (Mediterranean area)[a] (Thibaudon 2003)	Low	>10
	Moderate	>50
	High	>100
Artemisia		
Poland (Rapiejko et al. 2007)	First symptoms	30
	Symptoms present in all patients	55
Poland (Hofman et al. 1996)	?	20
Finland (Rantio-Lehtimäki et al. 1991)	Low	<10
	Moderate	10–30
	Abundant	>30
Israel[b] (Waisel et al. 2004)	Initiation of symptoms	>4–5
France[a] (Thibaudon 2003)	Low	5–19
	Moderate	20–69
	High	>69
Betula		
Finland (Viander and Koivikko 1978)	Highly sensitive patients develop symptoms	30
	90% of patients had symptoms	80
Poland (Rapiejko et al. 2007)	First symptoms	20
	Symptoms present in all patients	75
Poland (Hofman et al. 1996)	?	100

(continued)

Table 6.1 (continued)

Country(city) (Reference)	Criteria	Pollen threshold (grains/m³,24 h)
Finland (Rantio-Lehtimäki et al. 1991)	Low	<10
	Moderate	10–100
	Abundant	>100
France(North and Mid-France)[a] (Thibaudon 2003)	Low	5–20
	Moderate	20–99
	High	>99
France (Mediterranean area)[a] (Thibaudon 2003)	Low	5–99
	Moderate	100–299
	Low	>99
Castanea		
France[2] (Thibaudon 2003)	Low	>99
Corylus		
Poland (Rapiejko et al. 2007)	First symptoms	35
	Symptoms present in all patients	80
Poland (Hofman et al. 1996)	?	6–12
France[a] (Thibaudon 2003)	Low	6–13
	Moderate	14–227
	High	>227
Cupressaceae		
Israel[b] (Waisel et al. 2004)	Initiation of symptoms	10–50
France(North and Mid-France)[a] (Thibaudon 2003)	Low	70–141
	Moderate	>141
France (Mediterranean area)[a] (Thibaudon 2003)	Low	7–13
	Moderate	14–141
	High	>141
Olea		
Spain (Jaen) (Florido et al. 1999)	Regression analysis on mean symptoms score "development of mild symptoms"	>400
Israel[b] (Waisel et al. 2004)	Initiation of symptoms	>4
France (North and Mid-France)[a] (Thibaudon 2003)	Low	>60
France[a] (Thibaudon 2003)	Low	6–13
	Moderate	14–227
	High	>227
Parietaria		
Italy (Negrini et al., 1992)	% of patients with symptom score >2	
	31%	10
	47%	15
	67%	40
	90%	80
Italy (D'Amato et al. 1991)	?	30
Israel[b] (Waisel et al. 2004)	Initiation of symptoms	5
France (North and Mid-France)[a] (Thibaudon 2003)	Low	20–49
	Moderate	>49

(continued)

Table 6.1 (continued)

Country(city) (Reference)	Criteria	Pollen threshold (grains/m³,24 h)
France (Mediterranean area)[a] (Thibaudon 2003)	Low	5–19
	Moderate	20–69
	High	>69
Platanus		
Spain (Cordoba) (Alcázar et al. 2004)	Most patients had symptoms	>50
Spain (Madrid) (Camacho et al. 2008)	Combination of medical and statistical criteria (assures allergic reaction and "no-back")	30
France[a] (Thibaudon 2003)	Low	6–13
	Moderate	14–227
	High	>227
Poaceae		
London (Davies and Smith 1973)	All patients experience symptoms	50
Poland (Lipiec et al. 2005)	¼ of patients showed allergic reaction (eyes)	22
	All patients symptoms score 2 (eyes)	52
Poland (Rapiejko et al. 2007)	First symptoms	20
	Symptoms present in all patients	50
Poland (Hofman et al. 1996)	?	20
Spain (Antepara et al. 1995)	All patients experience moderate or severe symptoms	37
Finland (Rantio-Lehtimäki et al. 1991)	Low	<10
	Moderate	10–30
	Abundant	>30
Israel[b] (Waisel et al. 2004)	Initiation of symptoms	>4
France[a] (Thibaudon 2003)	Low	>1
	Moderate	>4
	High	>35
Quercus		
France[a] (Thibaudon 2003)	Low	6–13
	Moderate	14–227
	High	>227

The threshold levels were determined in different countries and at different criteria. Pollen levels are determined in ambient air at roof top level using a Hirst type pollen sampler unless otherwise indicated
[a]The scale of 3 levels is for the public, there is another scale (5 levels) for doctor
[b]Pollen levels are determined by a Rotorrod sampler at a sampling height of 80 cm

6.3.2 Ambrosia, Asteraceae (Compositae) Family

Although this genus comprises up to 45 species worldwide, the species that is spreading in Europe and Eastern North-America is essentially *Ambrosia artemisiifolia* or common ragweed. Ragweed established itself in Europe at the beginning of

Table 6.2 Prevalence of sensitization to allergenic pollen in Europe

Article	Country	Prevalence (%)											
		Cor	Aln	Bet	Pla	Que	Poa	Par	Art	Amb	Ole	Cup	Che
Pereira et al. (2006)	Spain			n:208	n:493		n:1,111	n:325	n:425		n:879	n:449	n:442
	Canary islands			2	4		18	9	18		4	3	8
	Centre			15	35		82	19	29		59	45	32
	Mediterranean			4	22		35	23	12		36	14	18
	North			10	17		52	9	15		25	15	11
	South			14	28		59	15	25		63	16	29
	Portugal:			n:150	n:182		n:588	n:294	n:265		n:243	n:117	n:161
	North			11	20		51	20	19		27	14	17
	Centre			19	13		53	33	30		13	5	12
	South			7	12		41	20	18		20	9	11
Zureik et al. (2002)	Northern Europe n: 264 (Iceland, Norway, Sweden)			39.8			39.4	1.9		3.8	6.1		
	Central Europe n:139 (Belgium, France, Netherlands)			18			38.1	5		3.6	15.1		
	Southern Europe n: 150 (Spain, Italy)			10			34.0	12		2.7	18.0		
Stach et al. (2007)	Poznan Poland n: 721								11.6				
Gioulekas et al. (2004)	Greece n: 1,311	9.6		6.8	8.2	7.6	40.4	15.3			31.8	12.7	18.3
Kilpeläinen et al. (2002)	Finland (Turku)												
	Farm n: 28			8.3			12.6		11.0				
	Nonfarm n: 253			24.2			30.3		5.7				
Belchi-Hernandez et al. (2001)	Spain (Murcia)												
	Polysensitized. n: 1,573				1.9		51.4	37.1	28.1		67.4	3.1	65.4
	Monosensitized n: 163				1.2		11.0	29.5	9.2		24.0	0.6	17.8

6 Impact of Pollen

Study	Location/Group											
Gonianakis et al. (2006)	Creta (Herakion) n: 576	21.5	16.9	24.6	12.3	16.9	21.5	64.02	40.96	16.13	11.63	3.11
Erkara et al. (2009)	Turkey (Eskisehir) n: 130											
	Sensitization severe	7.7	13.8	7.7	12.3	21.5	53.8	3.1	3.1	1.5		0.6
	Sensitization moderate					3.1		3.1	3.1			
	Sensitization mild	1.5	1.5	1.5	3.1	15.9 (11.9)	24.6					--
Heinzerling et al. (2009),	UK n: 3,034	15.9 (10.3)[b]	16.7 (11.1)	19.0 (11.9)	15.9 (11.9)		54.0 (50.8)	17.5 (16.7)	5.6 (3.2)	7.9 (7.1)	15.1 (12.7)	11.1 (8.7)
Burbach et al. (2009)	Austria	22.7 (13.3)	21.8 (12.4)	19.4 (9.5)	4.1 (0.0)		30.8 (20.2)	2.0 (1.0)	10.6 (6.9)	8.5 (5.3)	13.3 (5.9)	1.5 (0.3)
	Belgium	16.6 (13.5)	16.1 (14.3)	17.6 (13.9)	1.2 (0.9)		25.1 (24.5)	1.4 (0.7)	4.7 (4.7)	3.0 (3.0)	4.0 (3.4)	2.0 (1.2)
	Greece	10.3 (6.5)	8.4 (6.5)	9.8 (5.1)	6.5 (4.2)		49.5 (42.3)	24.8 (20.6)	16.9 (11.7)	11.7 (5.1)	35.0 (29.9)	5.6 (3.7)
	Denmark	49.4 (37.8)	47.0 (36.2)	57.4 (49.1)	8.9 (6.2)		69.9 (64.0)	5.6 (4.9)	28.3 (23.8)	17.1 (14.3)	9.1 (9.1)	5.8 (3.7)
	Germany	35.9 (32.4)	34.8 (31.7)	37.6 (34.1)	5.3 (2.3)		37.9 (34.1)	6.9 (3.9)	22.5 (19.1)	14.4 (9.3)	9.7 (4.9)	2.8 (0.8)
	Poland	22.3 (13.3)	22.8 (13.6)	27.7 (19.6)	4.0 (2.4)		38.0 (30.8)	3.0 (2.2)	26.2 (14.9)	10.8 (5.4)	2.7 (2.0)	1.2 (0.8)
	Portugal	7.4 (3.9)	6.8 (4.4)	6.8 (4.4)	7.0 (6.4)		34.4 (31.6)	17.5 (14.7)	16.3 (14.6)	12.4 (10.8)	21.3 (17.9)	5.1 (2.8)
	France	11.9 (7.4)	10.4 (4.8)	8.4 (4.0)	7.6 (2.7)		26.4 (19.3)	6.5 (3.6)	10.7 (3.5)	9.0 (4.5)	18.2 (8.9)	8.7 (5.3)
	Italy	9.3 (7.2	3.1 (2.3)	9.4 (7.7)	3.2 (2.5)		19.5 (18.6)	33.2 (30.7)	6.7 (5.8)	3.5 (3.1)	23.2 23.0)	8.1 (6.0)
	Hungary	20.2 (15.9)	16.0 (12.3)	20.1 (16.2)	7.0 (5.2)		40.6 (37.3)	3.1 (2.0)	44.3 (38.8)	53.8 (49.7)	14.4 (12.1)	2.9 (2.9)
	Finland	24.7 (22.9)	26.3 (24.6)	34.0 (30.0)	0.9 (0.9)		23.8 (18.5)	0.9 (0.9)	17.6 (13.0)	2.3 (1.4)	2.0 (1.0)	0.0 (0.0)
	Netherlands	24.8 (24.4)	24.5 (24.2)	26.9 (26.5)	4.7 (3.8)		35.5 (34.4)	8.7 (8.7)	6.2 (5.8)	18.6 (16.7)	12.3 (11.9)	1.5 (1.5)
	Switzerland	51.7 (24.8)	44.1 (22.8)	50.3 (43.4)	3.4 (2.1)		78.6 (71.0)	2.1 (0.7)	17.2 (6.2)	18.6 (9.7)	45.5 (32.4)	1.4 (1.4)

The table summarizes studies from the last 10 years (2001–2010) in which sensitization rates (%) are determined by skin prick tests in groups of either symptomatic patients or patients visiting the outpatient clinic

[a] n gives the size of the patient group for each study
[b] In brackets the clinically relevant sensitization rate (%) is given

the twentieth century, resulting in highly infested areas like the Balkan area, the Rhône valley in France and Northern Italy. In areas heavily infested with these plants ragweed pollen is considered as the major pollen allergen (e.g. Vitányi et al. 2003).

Ragweed pollen is often nearly the only pollen type in the air except for a very short period in the beginning of the flowering season when mugwort pollen (*Artemisia vulgaris*) is also airborne. Although the allergens present in mugwort and ragweed pollen show cross reactivity (Hirschwehr et al. 1998; Asero et al. 2006), this does not interfere with the diagnosis for ragweed pollinosis, since the flowering season of these two species only partly overlaps. If sufferers keep a daily symptom diary it is easy to know if symptoms correlate with the presence of either ragweed or mugwort pollen in the air.

European studies on the ragweed pollen threshold for symptom development are rare. This is most likely to be due to the fact that ragweed is an invasive weed that was not known in Europe two centuries ago and was introduced from the new World. Its spread is now intensive due to several factors such as (i) changes in lifestyle and land use, (ii) increases in suburbs resulting in the decrease of arable lands (iii) some bylaws of Common Agricultural Policy established in 1993–1994 (Déchamp et al. 2001) and recently (iv) possible effects of climate change.

A relevant paper on ragweed thresholds comes from Quebec, in the Eastern part of Canada (Comtois and Gagnon 1988). This study, which was conducted over two pollen seasons showed that symptoms appear at 1–3 pollen/m^3 of air. In France, many ragweed pollen allergic patients ask for treatment when the first pollen grains appear in the pollen monitoring trap (Déchamp et al. 1995).

Ragweed pollen is prevalent between 42° and 50° Northern latitude and rare in most countries around the Mediterranean coast and Portugal. Recent studies on the prevalence of ragweed sensitization in Europe show an increasing trend (Burbach et al. 2009). In contrast to other pollen types, ragweed pollinosis tends to be more prevalent in suburbs than in towns (Harf and Déchamp 2001).

6.3.3 *Artemisia, Asteraceae (Compositae) Family*

Mugwort is an unsuspicious weed, which looks very similar to *Ambrosia*. The major allergen of the mugwort pollen, Art v 1, is homologous to Amb a 4 of *Ambrosia* pollen and may cause cross reactivity in patients (Wopfner et al. 2005).

The weed is spread all over Europe. The most common species in Europe is *A. vulgaris* and it flowers in late summer (at the end of July, August and September). Two species which are common in Southern Europe *A. annua* and *A. verlotorum* flower even later (late summer to autumn). For this pollen type it was shown that street level counts are usually higher than the counts from rooftop samplers (Rantio-Lehtimäki et al. 1991). This may also hold for other pollen types due to the fact that these pollen grains are released during the early morning, when there is hardly any convection or turbulence in the air, preventing them from being lifted up to rooftop levels.

6.3.4 *Betula*, Betulaceae Family

Birch is the major pollen allergen producing tree in Northern Europe. The flowering season varies depending on the area. In Western Europe flowering starts by the end of March while in Central and Eastern Europe pollen grains appear in the air from the beginning to mid-April. Northward flowering starts, depending on the latitude, from late April to late May. Towards the South the number of birch trees is reduced and in Spain they are found mostly in mountainous areas and in zones in the interior of the North of the Peninsula with maximum values being detected during February and March (Hernandez et al. 1998).

Some allergens of these pollen grains show cross reactivity with other species belonging to this family. For example, the major allergen of birch Bet v 1 and Bet v 2 share IgE epitopes with the allergens in hazel and alder (Niederberger et al. 1998). Due to this cross reactivity the sequential flowering of the Betulaceae species may cause a prolonged hay fever season for sensitive patients.

6.3.5 *Castanea*, Fagaceae Family

There are no threshold studies published for *Castanea* pollen. This may be due to the fact that chestnut allergy (a) is strongly associated with latex-fruit syndrome and some other fruits (Raulf-Heimsoth et al. 2007), (b) has cross-reactivity with many other tree species like birch, walnut or *Quercus* (Cosmes et al. 2005; Teuber et al. 2003), (c) scores very low in monosensitization in positive skin prick-tested patients to *Castanea* (Lee et al. 2005; Laurent et al. 1993). Flowers bloom in late spring.

6.3.6 *Casuarina*, Casuarinaceae Family

The genus *Casuarina* (Australian pine) comprises nearly 60 species mainly natives to Australia. Some species are represented in warm regions forming part of the green urban spaces. It is an anemophilous tree that flowers during autumn, from September to November depending on the climatic characteristic of the region. *Casuarina* pollen allergenicity has been cited by various researchers (i.e. Agashe et al. 1994; García et al. 1997), but there are no studies on the pollen threshold for sensitized patients. Trigo et al. (1999) describes the presence of this pollen type in Spain. Although in most locations the daily average concentration of *Casuarina* pollen did not exceed 5 pollen/m^3 of air, in some locations in Andalusia Region and especially in Malaga (Sun Coast), 30 and 500 pollen/m^3 of air were reached, respectively. For this reason it is important to consider this tree as responsible of autumnal allergies in the Mediterranean Region.

6.3.7 Corylus, Betulaceae Family

Hazel is widely distributed in Europe and it typically has a number of shoots or trunks branching out at ground level. Due to this growth habit, it has been referred to as a bush rather than a tree. Flowering occurs from winter to early spring, depending on the climatic characteristics of the region.

The major allergen of hazel pollen, Cor a 1, is cross reactive with Bet v 1. Mari et al. (2003) showed that exposure to a high amount of hazel pollen in an intensively cultivated hazel area north of Rome (Italy) seems to lead to an IgE sensitization to allergenic molecules that are present only in hazel or hazel-related pollen.

6.3.8 Cupressaceae Family

Cupressaceae are the most widely distributed conifer family with a global range of nearly all continents except for Antarctica. In Europe, trees of the genus *Cupressus* are spread widely over the East Mediterranean e.g. *C. sempervirens* (native) and *C. arizonica* (introduced). Many species or genera of this family (e.g. *Cupressus, Juniperus, Thuya*) release pollen almost all year round and their pollen cannot be distinguished morphologically from each other. Daily concentrations can reach record peaks in the Mediterranean region during winter or early spring (Hidalgo et al. 2003; Charpin et al. 2005; Perez-Badia et al. 2010) although the pollen production shows a large variation from year to year. A gradual increase in the annual amounts of Cupressaceae pollen in the Mediterranean was observed since the 1980s, which in some cases was attributed to the increased planting of these trees (Calleja and Farrera 2000; Damialis et al. 2007).

Since reliable extracts to test the sensitization to Cupressaceae allergy have only been available since the last decade, the epidemiology of this allergy is poorly understood (Charpin et al. 2005). In Europe, allergy to Cupressaceae pollen was considered a rarity until 1975. Since then, considerable increases in the number of cases has been observed in Spain, France, Italy and Israel (Guerra et al. 1996; Papa et al. 2001; Charpin et al. 2005). An Italian multicentre study showed that 14.7% of the Cupressaceae allergic patients were monosensitized and that their average age was higher than that of the polysensitized patients (Italian Association of Aerobiology 2002). This may suggest that at least in some of the Cupressaceae allergic patients the sensitization may be caused by a long lasting exposure to the pollen, without being a typical atopic patient (Charpin et al. 2005).

6.3.9 Fagus, Fagaceae Family

Fagus is a tree of temperate South-West and Central Europe, extending to South Sweden. No threshold values for *Fagus* (beech) pollen can be found in the literature.

Also studies on prevalence rates are not very conclusive. In Germany, 12% of all hay fever patients react to it (Horak et al. 1979). Reactions against beech pollen allergen occurred in 87% of all persons allergic to birch pollen in a Swedish multi-centre study (Eriksson et al. 1984). The rate was the same in all Sweden, even in parts of the country where *Fagus* does not occur. It was concluded that the reactions against *Fagus* allergens resulted from of primary sensitization to *Betula* pollen. According to Horak et al. (1979), approximately 50% of all beech pollen allergies are due to cross-reactions with pollen from other species of Fagales. In a study from Swiss and Austrian birch pollen allergic patients, sera exhibited similar IgE reactivity profiles to birch, beech and oak pollen extracts due to beech and oak allergens that cross-react with the birch pollen allergens Bet v 1, 2 and 4 (Egger et al. 2008). In Sweden, the correlation coefficient between reactions to birch and beech pollen was 0.75 (Eriksson et al. 1987). The beech-specific IgE-levels were 0.063 in patients from Basel and 0.081 in patients from Vienna, respectively, as compared to 0.481 and 0.668 for birch (Egger et al. 2008). Beech is considered to be an early spring flowering tree.

6.3.10 *Fraxinus, Oleaceae Family*

In central Europe, *Fraxinus* pollen concentrations are considerably high. In Switzerland, for example, the concentrations are comparable with the amounts of birch pollen. Most of these pollen grains originate from *Fraxinus excelsior*, while other species with a more southern European distribution like *Fraxinus ornus* or *Fraxinus angustifolia* are represented with distinctly less pollen in the pollen traps (Guerra et al. 1996). Several studies show that ash pollen should be considered as a relevant allergen of spring pollinosis (Schmid-Grendelmeier et al. 1994; Hemmer et al. 2000). Sensitization to *Fraxinus* pollen is common in Switzerland and is also of clinical relevance as shown by nasal provocation tests (Schmid-Grendelmeier et al. 1994). The prevalence of *Fraxinus* sensitization in groups of pollen allergic patients varied from 17.6% in Vienna (Hemmer et al. 2000), 51.8% in Locarno and Lugano, Switzerland (Colombo 2010) to 56% in Zurich, Switzerland (Schmid-Grendelmeier et al. 1994).

The main ash allergen Fra e 1 is a homolog of Ole e 1, the highly allergenic olive pollen allergen. Fra e 1 has only a limited cross reactivity to birch pollen. Additionally this cross reactivity is confined to minor birch allergens (Wahl et al. 1996). Concerning thresholds for the development of symptoms, the study of Horak et al. (1979) gives a level of 167 pollen/m^3. In this study patients' symptoms were compared with the pollen measurements of the pollen trap in Vienna. The threshold is defined as the value for causing symptoms in the average patient. Thresholds for high pollen concentrations in Switzerland and Austria are 100 and 90 pollen/m^3 respectively, but these values are not based on symptom studies. *Fraxinus* is considered to be a late winter or spring flowering tree, depending on the species and the geographical area.

6.3.11 Ligustrum, Oleaeceae Family

The genus *Ligustrum* (privet) comprises about 20 species, subspecies and varieties. They are frequently found in urban environments in the Mediterranean Region, both as ornamental trees or hedges in parks and gardens. Privet is considered to be an entomophilous plant with large and heavy pollen grains which are not well collected by samplers located on the building roof tops. However, Cariñanos et al. (2002) have observed higher pollen counts in samplers situated at human height, where this plant is observed frequently in the urban green space.

Being a member of the Oleaceae family, privet pollen carries a homologue of Ole e 1. Since the flowering period occurs at the end of spring, coinciding with the post-peak olive pollen curve, for people suffering from olive pollen allergy the period with symptoms can be extended. For this reason privet pollen should be considered as a potential causative agent of local allergy problems (Cariñanos et al. 2002).

6.3.12 Olea Europaea, Oleaeceae Family

Olive is a cultivated evergreen tree native to the Mediterranean region. It is an ambophilous species, i.e. pollinated by both insects and the wind. Due to the selection through time of varieties with high flower and pollen production, nowadays the anemophilous character is more pronounced. Although the olive tree showed an entomophilous character in origin, today it is considered to be an anemophilous species due to the selection through time of varieties with high flower and pollen production. Although Ole e 1 was the first major allergen described and characterized, it has been demonstrated since that there are at least, 20 proteins present with allergic activity. Ten of these have been characterized, cloned and expressed (Cárdaba et al. 2007) (see paragraph 6.4.2). This tree flowers during the spring so that the olive and grass pollen seasons usually coincide. Although many studies support the role of olive pollen as one of the main causes of allergic disease in the Mediterranean area, only a few studies cite a specific risk threshold for sensitized patients (Table 6.1). As it has been mentioned before, the methodology used for pollen monitoring is an important issue to take into consideration when trying to define the pollen risk threshold. In these two studies (Table 6.1) a Hirst designed spore trap, located on a roof top, was used in Spain, while a Rotorod sampler located at a height of 80 cm was used in Israel.

6.3.13 Parietaria, Urticaceae Family

Pellitory-of-the-wall (*Parietaria*) is a typical allergenic pollen producing plant in the Mediterranean area. Par j 1 and Par j 2 are the two major allergens, with similar IgE epitopes, in *Parietaria judaica* (Asturias et al. 2003). *Parietaria* belongs to the

Urticaceae family together with *Urtica* (nettle) which is allergenically of little importance. The pollen of these two genera cannot be distinguished microscopically. Only the species *Urtica membranaceae* has a distinct pollen type, allowing the study of its individual pollen behaviour in the air (Galán et al. 2000). In the Mediterranean area the pollen season extends from spring to early summer and even autumn (Fotiou et al. 2011), depending on the species and the geographical area.

6.3.14 *Plantago*, Plantaginaceae Family

Plantain (*Plantago*) pollen is considered to be a minor cause of pollinosis in Europe (D'Amato et al. 1998). Mono allergy to plantain is unusual and often patients that are positive in a skin prick test for plantain pollen extract are also sensitized for grass pollen (Watson and Constable 1991; D'Amato et al. 1998). This is probably due to cross reactivity among these pollen species. Group 5 grass pollen allergen appeared to be responsible for most grass/plantain cross reactivity (Asero et al. 2000). The flowering season of plantain largely overlaps with the grass flowering period. Therefore the symptoms caused by *Plantago* cannot always be attributed to *Plantago* pollen.

6.3.15 *Platanus*, Platanaceae Family

Plane trees are widely grown as street and shade trees in European cities, especially in France, Spain, Italy and Greece. London Plane tree (*P. acerifolia* or *P. hybrida*) is the most common cultivated species. A high prevalence of positive skin prick tests has been recorded for *Platanus hybrida* in Madrid (Varela et al. 1997). In Córdoba (South Spain) 17% of the population is sensitive to *Platanus* pollen and patients presented symptoms almost immediately after the *Platanus* pollen season started, as this pollen type appears abruptly in high concentrations in the air. Patients suffer from symptoms at pollen concentrations of 51–200 pollen/m^3 during less than a month (Alcázar et al. 2004). *Platanus* pollen is produced at high levels in the southern European countries, where it is a relevant allergenic pollen species. Plane trees flower during early spring.

6.3.16 Poaceae Family

Poaceae, grasses, comprise a high number of species flowering from winter to the end of summer, depending on the species and on the geographical area. Although knowledge about the differences in the allergenic potency of the various species is scarce, most species are considered to be allergenic. A recent study showed that within the Pooidae subfamily 12 grass species which are common in the temperate and subtropical zones appeared to share similar IgE determinants (Andersson and Lidholm 2003). Threshold studies discussed here deal with the mixture of grass pollen that is produced in the particular area under study. Most threshold studies in ambient air determine pollen levels using a pollen sampler on a roof top, except for a study from Israel, but the criteria at which the threshold levels are determined differ among the different studies (Table 6.1). Recent European studies show that the prevalence of sensitization to grass pollen is significantly higher than that to other pollen species (Bauchau and Durham 2004; Bousquet et al. 2007) except for ragweed pollen in highly infested regions (Vitányi et al. 2003). The prevalence of grass pollen allergy differs among different European countries and among different studies (Table 6.2) and appears to be inversely associated with pollen exposure (Burr et al. 2003). This inverse association is very weak, but this study cautions us to conclude that high exposure to pollen increases the risk of rhinoconjunctivitis or/and asthma.

6.3.17 *Rumex*, Polygonaceae Family

Sorrel flowers from late May until August, which largely overlaps with the flowering seasons of *Plantago* and grass pollen. The clinical relevance of sorrel pollen cannot easily be defined. Allergy to sorrel may be overlooked, because it is not thought of during the grass pollen season. From a group of 96 grass pollen allergic patients in Austria, 70% of randomly selected sera tested positively for sorrel specific IgE. Relevant cross-reactivity between grass and sorrel pollen was excluded (Frank et al. 1991).

6.3.18 *Salix*, Salicaceae Family

Salix species (willow, sallow) are both insect- and wind pollinated, but the proportions of the two pollination modes differ between species. Pollen grains often tend to stick together (Dobson 1988; Karrenberg et al. 2002) and regional levels of airborne pollen are often low (Jones and Harrison 2004; Zeghnoun et al. 2005). No published threshold values of *Salix* are available from Europe and a Canadian study showed that the impact of an interquartile increase in daily *Salix* pollen levels upon asthma admissions was insignificant (Dales et al. 2008). The prevalence of *Salix* sensitization has been addressed in a Swedish multi-centre study (Eriksson et al. 1984). Positive skin prick reactions against *Salix* allergens occurred in 37% of people with birch pollen allergy, but they were strong (>2+) in only 12%. In patients reacting to other pollen types, only 6% showed mild positive results in prick tests (SPT=1+), and no one had a stronger reaction. Positive IgE tests against *Salix* allergens were found in 40% of the birch pollen sensitive people, but only 6% reacted with RAST class>2. In Eskisehir, Turkey, 24.9% of the hay fever patients reacted to *Salix* pollen. Of these, the allergy was considered severe in 16.9% (Egger et al. 2008).

6.3.19 Spore Types: *Alternaria* and *Cladosporium*

Epidemiological studies on mould allergy in Europe show that *Alternaria* and *Cladosporium* are the most important sources of fungal allergens; *Cladosporium* seems to be most prevalent in Northern Europe and *Alternaria*, especially in children, is most common in Southern Europe (D'Amato et al. 1997). A multicenter study in Europe showed that the prevalence of positive skin prick tests to these moulds varies from 3% in Portugal to 20% in Spain (D'Amato et al. 1997). Studies on the threshold level that induces symptoms are scarce for these spores. A study from 1979 mentions 100 spores/m^3 for *Alternaria* and 3,000 spores/m^3 for *Cladosporium* (Gravesen 1979) but no details on the determination are given.

In Europe, the main season for the occurrence of *Alternaria* spores is from June until October (Kasprzyk 2008), thus overlapping with the grass pollen season. The variation in annual totals is large, for instance concentrations from 10,000 to 25,000 spores/m^3/year have been described (Thibaudon and Lachasse 2006).

6.4 Factors Affecting the Threshold

6.4.1 Differences Among Individuals

6.4.1.1 Nonspecific and Allergen-Specific Hyperreactivity as Basis for Symptoms

The physiological background for symptom development in allergic individuals is the increased sensitivity of their mucosa after allergen exposure. This may result in an hyperreactivity (or hyperresponsiveness) of the mucosa of eyes, nose or bronchi. There are two types of hyperreactivity (or hyperresponsiveness) of the mucosa in eyes, noses, and bronchi which determine the severity of the clinical response (symptoms) in pollen allergic individuals following exposure to pollen allergens: the non-specific and the allergen-specific hyperreactivity. Since there are more data on the hyperreactivity of the airways than for nose and eyes, this part will focus primarily on bronchial hyperresponsiveness.

Airway or bronchial hyperresponsiveness (BHR) is defined as excessive bronchial narrowing and manifests itself as an exaggerated bronchoconstrictor response of the airways to various inhaled stimuli, including pollen (Sterk et al. 1993). BHR is determined at least in part by airway inflammation and airway structural remodeling (Busse 2010). The severity of BHR in non-asthmatic patients changes with pollen-induced rhinitis during the season and evidence suggests that during the pollen season, and even after the cessation of exposure to pollen, BHR is increased for weeks (Prieto et al. 1994).

The severity of allergic BHR is often measured as the provocative dose or concentration causing a 20% fall in the FEV_1 (=forced expiratory volume in 1 s) (Sterk et al.

1993). Non-specific BHR can be measured by using non-specific bronchoconstrictor stimuli such as histamine and methacholine that are well standardized and frequently used in clinical settings. Methacholine and histamine act directly on smooth muscle, while other stimuli can act indirectly by stimulating the release of inflammatory mediators and/or by stimulating neural pathways (Sterk et al. 1993).

6.4.1.2 Factors Influencing Hyperreactivity and Threshold

It is a well-known fact that the same amount of pollen does not cause the same intensity of complaints among individual patients with hay fever. Different factors or exposures have been implicated in temporary changes in nasal and airway reactivity. BHR may be variable over time because of several reasons.

6.4.1.3 Atopy

The most important and independent risk factors for BHR are lung function and atopy. Lower FEV_1, a lower Tiffeneau ratio (FEV_1 as a percentage of FVC), and the presence of atopy are associated with an increased occurrence of BHR (Britton et al. 1994).

The association between atopy and BHR is complex. Both skin prick test positivity as well as serum IgE are positively associated with BHR (Peat et al. 1996; Sunyer et al. 1995). Also it has been shown that BHR is frequently present in subjects without asthma but with allergic rhinitis. Moreover, BHR worsens during the allergen season in subjects with rhinitis who are atopic to pollen (Riccioni et al. 2002).

6.4.1.4 Gender and Age

Several studies have shown a relationship between BHR and gender. In children without asthma, BHR is approximately twice as frequent in boys as in girls (Ernst et al. 2002). On the other hand, the occurrence of BHR is more frequent among female than among male adults (Britton et al. 1994). This switch in prevalence among males and females seems to occur during and/or after puberty. The relationship between BHR and age is unclear: BHR is either not influenced by age or subjects of older age have a lower risk for BHR after correction for lung function or those of older age are more likely to have increased BHR (Hopp et al. 1985).

6.4.1.5 Influence of Smoking

Although smoking has been significantly associated with BHR, cigarette smoking appears less important than atopic status after correction for other risk factors (Britton et al. 1994). Smoking may increase BHR in part by increasing inflammatory cell numbers in the airways (Amin et al. 2003).

6.4.1.6 Allergen Exposure

Experimental exposure to allergen is associated with an increase in BHR the following day (Sterk et al. 1993). Seasonal variations in the allergen load and air pollution are paralleled by fluctuations in BHR, with higher allergen exposure levels being associated with increased responsiveness (van der Heide et al. 1994).

In subjects with asthma, a worsening in indices of BHR has been observed after respiratory infections (Cheung et al. 1995).

6.4.1.7 Effects of Lifestyle

Pollen counts show the pollen concentrations in the outside environment. However, most people spend most of their time indoors. Farrera et al. (2002), using a Cour's pollen trap inside and outside a house in the Rhône Valley (France), showed that 60% of the taxa found in the environment close to the house were also present inside the house during the same period. Ragweed pollen was the most prevalent species. It would be interesting to investigate the factors that are responsible for transfer the pollen grains inside the house.

Furthermore, the patient's personal agenda is relevant for the exposure to pollen. A person can sit inside a house all day or cycle through meadows. Also the traveling history of a person is relevant for the specific exposure to pollen. This was nicely demonstrated by Penel et al. (2009) who collected the hairwashing of volunteers in Valence and subsequently identified the pollen in these washings. In a period without any *Pinus* pollen on the trap he found that 36% of the pollen from the hair washing was *Pinus* pollen. This person arrived from Spain where *Pinus* was flowering. This study showed that hair can function as a pollen trap that is indicative for the site of life for a person.

6.4.1.8 Effects of Viral and Bacterial Infections

Transmissible viral infections occurring in the general population are also an environmental factor which can potentially modify the allergic response and asthma attacks. Viral infections and immunodeficiency (e.g. HIV) might be problematic, for instance, in immunotherapy. Depending on the virus and the person's state of health, various viruses can infect almost any type of body tissue, from the brain to the skin. Viral infections cannot be treated with antibiotics.

Busse (1989) summarizes the important role that viral illnesses could play in the allergic diathesis. Respiratory infections precipitate wheezing in many patients with asthma. A number of important conclusions in this relationship has come from detailed epidemiological studies. First, viral, not bacterial, respiratory infections provoke asthma in children and adults. Second, although many respiratory viruses trigger asthma, the prevalent organism responsible for increased wheezing varies with the age of the affected patient. For example, respiratory syncytial virus (RSV) infections predominate in infants and young children, whereas rhinovirus, influenza and para-influenza emerge with greater frequency as age increases. It is, however, unlikely that mechanisms of virus-induced wheezing are different for each respiratory virus. Rather, a common mode of asthma exacerbation should prove to be the rule. Third, the likelihood of increased wheezing with colds is greater when patients have severe symptoms such as malaises, fever, coryza, etc. Greiff et al. (2002) suggest that common colds in part through stimulation of granulocyte activity potentiate the airway inflammation in allergic diseases.

6.4.2 *Differences Among Different Ethnic Populations*

Different studies have shown that the pollen threshold levels for symptom development depend not only on allergy potency, but also on other environmental, genetic and socio-economic characteristics, e.g. ethnic population. This will be illustrated by studies on olive pollen, one of the main causes of allergic disease in the Mediterranean area. Olive allergy represents a unique allergy model where a sharp exposure gradient leads to different clinical allergy profiles (Barber et al. 2007). First of all, it is important to present the different proteins which have allergenic activity. Taking into account that major allergens are defined as those recognized by more than 50% of the patients allergic to a particular source (King et al. 1994), Ole e 1 was the first major allergen described and characterized (Blanca et al. 1983; Lauzurica et al. 1988a, b), being recognized by more than 70% of patients. Nowadays the presence of, at least, 20 proteins with allergenic activity has been demonstrated, and 10 of them have been characterized, cloned and expressed (Cárdaba et al. 2007), including Ole e 1 as the major allergen and other minor allergens (Lauzurica et al. 1988a, b; Villalba et al. 1994; Asturias et al. 1997; Ledesma et al. 1998; Barral et al. 2004, 2006). Recently, the 11th has been described (R. Rodriguez, personal communication).

In the case of olive pollen exposure, it has been demonstrated in several studies that high levels of pollen exposure can cause minor allergens to become major allergens for allergic patients that live in areas with high exposure to the respective pollen. For example, some minor allergens have been defined as major allergens, i.e. Ole e 4 and Ole e 7, by Rodríguez et al. (2001). Cárdaba et al. (2007) described a more complex IgE response to the various olive allergens in allergic patients from areas with extremely high pollen counts. They showed that not only Ole e 1 but also 2, 7 and 10 were major allergens in highly exposed populations, but not in populations with a lower antigenic exposure. On the other hand, Barber et al. (2007, 2008) defined next to Ole e 1, also Ole e 7 and Ole e 9 to be major allergens. Ole e 7 sensitization has been described with an average prevalence of about 47% in patients with olive allergy by Tejera et al. (1999). These authors show that high exposure to olive pollen can lead to different sensitization rates to this "minor allergen" Ole 7. In Jaen (South of Spain) up to 60% of the patients were sensitized to Ole 7, compared to less than 20% of the patients in the Madrid area (Central Spain) (Tejera et al. 1999).

Moreover, in a high olive pollen exposed area in the Andalusia Region, Spain (Jaen, Córdoba, Granada and Malaga provinces), it has been observed that 50% of the patients reacted only to Ole e 1, which is normal in areas with lower pollen counts. On the other hand, Ole e 9 sensitization always coincided with Ole e 1; however, Ole e 7 cases were Ole e 1 independent, 40% of Ole e 1-negative patients were sensitized to Ole e 7 (Barber et al. 2007, 2008). In this well exposed area of Andalusia it is important to consider not only Ole e 1, but also Ole e 9 and, particularly, Ole e 7. In other areas which are less exposed to olive pollen, Ole e 7 sensitization always coincided with Ole e 1. Clearly, new approaches to allergen standardization, diagnosis, and vaccination are necessary in different geographical areas.

Although it is generally believed that allergic sensitization to aeroallergens reflects the degree of exposure, and this also applies to pollen sensitivity, some studies show different results (Cárdaba et al. 2007). Geller-Bernstein et al. (1996) have observed different sensitization prevalences to olive pollen, in ethnically different populations of Israel. Geller-Bernstein et al. (2002) have observed very low prevalence of sensitization to olive pollen in the Druse Arab, (12%), Christian Arab (27%) and Muslim Arab (35%) populations, as compared to the Jewish population (66%). The Druse, the Christian and part of the Muslim Arabs have lived in the study region for several centuries. In this area, olive trees have been cultivated as a crop for centuries, so that the residents are massively exposed to olive pollen from birth. However their sensitization prevalence to olive pollen is lower than that of the Jewish population that has lived in the same region only for about 150 years. This interesting observation might indicate that a population exposed to a type of pollen (allergen) from birth onwards, for centuries, might lead to tolerance, rather than to sensitization – possibly another aspect of the "hygiene hypothesis" (Geller-Bernstein et al. 2002; Cárdaba et al. 2007).

Geller-Bernstein et al. (2002) compared the Israeli Druse population and the Andalusia population. Both populations are exposed to massive amounts of olive pollen in the air from birth. Such a continuous exposure has apparently lowered their possible sensitization to olive allergens. Finally, in studies comparing pollen thresholds for symptom development in different geographical areas, only a few papers focus on olive pollen, i.e. Jaen (Andalusia Region, Spain) (Florido et al. 1999) and Netzer Sireni (Coastal Plain of Israel, Israel) (Waisel et al. 2004). Florido et al. (1999) have published a pollen risk threshold of 400 olive pollen/m^3 for Jaen patients, the province with the biggest number of olive groves in the world. However, lower pollen counts in the air (2–4 pollen/m^3) seem to cause allergy among the Israeli population living in Netzer Sireni, with a smaller number of olive trees. The olive pollen threshold level for symptom development may differ in different geographical areas, depending upon amount of exposure and/or ethnic origins.

In general, different studies on migration offer information about the role of the environmental and genetic factors in the development of atopy and asthma under different scenarios (Hjern et al. 1999; Brabäck et al. 2004). These studies show that both the way of life and the environmental factors of industrialized countries play a role in the case of atopy and asthma. Furthermore, it appeared that in the beginning of respiratory allergies environmental factors play a more important role than genetic ones (Hjern et al. 2000; Rottem et al. 2005; Lombardi et al. 2008). The epidemiological information reveals that the prevalence of asthma and atopy depends on the age in which the migration occurs and is increased by the duration of the residence in the countries of destination.

All these studies demonstrate how difficult it is to define risk thresholds for sensitization to olive pollen in a particular geographical area.

6.4.3 Seasonal Variations in Threshold Level

Regarding the seasonal variations in threshold level, lower pre- and post-season exposure to relatively smaller pollen concentrations may cause allergic responses at a lower threshold. In the case of olive, Ole e 1 is also present in other trees belonging to the Oleaceae family, such as *Fraxinus* and *Ligustrum*. In the Andalusia region (Spain) different *Fraxinus* species flower at different periods of time. *F. angustifolia*, flowering in early spring, coincides with pre-season olive pollen exposure and *F. ornus*, flowering in late spring, coincides with high olive pollen exposure (Guerra et al. 1996). On the other hand, privet (*Ligustrum*) is an ornamental tree that flowers later, in June, coinciding with post-season olive exposure. This pollen appears in the air after the patients have been suffering from olive pollen allergy for a long time (Cariñanos et al. 2002). Privet pollen grains are very heavy and they are not transported easily by air. However, it has been demonstrated that people living near large areas with this ornamental tree suffer from allergy. For this reason, these homologous pollen types can be responsible in a percentage of olive allergic responses at the beginning and end of the olive pollen season.

It is also important to take into consideration pan-allergens (i.e. profilins, lipid-transfer, proteins, polcalcins, etc.) when trying to understand the seasonal variations in threshold levels (Barber et al. 2008; Quiralte et al. 2007). Homologous allergens in Oleaceae and non-Oleaceae species could account for cross-reactivity between different pollen (Quiralte et al. 2007). This is important especially in patients living in south Europe, where olive and grass pollen seasons usually coincide. Profilin sensitizations are mainly associated with grasses. Moreover other different allergenic pollen seasons usually overlap, i.e. *Rumex*, *Plantago*, *Artemisia*, *Ambrosia* or Amaranthaceae/Chenopodiaceae (nowadays both of them are included in the Amaranthaceae family) (Barber et al. 2008).

Another variation in symptom development during the season is observed in the grass pollen season in a Dutch grass pollen sensitive patient group. Patients seem to respond with more severe symptoms to the relatively low grass pollen concentrations in the early part of the season, while at the end of the season when the grass pollen concentrations were still relatively high, symptom scores were reduced (de Weger et al. 2008).

6.4.4 Pollen Allergens in the Air

Most pollen allergens in the air are carried by pollen grains, but the presence of airborne small respirable particles containing pollen allergens has also been reported (Spieksma et al. 1995a; Schäppi et al. 1997). The total amount of allergen in the air we breathe needs to be considered in the assessment of thresholds for symptom development.

6.4.4.1 Small Respirable Particles

Pollen is the carrier of the airborne allergens. Although the existence of non-pollen, small respirable particles that contain allergen is still debated, for birch for instance the only source of the main allergen Bet v 1 is the pollen. Even anatomically closely situated parts of the catkin to pollen, such as orbicules (anatomic part inside anthers to which developing pollen are attached) do not contain allergen Bet v 1 (Vinckier et al. 2006). At first this may seem surprising, but closer investigation showed that no Bet v 1 allergen is present in birch pollen until about 1 week before pollination (Buters et al. 2010). The Bet v 1 allergen was found only inside the pollen. Because orbicules were shown to be devoid of this allergen, there is apparently no transfer of Bet v 1 inside catkins, and separate parts of a birch tree including catkins must produce their own Bet v 1. So far Bet v1 has not been found in any plant parts studied (except one report on the presence of low levels of Bet v 1 in birch leaves) (Grote and Fromme 1986). Even if other parts of the tree such as leaves, bark or catkins contain some Bet v 1, they must become airborne which is unlikely. Thus, small respirable particles containing allergen must come from pollen from processes during or after pollen release, for example, rupture of pollen from birch (Schäppi et al. 1999b; Grote et al. 2003; Taylor et al. 2004; Namork et al. 2006) or grass (Rantio-Lehtimäki et al. 1994; Schumacher et al. 1988; Motta et al. 2006; Namork et al. 2006), or transfer of allergen after extraction by rain water to other environmental particles (Behrendt and Becker 2001). In Munich, Germany, during the last 5 years of continuous allergen monitoring in ambient air during the pollen season, Buters et al. (2010) did not find any birch allergen in the respirable fraction containing PM < 2.5 μm. The discrepancy between their results and other authors could be due either to specific climatic conditions not present in Munich, Germany, or to the difference between grass and birch pollen (Namork et al. 2006). Finding these small respirable particles is challenging as they occur in the same respirable fraction PM < 2.5 μm as diesel soot particles. Diesel soot adsorbs the allergen Bet v 1 and consequently no allergen can be detected anymore in the sampled fraction. It is also unclear if allergen-coated diesel particles release the allergen once inhaled, or if that the allergen is no longer bioavailable for interactions with the immune system (see Fig. 6.2).

Pollen releases several substances, such as lipid mediators (PALMS) and larger quantities of adenosine; factors which seem to influence the sensitization reaction

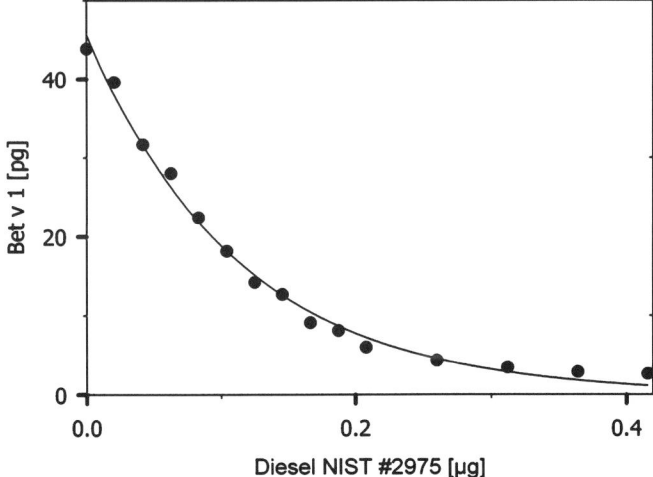

Fig. 6.2 Absorption of the allergen Bet v 1 to diesel soot particles. Allergen was incubated with diesel particles (NIST#2975) and the remaining allergen in the supernatant was determined by ELISA (data from J. Buters)

of the immune system. Once sensitized, the allergen seems to govern the observed patient symptoms (environmental pollutants influence the extent of the immune reaction too (Lubitz et al. 2010), but these compounds do not originate from pollen). The discussion concerning the threshold value then becomes the question of whether pollen releases a constant amount of allergen. If the pollen were constant in allergen release, then the pollen count would be a good surrogate marker for allergen exposure and a pollen-symptom threshold could be determined. If pollen release a variable amount of allergen, then an allergen-symptom threshold is perhaps possible.

6.4.4.2 Regional Variation in Allergen Release

There are very few studies concerning the regional variability of allergen release from the same amount of pollen. Buters et al. (2008) showed that two regions in Germany, about 600 km apart, varied three-fold in allergen release from the same amount of pollen of the same species, and that this difference remained similar during the 3 years studied (see Fig. 6.3). Also, *Ambrosia artimisiifolia* plants exposed to higher CO_2 values gave rise to higher pollen production but less allergen release per pollen (Ziska et al. 2003; Rogers et al. 2006). CO_2 varies among regions, although the differences are small, except for inner cities (Ziska et al. 2003). Interestingly, more individuals living in inner cities with a higher environmental pollution are sensitized than city-inhabitants that live in less polluted areas (Morgenstern et al. 2008).

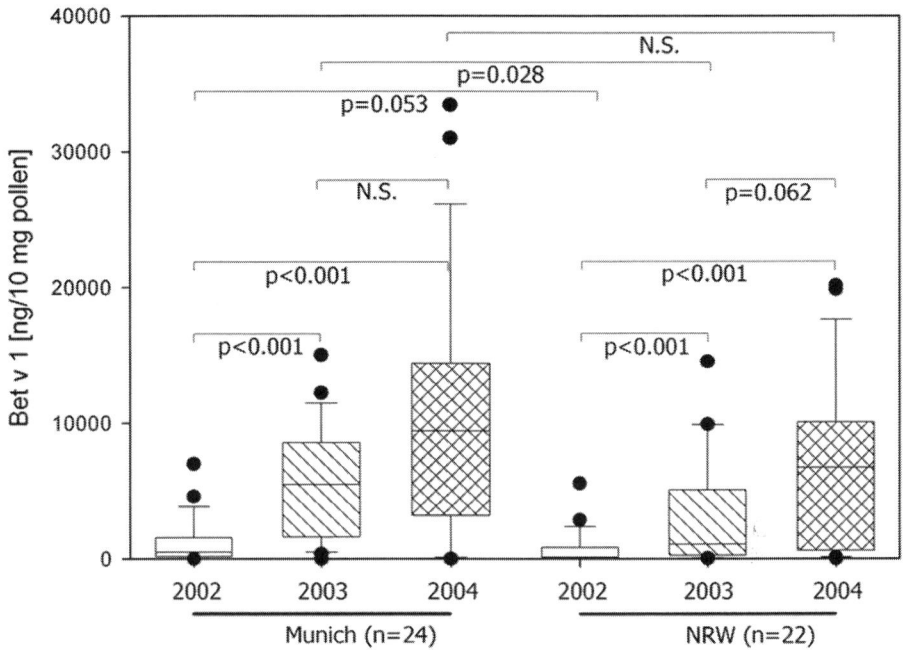

Fig. 6.3 Differences in Bet v 1 release from the same amount of pollen collected directly from trees in Munich and Nordrheinwestphalia (NRW) and from different years. The figure shows the median and the 90 and 75% confidence interval (with permission from Buters and Behrendt (2008))

6.4.4.3 Day-to-Day Variability in Allergen Release

For birch, several authors monitored the allergen release from pollen collected from ambient air (see Fig. 6.4) (Buters et al. 2010).

After rainfall (recognized by a very high humidity) both pollen and the allergen are greatly reduced. In the 1990s Rantio-Lehtimäki in Finland (Pehkonen and Rantio-Lehtimaki 1994; Jensen et al. 1989) had shown that allergen and pollen flight were not congruent, suggesting that the release of allergen per pollen is not constant and varied per day. Schäppi and colleagues, first in Switzerland and later also in Australia, showed the same results (Schäppi et al. 1996, 1997, 1999a,b). Riediker et al. (2000) confirmed this finding but also cautioned that allergen present in other fractions of ambient air might be due to sampling artefacts. The same was described for olive pollen (De Linares et al. 2007) although surprisingly most olive allergen was detected in those fractions that do not contain pollen. For grass pollen the differences in release is to be expected as the pollen from different grass species cannot be discriminated in a Hirst-type pollen trap, as microscopically the main species of grass pollen (such as *Lolium perenne*, *Dactylis glomerata* or *Phleum pratense*) show the same morphology (Winkler et al. 2001).

The allergen release per fixed amount of grass pollen varied (Schumacher et al. 1988). In addition, it can be demonstrated that, Phl p 5 was only partly congruent to grass pollen counts over Europe (Buters et al. unpublished observation).

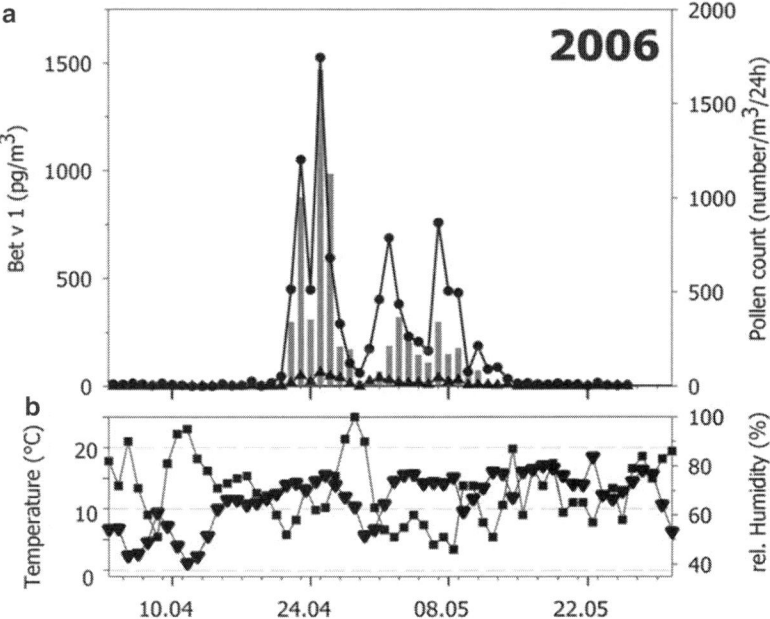

Fig. 6.4 (**a**) Daily variation in birch pollen allergen in different fractions of ambient air (-●- PM>10 μm, -▲- 10 μm>PM>2.5 μm: not shown 2.5 μm>PM>0.12 μm, since all fractions resulted in baseline values) and pollen (*grey bars*) in ambient air from Munich, Germany in 2006. At the end of the birch pollen season pollen counts decline, but Bet v 1 does not. (**b**) Concomitant humidity (■) and temperature (▼) at the sampling site. Adapted from Buters et al. (2010).

6.4.4.4 Year-to-Year Variability in Allergen Release

Currently, to our knowledge only one study has examined the effect of different seasons on the allergen release of birch pollen. The authors reported an up to five-fold difference in allergen release between the same amount of pollen sampled from the same trees in different years (Buters et al. 2008). Although day of sampling will influence the allergenicity of pollen due to the rapid increase of allergen just before pollination, the year-to-year differences remained between all years studied. Indeed, pollen sampled from ambient air from the same location varied up to ten-fold in allergen release potency between different years (Buters et al. 2010). Thus allergen release from the same amount of birch pollen is variable. Also, the high variability among individual birch trees was not constant between years, although a few trees constantly remained in the lower percentile (Buters et al. 2008) A Europe-wide project was initiated that studies these questions in detail for birch, grass and olive pollen (see www.hialine.com).

6.4.4.5 Allergen Release from Different Pollen Species

Differences in allergen release from the same amount of pollen was best studied for birch (see above) but was also reported for grass, olive and ragweed pollen (Namork et al. 2006; Buters et al. 2008, 2010). The phenomenon seems to be present in all the pollen species studied. This phenomenon is likely to be a general natural characteristic of plants since it has been known for a long time that the specific contents of certain ingredients vary also in non-allergenic species (Stahl and Jork 1964; Filipowicz et al. 2006). Pollen is thus no exception.

Although previous findings suggest that airborne allergen measurements correlate with pollen counts (Marsh et al. 1987; Spieksma et al. 1995a), the recent studies discussed above suggest that it is likely that pollen contain different amounts of allergen per pollen grain. The question remains, if the differences observed result in differences in patient symptoms. An additional advantage of allergen monitoring could be that automation and on-line monitoring of allergen is less complicated than pollen counting, as the current techniques have difficulties discriminating optically similar but allergenically different pollen species (Delaunay et al. 2007).

Pollen grains release more than allergen alone (Traidl-Hoffmann et al. 2005) and the allergens themselves are a mixture of several different proteins, each protein again consisting of a mix of different isoforms. Thus birch pollen releases a mixture consisting of at least PALMs, adenosine, Bet v 1, Bet v 2, and Bet v 4. Bet v 1 consists of 37 reported isoforms, of which 6 are commonly present in European samples.

The major isoform of Bet v 1 is Bet v 1.0101. About >90% of European sensitized individuals are sensitized to Bet v 1 (Moverare et al. 2002), and >50% of Bet v 1 is made up by the isoform Bet v 1.0101 (Swoboda et al. 1995). The other isoforms of Bet v 1 are close homologues of Bet v 1.0101 and birch pollen sensitized individuals react to all these isoforms equally, except for the hypoallergenic isoform Bet v 1.0401 (Bet v 1d), which is a minor isoform (Hartl et al. 1999). Patients sensitized to the minor allergen Bet v 2 or Bet v 4 are often simultaneously sensitized to Bet v 1 (Moverare et al. 2002).

The major allergen in grass pollen is Phl p 5. Phl p 1 is a major allergen too and often sensitization to both occurs simultaneously (Moverare et al. 2006). ELISA, the technique used to determine the allergen content of ambient air in all published studies, is mostly designed to recognize only one allergen, of which it then recognizes the isoform mix relevant for patients. For Bet v 1 the antibodies used in the ELISA recognized all isoforms (except the hypoallergenic Bet v 1.0104) of Bet v 1, and the ELISA used predicted patient reactivity. Therefore, this technique is suitable for determining allergen exposure.

6.4.4.6 Allergen Release from Other Plant Parts

Allergen specific IgE binding activity appeared to be present in other plant parts than pollen of e.g. ragweed (Agarwal et al. 1984) and birch (Fountain et al. 1992). It is not known whether leaves, stems, seeds and roots can release relevant amounts of allergens into the air under general conditions. However, release of allergens from other plant parts than pollen may be relevant during the annual campaigns to draw ragweed in the Rhône valley. During these campaigns which take place before the flowering season of ragweed, plants are being broken and snapped and during these actions some of the voluntary workers suffered from sudden attacks of asthma, conjunctivitis and/or urticaria (C. Déchamp, personal communication).

6.4.5 Environmental Aspects

6.4.5.1 Temperature and CO_2

Temperature is the primary determinant of the metabolic rate of plants, and therefore it is an important determinant of phenology including that of allergenic plants (Linderholm 2006). Higher CO_2 concentrations stimulate photosynthesis in many plants especially when enough nutrients such as nitrogen are available (Bazzaz 1990). Therefore increasing temperature and high CO_2 may affect the pollen load in a region.

Spieksma et al. (1995b) examined atmospheric birch (*Betula*) pollen data from five European stations (Basel, Vienna, London, Leiden, and Stockholm) between 1961 and 1993 (records from 18 to 30 years). They found weakly rising trends of annual sums of daily concentrations at all five stations. More recent European studies, in Denmark (Rasmussen 2002) and Switzerland (Frei 1998) found trends of increasing amounts of pollen over the latter decades of the 1900s that were related to climate change. Teranishi et al. (2000) studied the association between Japanese cedar (*Cryptomeria japonica*) pollen and temperature from 1983 to 1998 in Japan. They found a significant positive correlation between total pollen count in a year and temperature in July the previous year. In North America, Levetin (2001) has found statistically significant increases in a number of taxa, including *Juniperus*, *Quercus*, *Carya*, and *Betula*, since 1987. There is some evidence that these increases have been associated with increases in average winter temperature.

Rising CO_2 concentrations may affect the amount of pollen produced per plant as well as the allergenic content of the pollen grains itself. Greenhouse experiments with ragweed plants showed that plants exposed to elevated levels of CO_2 increased their ragweed pollen production (Ziska and Caulfield, 2000) as well as the Amb a 1 content per pollen grain (Singer et al. 2005).

6.4.5.2 Atmospheric Pressure

Although only a few studies on the importance of atmospheric pressure have been carried out, the effects of pressure on the threshold levels can be very diverse. (i) In connection with the NAO (North Atlantic Oscillation)-studies it has been shown that that pressure is a much more important factor causing changes in pollen concentrations than has been realized thus far (e.g. Smith et al. 2009). The role of pressure is revealed in long-distance pollen transport. Emberlin (1994) suggested that an increase in the strength of the Atlantic westerlies over north-west Europe would enhance the long distance transport of pollen from northern and central Europe to Scandinavia. This aspect should be taken into account for pollen forecast, since it may affect the expected pollen load for the coming day(s) considerably. (ii) Although the role of climatic factors in exacerbating asthma is poorly understood (D'Amato et al. 2001), high and low atmospheric pressures have been linked to asthma attacks (Garty et al. 1998; Celenza et al. 1996). There is evidence that changes in temperature, barometric pressure and relative humidity have some influence on the worsening of asthmatic symptoms (e.g. Hashimoto et al. 2004). (iii) Differences in atmospheric pressure can cause remarkable error in the function of pollen trap; for instance, in Switzerland at an altitude of about 2,300 m at Gütsch where the mean air pressure is 770 hPa this causes a 10% error compared to sampling in Basel where the air pressure is 979 hPa on an average (Gehrig and Peeters 2000).

6.4.5.3 Thunderstorms

Thunderstorms have been linked to asthma epidemics, especially during the pollen seasons, and there are descriptions of asthma outbreaks associated with thunderstorms. This is mainly because pollen grains release allergenic particles when they get in to contact with rainwater in combination with an electric field.

Grass pollen has been responsible for triggering allergic asthma, gaining impetus as a result of the 1987/1989 Melbourne and 1994 London thunderstorm-associated asthma epidemics (Suphioglu 1998). Ryegrass pollen ruptures upon contact with water releasing about 700 starch granules which not only contain the major allergen Lol p 5, but have been shown to trigger both *in vitro* and *in vivo* IgE-mediated responses. Starch granules have been isolated in Melbourne atmosphere with 50-fold increase following rainfall. Free grass pollen allergen molecules have been shown to interact with other particles including diesel exhaust carbon particles, providing a further transport mechanism for allergens to gain access into lower airways (Suphioglu 1998). Taylor et al. (2007) also suggested that thunderstorm asthma epidemics may be triggered by grass pollen rupture in the atmosphere and the entrainment of respirable-sized particles in the outflows of air masses at ground level. Pollen contains nicotinamide adenine dinucleotide phosphate (reduced) oxidases and bioactive lipid mediators which likely contribute to the inflammatory response.

Vaidyanathan et al. (2006) decided to determine whether the potential for pollen fragmentation is increased during thunderstorms by exploring the effects of electric fields, with magnitude as found in the outdoor environment. Fresh pollen grains were collected from Bermuda grass flowers. Water was added to test for pollen rupture and to assess pollen viability. Bermuda grass pollen did not rupture within 1 h of contact with water. Only after exposure to an electric field the Bermuda grass pollen showed a considerable amount of rupturing immediately upon immersion in water. The higher the voltage the pollen is exposed to before coming into contact with water, the higher the percentage of rupture of the pollen. Electric fields, generated in the laboratory and of magnitude found during thunderstorms, affected the pollen after as little as a 5 s exposure. Thunderstorms regularly generate electric fields up to 5 kV/m in strength, can reach 10 kV/m and cover several km in distance.

6.4.5.4 Ozone

Exposure to increased atmospheric levels of O_3 causes increased airway reactivity to pollen and is related to an increased risk of asthma exacerbation (Chen et al. 2004). It has been observed that O_3 exposure has a priming effect on allergen-induced responses as well as an intrinsic inflammatory effect in the airways of allergic asthmatics (Molfino et al. 1991; Kehrl et al. 1999). Exposure to O_3 may increase the risk of allergic sensitization in predisposed subjects. Indeed, by lowering the threshold concentration of allergen able to induce clinical symptoms, O_3 can enhance the airway responsiveness of sensitized subjects. Molfino et al. (1991) reported that a 1-h exposure to 0.12 ppm O_3 while at rest caused a two-fold reduction in the provocation concentration of inhaled antigen required to cause early bronchoconstriction in specifically sensitized asthmatic subjects. The mean provocation dose of ragweed pollen necessary to reduce FEV_1 by 20% in sensitized asthmatic subjects was significantly reduced to about one-half the dose of ragweed allergen when the patients were pre-exposed to 0.12 ppm O_3 for 1 h *versus* pre-exposure to air. Jörres et al. (1996) using a higher effective dose (0.25 ppm inhaled through a mouthpiece with intermittent exercise) and a longer duration of exposure (3 h), found that 23 of 24 mild asthmatic subjects required a lower provocation dose of allergen to cause a 20% decrease in FEV_1 (PD_{20}) after O_3 exposure. The previous exposure to O_3 (and NO_2) on subsequent pollen allergen significantly increases the allergen-induced release of eosinophil cationic protein in nasal lavage of patients with seasonal allergic rhinitis (Devalia et al. 1998).

Rogerieux et al. (2007) exposed timothy grass pollen to ozone (O_3), nitrogen dioxide (NO_2) and sulphur dioxide (SO_2) alone or in combination. For O_3-treated pollen, tests with patients' sera showed an acidification of allergens Phl p 1b, Phl p 4, Phl p 5 and Phl p 6. Moreover, samples treated with a mix of NO_2/O_3 or NO_2/SO_2 showed a higher decrease in allergen content, compared with samples treated with only one pollutant.

6.4.5.5 Nitrogen Dioxide (NO_2) and Sulphur Dioxide (SO_2)

Automobile exhaust is the most significant source of NO_2 which is an oxidant pollutant. It is less chemically reactive than O_3 and thus the inhalation of NO_2 is usually not associated with significant changes in bronchial function of asthmatic patients but it has an obstructive influence on bronchi during exercise (Roger et al. 1990). Tunnicliffe et al. (1994) demonstrated a greater allergen-induced early bronchoconstriction by a combined inhalation of allergen and NO_2 for 1 h, while at rest, than after sham exposure. SO_2 is generated primarily from burning of sulphur-containing fossil fuel. It has been shown to induce bronchial obstruction very rapidly (within 2–5 min) in asthmatic patients at low concentrations (Riedel et al. 1988). O_3, NO_2 and SO_2 have been found to interact with allergenic pollen in the ambient air to aggravate symptoms of asthma (Feo Brito et al. 2007).

Rogerieux et al. (2007) showed that exposure of grass pollen to nitrogen dioxide (NO_2) and sulphur dioxide (SO_2) induced a decrease of grass allergen recognition by patients' sera. This decrease could be due to a mechanical loss of allergens from the altered pollen grains and/or post-translational modifications affecting allergen recognition by IgE.

6.4.5.6 Particulate Matter

Particulates, alternatively referred to as particulate matter (PM) or fine particles, are tiny particles of solid or liquid suspended in a gas. Sources of particulate matter can be man made or natural. Some particulates occur naturally, originating from volcanoes, dust storms, forest and grassland fires, living vegetation, and sea spray. Human activities, such as the burning of fossil fuels in vehicles, power plants and various industrial processes also generate significant amounts of aerosols. Increased levels of fine particles in the air are linked to health hazards such as heart disease, altered lung function and lung cancer. Fernvik et al. (2002) studied health effects due to air pollution arising from motor vehicles, a major public and political concern worldwide. Epidemiological studies have shown that the manifestations of asthma are increased by air pollution in individuals who are already affected, e.g., the interaction between ambient PM_{10} and pollen in studies of short-term effects (Lierl and Hornung 2003; Feo Brito et al. 2007) To investigate the potential role of traffic particulate matter, TPM, or pure carbon core particles in the initiation and persistence of experimental allergic inflammation, mice were immunized with birch pollen alone, or pollen together with TPM, or with birch pollen and $Al(OH)_3$, or with birch pollen and carbon core particles. Mice immunized with birch pollen alone and challenged intranasally with pollen or a mixture of pollen and TPM demonstrated the highest levels of IL-4 and IL-5 (Fernvik et al. 2002). These results highlight the importance of the exposure to a combination of particulate matters and pollen allergens. The results strongly suggest that it is most likely to be the organic phase bound to the carbon core of the diesel exhaust particles (DEP) that might have an important adjuvant effect in the induction of experimental allergy.

Diaz-Sanchez et al. (1996) studied the effect of DEP on antigen in ragweed-sensitive subjects challenged (nasal provocation test) with DEP, Amb a1 (the major ragweed allergen) and a combination of DEP and Amb a 1. Provocation with ragweed led to an increase in both total and ragweed-specific IgE in nasal lavage fluid measured at 18 h, 4 and 8 days post challenge. The DEP challenge increased the concentration of ragweed-specific IgE 16-fold *versus* concentrations observed after challenge with ragweed allergen alone. Combination of DEP and ragweed allergen challenge markedly enhances human *in vivo* nasal ragweed-specific IgE and skews cytokine production to a T-helper cell type-2 (Th2) pattern. These results indicate that DEP plays a role in the enhanced allergic inflammatory response thus decreasing the threshold value (Rudell et al. 1994).

6.5 Impact of Allergenic Pollen on Non-allergic Diseases

Recent observations draw attention to rather consistent relationships between daily variations in both pollen counts and the incidence or prevalence of various non-allergic diseases. This applies especially to three groups of diseases: non-allergic respiratory diseases, cardio- and cerebrovascular diseases, and psychiatric diseases including suicide and suicide attempt.

6.5.1 Effects on Non-allergic Respiratory Diseases

An impressive amount of data supports the effects of air pollution on respiratory diseases both in terms of mortality and morbidity (Schwela 2000; Baldacci and Viegi 2002; Pascal 2009). However, pollen and spores have been included rarely among pollutants and their possible effects are often neglected. Brunekreef et al. (2000) were the first in 2000 to report an association between pollen concentrations and mortality. The authors evaluated daily non-accidental deaths due to cardiovascular and respiratory diseases and pneumonia between 1986 and 1994 in the Netherlands, and daily pollen levels for Poaceae, *Betula, Quercus, Fraxinus, Artemisia*, and *Rumex* from two stations (Leiden, west and Helmond, south). Results were also adjusted for long-term and seasonal trends, temperature, relative humidity, influenza morbidity, day of the week and holidays. Furthermore, lag (−7, −2, −1, 0) effect was evaluated. The main findings were that (i) Weekly concentrations of Poaceae were associated particularly with daily deaths due to chronic obstructive pulmonary disease (COPD) and pneumonia (ii) there was a consistent dose–response relationship and (iii) *Betula* and *Rumex* were positively associated with mortality. None of these findings were confounded by air pollution.

More recently, several papers confirmed and clarified Brunekreef's (Brunekreef et al. 2000) study. A symmetric case-crossover analysis, including pollen levels of Asteraceae, Betulaceae, Cupressaceae, Chenopodiaceae, Ericaceae, Fagaceae, Myrtaceae, Oleaceae, Pinaceae, Plantaginaceae, Platanaceae, Poaceae, Polygonaceae, Salicaceae and Urticaceae, performed in the city of Vigo (Spain) provided confirmation of a short-term relationship between pollen levels of *Betula*, Chenopodiaceae, *Corylus* and *Alnus* and the medical emergency calls for respiratory and cardiovascular causes in the period 1996–1999, even though there was no evident interaction between pollen and chemical air pollutants (Carracedo-Martinez et al. 2008). Interestingly, this association was statistically significant at lag 0 for Chenopodiaceae, at lag 2 for *Corylus* and *Alnus* and at lag 3 for *Betula*, whereas no relationships were found at lag 1.

In Darwin (Australia), positive linear associations were shown between hospital admissions for total respiratory diseases and COPD in the period April 2004–November 2005 and total pollen counts, without lag effect. But no convincing evidence was found for relationships between total pollen and asthma or respiratory infections. The finding of an association between pollen concentrations and respiratory hospital admissions that could not be explained by asthma admissions suggests that ambient airborne pollen grains might have a wider public health impact than previously recognized. When individual taxa were investigated, associations were the strongest in relation to Myrtaceae pollen (the dominant tree taxon in the region), while positive associations not attaining statistical significance were observed for Poaceae, Cyperaceae and Arecaceae. On the other hand, no associations were evident for any conditions with fungal spores (Hanigan and Johnston 2004).

Another study, published by Burney et al. (2008), shed new light on intrinsic (i.e. non-allergic) asthma. A cohort of 297 patients using bronchodilators aged 18–64 years was extensively evaluated with skin tests in London and they were asked to report any acute respiratory event over the coming months. Small particles with a mean aerodynamic diameter <10 μm (PM_{10}) were collected and the ability of airborne particles to bind IgE from the patients was compared with particles sampled on the weekend before their reported exacerbation and with particles sampled on the weekend 2 weeks before or after. A highly significant association ($p=0.00089$) between exacerbations and a 25% increase in IgE binding to particles collected on the previous weekend compared with the control weekends was found, even in patients with negative skin prick tests to grass or tree pollen. This observation indicates a role of unidentified allergens in asthma diagnosed as intrinsic, suggesting limits in current allergy diagnostics.

6.5.2 *Effects on Cardio- and Cerebrovascular Diseases*

The study in the Netherlands by Brunekreef et al. (2000) does not allow any conclusions on the possible effects on mortality from cardiovascular diseases to be drawn. The association between pollen and mortality was seen only for two taxa, *Betula* and *Rumex*, and the authors did not show results by disease type. On the other hand, high Poaceae pollen counts were not associated with mortality from cardiovascular diseases. Similarly, in the study with patients who made medical emergency calls in Vigo (Carracedo-Martinez et al. 2008), elevations in pollen levels did not increase medical emergency calls because of cardiovascular causes.

However, Low et al. (2006) analyzed the effects of numerous environmental variables, including pollen and spore counts, on hospital admissions for stroke in New York. Using a statistical time series modelling technique, the authors found a relatively small, but significant independent association ($p=0.0341$), between stroke events and grass pollen concentration. It is a pity that they did not provide any hypothesis supporting this observation, apart from the fact that alveolar inflammation is proposed as the mechanistic link.

Similarly, according to Matheson et al. (2008), hay fever seems to be a risk factor for stroke during a 4.4 year period in South Carolina. Patients with a history of allergic rhinitis had an unadjusted hazard ratio of 1.72 (95% CI, 1.08–2.27) for stroke *versus* patients without allergic rhinitis. Risk of stroke remained significant (hazard ratio, 1.87 [95% CI, 1.17–2.99]) after controlling for age, sex, race, smoking status, body mass index, diabetes, hypertension, alcohol use, and hyperlipidemia. The authors suggested the mechanism to be the systemic inflammation found in those with allergic rhinitis. Yet, Ng et al. (2008) put forward an additional explanation, i.e. obstructive sleep apnea (OSA), for the increased risk. It is now well accepted that allergic rhinitis is associated with an increased risk of OSA in adults and children and that adults with OSA have an increased risk of cerebrovascular disease independent of atherosclerotic risk factors.

6.5.3 *Effects on Psychiatric Diseases*

The seasonal trend of some psychiatric diseases has been reported consistently, but its causality is still poorly understood. In this view, several papers investigated the possible role of the pollen season which overlaps with the seasonality of some psychiatric diseases. An experimental study showed a regression in children with autistic spectrum disorders (ASD) or with attention deficit hyperactive disorder (ADHD) after nasal challenge with oak, timothy grass and ragweed pollen. During natural pollen exposure, 15 of 29 (52%) children with ASD and 10 of 18 (56%) children with ADHD demonstrated neurobehavioral regression. There was no correlation with the child's allergic status (IgE, skin tests and RAST) or allergy symptoms. The same effect was observed in allergic and non-allergic children (Boris and Goldblatt 2004). The authors suggested that, although they had not conducted an empirically test, an unknown non IgE-mediated mechanism underlined this relationship.

The most numerous studies are about suicide or suicide attempt, including those done at the University of Maryland School of Medicine at Baltimore. A strong and highly replicated seasonality of suicide in late spring and, less consistent, in late summer and early fall is well known (Hakko et al. 1998). For this reason, Postolache et al. (2005) performed an epidemiologic study analyzing 2,417 suicides in the tree pollen season and 1,811 in the ragweed season for the continental US and Canada in the period 1995–1998. They found a two-fold increase in the rate of non-violent suicides among younger females during the period with high tree pollen, in comparison to similar periods with low tree pollen (95% CI, 1.3–3.0). There was no difference between the post pollen period and the pre pollen period. The association was thus found only for non-violent suicides, only in females, and only for tree pollen (and not also for ragweed pollen). While these results may have been confounded by a number of biological and psychosocial factors (such as impact of feeling sick), acting during the allergy season on individuals and their social support systems, two possible explanations were proposed. These were the side-effect of anti-allergic drugs (pseudoephedrine, antihistamines and corticosteroids), which may worsen prosuicidal factors such as night-insomnia, day-somnolence, agitation, anxiety, impulsivity, and cognitive disturbance (Pretorius 2004); or a mechanism linked to an increased expression of cytokines during the immune activation. Both hypotheses appear rather weak. Nevertheless, three points require careful thought: (i) If there is insufficient data to prove any link between the other medications used to treat allergy and suicidality, systemic (but not topical) corticosteroids used for severe symptoms have been really associated with manic and depressive episodes and mixed mood states (Postolache et al. 2008). On June 12, 2009 the US Food and Drug Administration issued a warning raising concerns about the suicidality potential of montelukast (trade name Singulair®) and other leukotriene receptor antagonists (LTRA) (Manalai et al. 2009). (ii) Another mechanism linking allergic inflammation, depression, and suicide has been demonstrated. Reeves et al. (2007) outlined the role of alcohol, that is a well-established risk factor for adolescent suicide, in inducing phosphorylation and rearrangement of tight junction proteins of the blood–brain barrier (BBB) resulting in increased "leakiness", i.e. passage of cells and molecules. Many authors have proposed for a long time that allergic inflammation influences depression-related brain function via molecular and cellular mediators, but those mediators have a very limited access to the brain when the BBB is intact. Alcohol intake disrupts BBB, allowing increased brain exposure to cellular mediators of allergy, especially in youth when alcohol use starts. (iii) It has also been shown in Brown Norway rats that sensitization and exposure to aeroallergens induces anxiety-like and aggressive behaviours as well as allergy-related helper T-cell type 2 (Th2) cytokine gene expression in the prefrontal cortex. Thus, it is possible that sensitization and exposure to aeroallergens, which peak in spring, may be conducive to seasonal exacerbation of suicide risk factors such as anxiety, depression, hostility/aggression, and sleep disturbance. The recent report of Th2 (allergy-mediating) cytokine expression in the orbitofrontal cortex of suicide victims should lead to future studies to test the hypothesis that mediators of allergic inflammation in the nasal cavities may result in Th2 cytokine expression in the brain, influencing affect and behavioural modulation (Postolache et al. 2008).

Other studies showed a significant relationship ($p<0.001$) between high pollen season and self-reported non-winter seasonal affective disorder (SAD), distinguished by depression in summer and remission in fall and winter (Guzman et al. 2007). The survey was conducted among college students at local universities in Washington, DC. However, the sample included only African American and African students and, therefore, the results may not be applicable to the entire population. The study was based only on a questionnaire and patients were not tested for pollen sensitization. No direct measures of depression scores or pollen counts were collected. It is nevertheless a fact that the findings confirm certain results of the oldest study about this question that used an ethnically diverse sample (Bartko and Kasper 1989).

Postolache et al. (2008) stated that (i) the rates of depression, anxiety, and sleep disturbance (suicide risk factors) are greater in patients with allergic rhinitis than in the general population, (ii) that the rate of allergy is greater in patients with depression and (iii) finally that patients with a history of allergy may have an increased rate of suicide. However, it remains unknown if these associations are true or spurious, and if true, if they represent trait (vulnerability) – level relationship or if mood-states are involved, i.e. if there is adequate evidence to suggest that exposure to allergens might trigger mood changes in vulnerable individuals. In another study, confirming the relationship between tree pollen and changes in mood in patients sensitized to that pollen, Postolache et al. (2007) suggested a connection between allergy and depression decompensation rather than between pollen and depression.

To sum up, allergy can influence mood by several potential mechanisms. There may be somatic changes, such as discomfort caused by allergic inflammatory processes in the upper airway that may affect well being. This situation may lead to other possible mediators affecting mood, such as the use of medications like corticosteroids or LTRA, or disturbance of sleep caused by multiple factors including obstruction of the airways. The release of inflammatory mediators including cytokines is one likely mechanism that may promote the worsening in mood. This mood worsening can occur by either acting directly in the brain or through other pathways, such as interactions with the HPA-axis and/or the IDO enzyme. Cytokines have been shown to induce depression and anxiety. The findings of Postolache et al. (2007) relating allergic symptoms with depression scores may be explained in part by the release of cytokines during allergic reactions. *In vitro* studies have also shown that certain human polymorphisms increase the expression of cytokines including TNF-α, IL-13, and IFN-γ. In addition, certain cytokine polymorphisms increase susceptibility to allergic disease such as asthma. Moreover, allergic symptoms have been correlated with the amount of cytokines released during some allergic reactions. It is possible that depression scores in some individuals may reflect the increased amount of cytokine release during allergic processes. This issue is a matter to be clarified with future investigations.

6.6 Conclusion

Pollen has a substantial impact on human health. Besides the role of pollen and fungal spores in allergic diseases, recent research also suggests an effect of allergenic pollen on non-allergic diseases. Current knowledge is insufficient to draw a conclusion on these effects. Experimental studies showing a proinflammatory and immunomodulatory activity of pollen support the hypothesis of a non-IgE mediated effect on human health, even if with several limitations, some epidemiological observations are consistent with the experimental data. Other indications come from studies on respiratory and psychiatric diseases which present some diagnostic difficulties and limitations. However, all these findings have not been sufficiently reproduced so far. Consequently, further research is needed at clinical, epidemiological, animal and postmortem tissue levels to clarify the complex role of pollen on human health, which probably goes beyond allergic diseases. Experimental studies should identify the role of pollen-derived particles in airborne particulate matter (e.g. PM_{10}) and their effects on human health. Moreover, epidemiological studies should be specifically designed by a multidisciplinary research team, with the aim of confirming the association between pollen counts and diseases. The advances in the field of detection of airborne allergens could contribute to distinguishing the effects of allergens from those of other non-allergenic proteins of pollen (Thibaudon and Sindt 2008).

Most knowledge on the impact of pollen on human health deals with the allergic sensitization upon pollen exposure causing IgE-mediated responses in susceptible individuals. These responses, resulting in respiratory diseases and allergic eye irritation, have a marked impact not only on clinical parameters, but also on functional status as shown by a study demonstrating a reduction in productivity of 11–40% (Vandenplas et al. 2008). This impact of pollen on health status differs among patients, both on an individual as well as on a population level. It is dependent on weather conditions (e.g. temperature, thunderstorms), presence of air pollutants (e.g. O_3, NO_2, SO_2, particulate air pollution [PM]) and the impact may change during the season. Furthermore, the amount of allergen per pollen grain can differ which may affect the impact of pollen.

The diversity of factors that influence the health impact of pollen has hampered the definition of a straightforward relationship between pollen and the severity of symptoms. Several studies report pollen levels at which symptoms develop, but there is still no consensus on these threshold levels. The various factors discussed in this overview that influence the interactions between pollen and patients, are responsible for the fact that threshold values not only vary by pollen type, but also by region, population, year, time of the season, and landscape architecture (e.g. urban or rural). However, there is a need for definition of threshold pollen levels, not only for scientists and physicians, but also for patients that want to know at which pollen concentration they are at risk for symptom development, and for policymakers that want to create distinct rules for public information services. At this moment, science cannot formulate a validated solution. However, one first approach to tackle this problem could be to define preliminary threshold values for different regions. Whereas this is inevitably an oversimplification, such preliminary threshold values could be validated using symptoms scores of patients that can be entered by special websites (e.g. www.allergieradar.nl or www.polleninfo.org). Such websites may reach a wide audience and symptoms scores can be collected including GPS coordinates. The combination of geographical coordinates and symptoms scores will result in relevant new information which can teach us whether these preliminary threshold values are correct or whether they should be adjusted based on interacting mechanisms.

Acknowledgements The authors acknowledge the contributions made by these persons to this chapter:

Rui Brandao[1], Andreja Seliger[2], Regula Gehrig[3], Jean Emberlin[4], Ingrida Sauliene[5]

[1]Universidade de Evora, University of Evora, Life Sciences, Area Departamental Ciências Natureza e Ambiente Biology, Largo dos Colegiais, Centro de Estudos da Mitra, P-7000 Évora, Portugal
[2]Institute of public health of the republic of Slovenia, Bohoričeva 15, SI-1000 Ljubljana, Croatia
[3]MeteoSwiss, Kraehbuehlstrasse 58, 8044 Zurich, Switzerland
[4]University of Worcester, Health, Social care and Psychology, National Pollen and Aerobiology Research Unit, WR2 6AJ Worcester, UK
[5]Siauliai university, Environmental reseach,Vilniaus St. 88,LT-76285 Siauliai, Lithuania

References

Abelson, M. B., George, M. A., Schaefer, K., & Smith, L. M. (1994). Evaluation of the new ophthalmic antihistamine, 0.05% levocabastine, in the clinical allergen challenge model of allergic conjunctivitis. *Journal of Allergy and Clinical Immunology, 94*, 458–464.

Agarwal, M. K., Swanson, M. C., Reed, C. E., & Yunginger, J. W. (1984). Airborne ragweed allergens: Association with various particle sizes and short ragweed plant parts. *Journal of Allergy and Clinical Immunology, 74*, 687–693.

Agashe, S. N., Bapat, B. N., Bapat, H. N., & Philip, E. (1994). Aerobiology of *Casuarina* pollen and its significance as a potential aeroallergen. *Aerobiologia, 10*, 123–128.

Ahlholm, J. U., Helander, M. L., & Savolainen, J. (1998). Genetic and environmental factors affecting the allergenicity of birch (*Betula pubescens* ssp. czerepanovii [Orl.] Hamet-ahti) pollen. *Clinical and Experimental Allergy, 28*(11), 1384–1388.

Ahlstrom-Emanuelsson, C., Andersson, M., Persson, C., Schrewelius, C., & Greiff, L. (2004). Topical treatment with aqueous solutions of rofleponide palmitate and budesonide in a pollen-season model of allergic rhinitis. *Clinical and Experimental Allergy, 34*(5), 731–735.

Akerlund, A., Andersson, M., Leflein, J., Lildholdt, T., & Mygind, N. (2005). Clinical trial design, nasal allergen challenge models, and considerations of relevance to pediatrics, nasal polyposis, and different classes of medication. *Journal of Allergy and Clinical Immunology, 115*(3 Suppl. 1), 460–482.

Alcázar, P., Galán, C., Cariñanos, P., & Domínguez, E. (1998). Vertical variation in Urticaceae airborne pollen concentration. *Aerobiologia, 14*, 131–134.

Alcázar, P., Galán, C., Cariñanos, P., & Domínguez, E. (1999a). Diurnal variations of airborne pollen at two different heights. *Journal of Investigational Allergology and Clinical Immunology, 9*, 89–95.

Alcázar, P., Galán, C., Cariñanos, P., & Domínguez, E. (1999b). Effects of sampling height and climatic conditions in aerobiological studies. *Journal of Investigational Allergology and Clinical Immunology, 9*, 253–261.

Alcázar, P., Cariñanos, P., De Castro, C., Guerra, F., Moreno, C., Domínguez-Vilches, E., & Galán, C. (2004). Airborne plane-tree (*Platanus hispanica*) pollen distribution in the city of Córdoba, South-western Spain, and possible implications on pollen allergy. *Journal of Investigational Allergology and Clinical Immunology, 14*(3), 238–243.

Amin, K., Ekberg-Jansson, A., Lofdahl, C. G., & Venge, P. (2003). Relationship between inflammatory cells and structural changes in the lungs of asymptomatic and never smokers (a biopsy study). *Thorax, 58*, 135–142.

Andersson, K., & Lidholm, J. (2003). Characteristics and immunobiology of grass pollen allergens. *International Archives of Allergy and Immunology, 130*, 87–107.

Antepara, I., Fernandez, J. C., Gamboa, P., Jauregui, I., & Miguel, F. (1995). Pollen allergy in the Bilbao area (European Atlantic seaboard climate): Pollination forecasting methods. *Clinical and Experimental Allergy, 25*, 133–140.

Asero, R., Mistrello, G., Roncarolo, D., & Casarini, M. (2000). Detection of allergens in plantain (*Plantago lanceolata*) pollen. *Allergy, 55*, 1059–1062.

Asero, R., Wopfner, N., Gruber, P., Gadermaier, G., & Ferreira, F. (2006). *Artemisia* and *Ambrosia* hypersensitivity: Co-sensitization or co-recognition? *Clinical and Experimental Allergy, 36*, 658–665.

Asher, M. I., Stewart, A. W., Mallol, J., Montefort, S., Lai, C. K., Aït-Khaled, N., & Odhiambo, J. (2010). Which population level environmental factors are associated with asthma, rhinoconjunctivitis and eczema? Review of the ecological analyses of ISAAC Phase One. *Respiratory Research, 11*, 8.

Asturias, J. A., Arilla, M. C., Gomez-Bayon, N., Martinez, A., Martinez, J., & Palacios, R. (1997). Cloning and expression of the panallergen profilin and the major allergen (Ole e 1) from olive tree pollen. *Journal of Allergy and Clinical Immunology, 100*, 365–372.

Asturias, J. A., Gómez-Bayón, N., Eseverri, J. L., & Martínez, A. (2003). Par j 1 and Par j 2, the major allergens from *Parietaria judaica* pollen, have similar immunoglobulin E epitopes. *Clinical and Experimental Allergy, 33*, 518–524.

Baldacci, S., & Viegi, G. (2002). Respiratory effects of environmental pollution: Epidemiological data. *Monaldi Archives for Chest Disease, 57*(3–4), 156–160.

Banken, R., & Comtois, P. (1990). Concentration de l'herbe à poux et prévalence de la rhinite allergique dans deux municipalités des Laurentides. *L'Unión Médicale du Canada, 119*, 178–182.

Barber, D., Moreno, C., Ledesma, A., Serrano, P., Galán, A., Villalba, M., Guerra, F., Lombardero, M., & Rodríguez, R. (2007). Degree of olive pollen exposure and sensitization patterns. Clinical implications. *Journal of Investigational Allergology and Clinical Immunology, 17*(Suppl. 1), 63–68.

Barber, D., de la Torre, F., Feo, F., Florido, F., Guardia, P., Moreno, C., Quiralte, J., Lombardero, M., Villalba, M., & Salcedo, G. (2008). Understanding patient sensitization profiles in complex pollen areas: A molecular epidemiological study. *Allergy, 63*, 1550–1558.

Barral, P., Tejera, M. L., Treviño, M. A., Batanero, E., Villalba, M., Bruix, M., & Rodríguez, R. (2004). Recombinant expression of Ole e 6, a Cysenriched pollen allergen, in *Pichia pastoris* yeast. Detection of partial oxidation of methionine by NMR. *Protein Expression and Purification, 37*, 336–343.

Barral, P., Serrano, A. G., Batanero, E., Pérez-Gil, J., Villalba, M., & Rodríguez, R. (2006). A recombinant functional variant of the olive pollen allergen Ole e 10 expressed in baculovirus system. *Journal of Biotechnology, 121*, 402–409.

Bartko, J. J., & Kasper, S. (1989). Seasonal changes in mood and behavior: A cluster analytic approach. *Psychiatry Research, 28*(2), 227–239.

Bauchau, V., & Durham, S. R. (2004). Prevalence and rate of diagnosis of allergic rhinitis in Europe. *European Respiratory Journal, 24*(5), 758–764.

Bazzaz, F. A. (1990). The response of natural ecosystems to the rising global CO_2 levels. *Annual Review of Ecology and Systematics, 21*, 167–196.

Behrendt, H., & Becker, W. M. (2001). Localization, release and bioavailability of pollen allergens: The influence of environmental factors. *Current Opinion in Immunology, 13*, 709–715.

Belchi-Hernandez, J., Moreno-Grau, S., Bayo, J., Rendueles, B. E., Moreno, J., Angosto, J. M., Iniesta-Perez, B., & Gonzalez, A. M. (2001). Pollinosis related to Zygophyllum fabago in a Mediterranean area. *Aerobiologia, 17*, 241–246.

Blanca, M., Boulton, P., Brostoff, J., & Gonzalez-Reguera, I. (1983). Studies of the allergens of *Olea europea* pollen. *Clinical Allergy, 13*(5), 473–478.

Bonini, S., Bonini, S., Bucci, M. G., Berruto, A., Adriani, E., Balsano, F., & Allansmith, M. R. (1990). Allergen dose response and late symptoms in a human model of ocular allergy. *Journal of Allergy and Clinical Immunology, 86*, 869–876.

Boris, M., & Goldblatt, A. (2004). Pollen exposure as a cause for the deterioration of neurobehavioral function in children with autism and attention deficit hyperactive disorder: Nasal pollen challenge. *Journal of Nutritional and Environmental Medicine, 14*(1), 47–54.

Bousquet, P. J., Chinn, S., Janson, C., Kogevinas, M., Burney, P., & Jarvis, D. (2007). Geographical variation in the prevalence of positive skin tests to environmental aeroallergens in the European Community Respiratory Health Survey I. *Allergy, 62*(3), 301–309.

Brabäck, L., Hjernwz, A., & Rasmussen, F. (2004). Trends in asthma, allergic rhinitis and eczema among Swedish conscripts from farming and non-farming environments. A nationwide study over three decades. *Clinical and Experimental Allergy, 34*, 38–43.

Britton, J., Pavord, I., Richards, K., Knox, A., Wisniewski, A., Wahedna, I., Kinnear, W., Tattersfield, A., & Weiss, S. (1994). Factors influencing the occurrence of airway hyperreactivity in the general population: The importance of atopy and airway calibre. *European Respiratory Journal, 7*, 881–887.

Brunekreef, B., Hoek, G., Fischer, P., & Spieksma, F. T. (2000). Relation between airborne pollen concentrations and daily cardiovascular and respiratory-disease mortality. *The Lancet, 355*(9214), 1517–1518.

Burbach, G. J., Heinzerling, L. M., Edenharter, G., Bachert, C., Bindslev-Jensen, C., Bonini, S., Bousquet, J., Bousquet-Rouanet, L., Bousquet, P. J., & Bresciani, M. (2009). GA(2)LEN skin test study II: Clinical relevance of inhalant allergen sensitizations in Europe. *Allergy, 64*, 1507–1515.

Burney, P. G., Newson, R. B., Burrows, M. S., & Wheeler, D. M. (2008). The effects of allergens in outdoor air on both atopic and nonatopic subjects with airway disease. *Allergy, 63*(5), 542–546.

Burr, M. L., Emberlin, J. C., Treu, R., Cheng, S., & Pearce, N. E. (2003). Pollen counts in relation to the prevalence of allergic rhinoconjunctivitis, asthma and atopic eczema in the International Study of Asthma and Allergies in Childhood (ISAAC). *Clinical and Experimental Allergy, 33*(12), 1675–1680.

Busse, W. W. (1989). The relationship between viral infections and onset of allergic diseases and asthma. *Clinical and Experimenal Allergy, 19*, 1–9.

Busse, W. W. (2010). The relationship of airway hyperresponsiveness and airway inflammation: Airway hyperresponsiveness in asthma: Its measurement and clinical significance. *Chest, 138*, 4–10.

Buters, J. T., & Behrendt, H. (2008). *Kombinationswirkungen: Pollen und Schadstoffe. Rundgesprache der Kommission für Ökologie, Bd 38*. Bioaerosole und ihre Bedeutung für die Gesundheit (pp. 97–105). München: Verlag Dr. Friedrich Pfeil.

Buters, J. T., Kasche, A., Weichenmeier, I., Schober, W., Klaus, S., Traidl-Hoffmann, C., Menzel, A., Huss-Marp, J., Kramer, U., & Behrendt, H. (2008). Year-to-year variation in release of Bet v 1 allergen from birch pollen: Evidence for geographical differences between West and South Germany. *International Archives of Allergy and Immunology, 145*, 122–130.

Buters, J. T., Weichenmeier, I., Ochs, S., Pusch, G., Kreyling, W., Boere, A. J., Schober, W., & Behrendt, H. (2010). The allergen Bet v 1 in fractions of ambient air deviates from birch pollen counts. *Allergy, 65*, 850–858.

Calleja, M., & Farrera, I. (2000). Le Cyprès:un nouveau fléau pour la région Rhône-Alpes? *Allergy and Immunology, 32*, 125–127.

Camacho, J. L., Cano, D., Cervigón, P., Gutiérrez, A. M., & Subiza, J. (2008). Forecasting climate and health related events on plane tree pollen and grass pollen seasons in Madrid region (Spain). Climate prediction applications science workshop, Chapel Hill, NC March 4–7 (poster session).

Canonica, G. W., & Compalati, E. (2009). Minimal persistent inflammation in allergic rhinitis: Implications for current treatment strategies. *Clinical and Experimental Immunology, 156*, 260–271.

Cárdaba, B., Llanes, E., Chacártegui, M., Sastre, B., López, E., Mollá, R., del Pozo, V., Florido, F., Quiralte, J., Palomino, P., & Lahoz, C. (2007). Modulation of allergic response by gene–environment interaction: Olive pollen Allergy. *Journal of Investigational Allergology and Clinical Immunology, 17*(Suppl 1), 83–87.

Cariñanos, P., Alcázar, P., Galán, C., & Domínguez, E. (2002). Privet pollen (Ligustrum sp.) as potential cause of pollinosis in the city of Cordoba, south-west Spain. *Allergy, 57*, 92–97.

Carracedo-Martinez, E., Sanchez, C., Taracido, M., Saez, M., Jato, V., & Figueiras, A. (2008). Effect of short-term exposure to air pollution and pollen on medical emergency calls: A case-crossover study in Spain. *Allergy, 63*(3), 347–353.

Celenza, A., Fothergill, J., Kupek, E., & Shaw, R. J. (1996). Thunderstorm associated asthma: A detailed analysis of environmental factors. *British Medical Journal, 312*, 604–607.

Charpin, D., Hughes, B., Mallea, M., Sutra, J. P., Balansard, G., & Vervloet, D. (1993). Seasonal allergic symptoms and their relation to pollen exposure in south-east France. *Clinical and Experimental Allergy, 23*, 435–439.

Charpin, D., Calleja, M., Lahoz, C., Pichot, C., & Waisel, Y. (2005). Allergy to cypress pollen. *Allergy, 60*, 293–301.

Chen, L. L., Tager, I. B., Peden, D. B., Christian, D. L., Ferrando, R. E., Welch, B. S., & Balmes, J. R. (2004). Effect of ozone exposure on airway responses to inhaled allergen in asthmatic subjects. *Chest, 125*(6), 2328–2335.

Cheung, D., Dick, E. C., Timmers, M. C., de Klerk, E. P., Spaan, W. J., & Sterk, P. J. (1995). Rhinovirus inhalation causes long-lasting excessive airway narrowing in response to methacholine in asthmatic subjects in vivo. *American Journal of Respiratory and Critical Care Medicine, 152*, 1490–1496.

Ciprandi, G., Filaci, G., Negrini, S., De, A. M., Fenoglio, D., & Marseglia, G. (2009). Serum leptin levels in patients with pollen-induced allergic rhinitis. *International Archives of Allergy and Immunology, 148*(3), 211–218.

Colombo, V. (2010). Zum Sensibilisierungsspektrum von Pollenallergikern im Kanton Tessin. Eine prospektive Studie in Locarno und Lugano 2009. Thesis. Zurich: University of Zurich.

Comtois, P., & Gagnon, L. (1988). Pollen concentration and frequency of pollinosis symptoms: Method of determination of the clinical threshold. *Revue Française d'Allergologie, 28*, 279–286.

Connell, J. T. (1968). Quantitative intranasal pollen challenge. II. Effect of daily pollen challenge, environmental pollen exposure, and placebo challenge on the nasal membrane. *Journal of Allergy, 41*(3), 123–139.

Cosmes, M. P., Moreno, A. A., Dominguez, N. C., Gutierrez, V. A., Belmonte, S. J., & Roure, N. J. (2005). Sensitization to *Castanea sativa* pollen and pollinosis in northern Extremadura (Spain). *Allergologia et immunopathologia, 33*(3), 145–150.

D'Amato, G., Ruffilli, A., & Ortolani, C. (1991). Allergenic significance of Parietaria. In G. D'Amato, F. Th. M. Spieksma, & S. Bonini (Eds.), *Allergenic pollen and pollinosis in Europe* (pp. 113–118). Oxford: Blackwell Scientific Publishers.

D'Amato, G., Chatzigeorgiou, G., Corsico, R., Gioulekas, D., Jager, L., Jager, S., Kontou-Fili, K., Kouridakis, S., Liccardi, G., Meriggi, A., Palma-Carlos, A., Palma-Carlos, M. L., Pagan Aleman, A., Parmiani, S., Puccinelli, P., Russo, M., Spieksma, F. T., Torricelli, R., & Wuthrich, B. (1997). Evaluation of the prevalence of skin prick test positivity to *Alternaria* and *Cladosporium* in patients with suspected respiratory allergy. A European multicenter study promoted by the Subcommittee on Aerobiology and Environmental Aspects of Inhalant Allergens of the European Academy of Allergology and Clinical Immunology. *Allergy, 52*(7), 711–716.

D'Amato, G., Spieksma, F., Liccardi, G., Jäger, S., Russo, M., Kontou-Fili, K., Nikkels, H., Wüthrich, B., & Bonini, S. (1998). Pollen-related allergy in Europe. *Allergy, 53*, 567–578.

D'Amato, G., Liccardi, M., & Cazzola, M. (2001). The role of outdoor air pollution and climatic changes on the rising trends in respiratory allergy. *Respiratory Medicine, 95*, 606–611.

Dales, R. E., Cakmak, S., Judek, S., & Coates, F. (2008). Tree pollen and hospitalization for asthma in urban Canada. *International Archives of Allergy and Immunology, 146*(3), 241–247.

Damialis, A., Halley, J. M., Giouleka, D., & Vokou, D. (2007). Long-term trends in atmospheric pollen levels in the city of Thessaloniki, Greece. *Atmospheric Environment, 41*, 7011–7021.

Davies, R. R., & Smith, L. P. (1973). Forecasting the start and severity of the hay fever season. *Clinical Allergy, 3*(3), 263–267.

Day, J. H., & Briscoe, M. P. (1999). Environmental exposure unit: A system to test anti-allergic treatment. *Annals of Allergy, Asthma & Immunology, 83*(2), 83–89.

Day, J. H., Briscoe, M. P., Rafeiro, E., Ellis, A. K., Pettersson, E., & Akerlund, A. (2000). Onset of action of intranasal budesonide (Rhinocort aqua) in seasonal allergic rhinitis studied in a controlled exposure model. *Journal of Allergy and Clinical Immunology, 105*(3), 489–494.

Day, J. H., Briscoe, M. P., Ratz, J. D., Ellis, A. K., Yao, R., & Danzig, M. (2009). Onset of action of loratadine/montelukast in seasonal allergic rhinitis subjects exposed to ragweed pollen in the Environmental Exposure Unit. *Allergy and Asthma Proceedings, 30*(3), 270–276.

Déchamp, C., Legal, M., & Deviller, P. (1995). Prevalence of ragweed pollinosis in the South and Eastern suburbs of Lyon in 1993. *Allergie et Immunologie, 27*, 320–325.

Déchamp, C., Rimet, M. L., Méon, H., & Deviller, P. (1997). Parameters of ragweed pollination in the LYON's area (France) from fourteen years of pollen counts. *Aerobiologia, 13*, 275–279.

Déchamp, C., Calleja, M., Deviller, P., Harf, R., & Méon, H. (2001). L'ambroisie dans le Rhône et la Politique Agricole Commune. Le rôle des jachères européennes et des cultures de tournesol sur la pollution biologique aéroportée par le pollen d'ambroisie. *Phytoma, 538*, 13–16.

Delaunay, J. J., Sasajima, H., Okamoto, Y., & Yokota, M. (2007). Side-by-side comparison of automatic pollen counters for use in pollen information systems. *Annals of Allergy, Asthma & Immunology, 98*, 553–558.

De Linares, C., Nieto-Lugilde, D., Alba, F., Diaz de la Guardia, C., Galán, C., & Trigo, M. M. (2007). Detection of airborne allergen (Ole e 1) in relation to *Olea europaea* pollen in S Spain. *Clinical and Experimental Allergy, 37*, 125–132.

Devalia, J. L., Rusznak, C., & Davies, R. J. (1998). Allergen/irritant interaction – Its role in sensitization and allergic disease. *Allergy, 53*, 335–345.

De Weger, L. A., Beerthuizen, T., Gast-Strookman, J. M., van der Plas, D. T., Hiemstra, P. S., Terreehorst, I., & Sont, J. K. (2008). Seasonal variability in the relationship between grass pollen counts and ICT-based daily hay fever symptom scores in patients with allergic rhinitis: Implications for hay fever forecasts. EAACI 2008, Barcelona, Spain.

Diaz-Sanchez, D., Tsien, A., Casillas, A., Dotson, A. R., & Saxon, A. (1996). Enhanced nasal cytokine production in human beings after in vivo challenge with diesel exhaust particles. *Journal of Allergy and Clinical Immunology, 98*(1), 114–123.

Diaz-Sanchez, D., Garcia, M. P., Wang, M., Jyrala, M., & Saxon, A. (1999). Nasal challenge with diesel exhaust particles can induce sensitization to a neoallergen in the human mucosa. *Journal of Allergy and Clinical Immunology, 104*, 1183–1188.

Dobson, H. E. M. (1988). Survey of pollen and pollenkitt lipids – Chemical cues to flower visitors? *American Journal of Botany, 75*(2), 170–182.

Egger, C. M., Focke, C., Bircher, A. J., Scherer, K., Mothes-Luksch, N., Horak, F., & Valenta, R. (2008). The allergen profile of beech and oak pollen. *Clinical and Experimental Allergy, 38*(10), 1688–1696.

Emberlin, J. (1994). The effects of patterns in climate and pollen abundance on allergy. *Allergy, 49*, 15–20.

Emberlin, J., Smith, M., Close, R., & Adams-Groom, B. (2007). Changes in the pollen seasons of the early flowering trees *Alnus* spp. and *Corylus* spp. in Worcester, United Kingdom, 1996–2005. *International Journal of Biometeorology, 51*, 181–191.

Eriksson, N. E., Wihl, J.-Å., Arrendal, H., & Strandhede, S.-O. (1984). Tree Pollen Allergy. *Allergy, 39*(8), 610–617.

Eriksson, N. E., Wihl, J.-Å., Arrendal, H., & Strandhede, S.-O. (1987). Tree pollen allergy. III. Cross reactions based on results from skin prick tests and the RAST in hay fever patients. A multi-centre study. *Allergy, 42*(3), 205–214.

Erkara, I. P., Cingi, C., Ayranci, U., Gurbuz, K. M., Pehlivan, S., & Tokur, S. (2009). Skin prick test reactivity in allergic rhinitis patients to airborne pollens. *Environmental Monitoring, 151*, 401–412.

Ernst, P., Ghezzo, H., & Becklake, M. R. (2002). Risk factors for bronchial hyper responsiveness in late childhood and early adolescence. *European Respiratory Journal, 20*, 635–639.

Farrera, I., Calleja, M., & Déchamp, C. (2002). Indoor and outdoor sedimentation in the Rhône-Alpes region during the ragweed pollen season. *Revue française d'allergologie et d'immunologie Clinique, 42*(7), 750–753.

Feo Brito, F., Mur Gimeno, P., Martinez, C., Tobias, A., Suarez, L., Guerra, F., Borja, J. M., & Alonso, A. M. (2007). Air pollution and seasonal asthma during the pollen season. A cohort study in Puertollano and Ciudad Real (Spain). *Allergy, 62*(10), 1152–1157.

Fernvik, E., Scharnweber, T., Knopp, D., Niessner, R., Vargaftig, B. B., & Peltre, G. (2002). Effects of fractions of traffic particulate matter on TH2-cytokines, IgE levels, and bronchial hyperresponsiveness in mice. *Journal of Toxicology and Environmental Health. Part A, 65*, 1025–1045.

Filipowicz, N., Piotrowski, A., Ochocka, J. R., & Asztemborska, M. (2006). The phytochemical and genetic survey of common and dwarf juniper (*Juniperus communis* and *Juniperus nana*) identifies chemical races and close taxonomic identity of the species. *Planta Medica, 72*, 850–853.

Fiorina, A., Scordamaglia, A., Mincarini, M., Fregonese, L., & Canonica, G. W. (1997). Aerobiologic particle sampling by a new personal collector (Partrap FA52) in comparison to the Hirst (Burkard) sampler. *Allergy, 52*, 1026–1030.

Florido, J. F., Delgado, P. G., de San Pedro, B. S., Quiralte, J., de Saavedra, J. M., Peralta, V., & Valenzuela, L. R. (1999). High levels of *Olea europaea* pollen and relation with clinical findings. *International Archives of Allergy and Immunology, 119*, 133–137.

Fotiou, C., Damialis, A., Krigas, N., & Vokou, D. (2011). *Parietaria judaica* flowering phenology, pollen production, viability and atmospheric circulation, and expansive ability in the urban environment: Impacts of environmental factors. *International Journal of Biometeorology, 55*(1), 35–50.

Fountain, D. W., Berggren, B., Nilsson, S., & Einarsson, R. (1992). Expression of birch pollen-specific IgE-binding activity in seeds and other plant parts of birch trees (*Betula verrucosa* Ehrh.). *International Archives of Allergy and Immunology, 98,* 370–376.

Frank, E., Leonhardt, L. G. W., & Jaeger, S. (1991). Allergenic significance of *Rumex* pollen. In G. D'Amato, F. Th. M. Spieksma, & S. Bonini (Eds.), *Allergenic pollen and pollinosis in Europe* (pp. 119–120). Oxford: Blackwell Scientific Publications.

Frei, T. (1998). The effects of climate change in Switzerland 1969–1996 on airborne pollen quantities from hazel, birch and grass. *Grana, 37,* 172–179.

Frei, T., & Gassner, E. (2008). Trends in prevalence of allergic rhinitis and correlation with pollen counts in Switzerland. *International Journal of Biometeorology, 52,* 841–847.

Galán, G., Alcázar, P., Domínguez, E., Villamandos, F., & Infante, F. (1995). Airborne pollen grains concentration at two different heights. *Aerobiologia, 11,* 105–109.

Galán, C., Alcázar, P., Cariñanos, P., Garcia, H., & Domínguez-Vilches, E. (2000). Meteorological factors affecting daily urticaceae pollen counts in southwest Spain. *International Journal of Biometeorology, 43,* 191–195.

García, J. J., Trigo, M. M., Cabezudo, B., Recio, M., Vega, J. M., Barber, D., Carmona, M. J., Cervera, J. A., Toro, F. J., & Miranda, A. (1997). Pollinosis due to Australian Pine (*Casuarina*): An aerobiologic and clinical study in southern Spain. *Allergy, 52,* 11–17.

Garty, B. Z., Kosman, E., Ganor, E., Berger, V., Garty, L., Wietzen, T., & Waisman, Y. (1998). Emergency room visits of asthmatic children, relation to air pollution, weather, and airborne allergens. *Annals of Allergy, Asthma & Immunology, 81,* 563–570.

Gehrig, R., & Peeters, A. (2000). Pollen distribution at elevations above 1000 m in Switzerland. *Aerobiologia, 16,* 69–74.

Geller-Bernstein, C., Arad, G., Keynan, N., Lahoz, C., Cardaba, B., & Waisel, Y. (1996). Hypersensitivity to pollen of *Olea europaea* in Israel. *Allergy, 51,* 356–359.

Geller-Bernstein, C., Lahoz, C., Cárdaba, B., Hassoun, G., Iancovici-Kidon, M., Kenett, R., & Waisel, Y. (2002). Is it "bad hygiene" to inhale pollen early life? *Allergy, 57,* 43–46.

Georgitis, J. W., Meltzer, E. O., Kaliner, M., Weiler, J., & Berkowitz, R. (2000). Onset-of-action for antihistamine and decongestant combinations during an outdoor challenge. *Annals of Allergy, Asthma & Immunology, 84*(4), 451–459.

Gilles, S., Fekete, A,. Zhang, X., Blume, C., Behrendt, H., Schmitt-Kopplin, P., et al. (2010). Pollen-derived adenosine modulates dendritic cell function resulting in inhibition of Th1 responses. 8th EAACI-Ga2len immunology winter school. Eibsee, Germany, 11–14 February 2010.

Gioulekas, D., Papakosta, D., Damialis, A., Spieksma, F., Giouleka, P., & Patakas, D. (2004). Allergenic pollen records (15 years) and sensitization in patients with respiratory allergy in Thessaloniki, Greece. *Allergy, 59,* 174–184.

Gonianakis, M., Neonakis, I., Gonianakis, I., Baritaki, M., Kypriotakis, Z., Potamias, G., Bouros, D., & Kontou-Fili, K. (2006). A 10-year aerobiological study (1994–2003) in the Mediterranean island of Crete, Greece: Grasses and other weeds, aerobiological data, and botanical and clinical correlations. *Allergy and Asthma Proceedings, 27,* 363–370.

Gravesen, S. (1979). Fungi as a cause of allergic disease. *Allergy, 34*(3), 135–154.

Greiff, L., Venge, P., Andersson, M., Enander, I., Linden, M., Myint, S., & Persson, C. G. (2002). Effects of rhinovirus-induced common colds on granulocyte activity in allergic rhinitis. *Journal of Infection, 45,* 227–232.

Grote, M., & Fromme, H. G. (1986). Cross-reactivity of birch anthers and leaves with birch pollen antigens and allergens. A fine-structural immunocytochemical study using the post-embedding protein-A-gold technique. *Journal of Histochemistry and Cytochemistry, 34,* 1459–1464.

Grote, M., Valenta, R., & Reichelt, R. (2003). Study of sensitivity of *Fraxinus* spp. (Oleaceae) in Córdoba, Spain. *Journal of Investigational Allergology and Clinical Immunology, 5,* 166–170.

Guerra, F., Daza, J. C., Miguel, R., Moreno, C., Galán, C., Domínguez, E., & Sanchez Guijo, P. (1996). Sensitivity to Cupressus, allergenic significance in Córdoba (Spain). *Journal of Investigational Allergology and Clinical Immunology, VI*(2), 117–120.

Guzman, A., Tonelli, L. H., Roberts, D., Stiller, J. W., Jackson, M. A., Soriano, J. J., Yousufi, S., Rohan, K. J., Komarow, H., & Postolache, T. T. (2007). Mood-worsening with high-pollen-counts and seasonality: A preliminary report. *Journal of Affective Disorders, 101*(1–3), 269–274.

Hakko, H., Räsänen, P., & Tiihonen, J. (1998). Seasonal variation in suicide occurrence in Finland. *Acta Psychiatrica Scandinavica, 98*(2), 92–97.

Hanigan, I. C., & Johnston, F. H. (2004). Respiratory hospital admissions were associated with ambient airborne pollen in Darwin, Australia, 2004–2005. *Clinical and Experimental Allergy, 37*(10), 1556–1565.

Harf, R., & Déchamp, C. (2001). Pollinosis and anti-allergic drug use: *Ambrosia* in the Rhône-Alpes Region. *Revue des Maladies Respiratoires, 18*, 517–522.

Hartl, A., Kiesslich, J., Weiss, R., Bernhaupt, A., Mostbock, S., Scheiblhofer, S., Flöckner, H., Sippl, M., Ebner, C., Ferreira, F., & Thalhamer, J. (1999). Isoforms of the major allergen of birch pollen induce different immune responses after genetic immunization. *International Archives of Allergy and Immunology, 120*, 17–29.

Hashimoto, M., Fukuda, T., Shimizu, T., Watanabe, S., Watanuki, S., Eto, Y., & Urashima, M. (2004). Influence of climate factors on emergency visits for childhood asthma attack. *Pediatrics International, 46*, 48–52.

Heinzerling, L. M., Burbach, G. J., Edenharter, G., Bachert, C., Bindslev-Jensen, C., Bonini, S., et al. (2009). GA(2)LEN skin test study I: GA(2)LEN harmonization of skin prick testing: Novel sensitization patterns for inhalant allergens in Europe. *Allergy, 64*, 1498–1506.

Hemmer, W., Focke, M., Wantke, F., Götz, M., Jarisch, R., & Jäger, S. (2000). Ash (*Fraxinus excelsior*)-pollen allergy in central Europe: Specific role of pollen panallergens and the major allergen of ash pollen, Frax e 1. *Allergy, 55*, 923–930.

Hernandez, P. M., Lorente, T. F., Romo, C. A., Davila, G. I., Laffond, Y. E., & Calvo, B. A. (1998). Pollen calendar of the city of Salamanca (Spain). Aeropalynological analysis for 1981–1982 and 1991–1992. *Allergologia et immunopathologia, 26*, 209–222.

Hidalgo, P. J., Galán, C., & Domínguez, E. (2003). Male phenology of three species of *Cupressus*: Correlation with airborne pollen. *Trees, 17*, 336–344.

Hirschwehr, R., Heppner, C., Spitzauer, S., Sperr, W. R., Valent, P., Berger, U., Horak, F., Jäger, S., Kraft, D., & Valenta, R. (1998). Identification of common allergenic structures in mugwort and ragweed pollen. *Journal of Allergy and Clinical Immunology, 101*, 196–206.

Hjern, A., Rasmussen, F., Johansson, M., & Aberg, N. (1999). Migration and atopic disorder in Swedish Conscripts. *Pediatric Allergy and Immunology, 10*, 209–215.

Hjern, A., Haglund, B., & Hedlin, G. (2000). Ethnicity, childhood environment and atopic disorder. *Clinical and Experimental Allergy, 30*, 521–528.

Hofman, T., Wykretowicz, G., Stach, A., Springer, E. K. B., & Ossowski, M. (1996). A multicentre analysis of a population of patients with newly-diagnosed pollinosis in Poznan, Poland, in the year 1995. *Annals of Agricultural and Environmental Medicine, 3*, 171–177.

Hopp, R. J., Bewtra, A., Nair, N. M., & Townley, R. G. (1985). The effect of age on methacholine response. *Journal of Allergy and Clinical Immunology, 76*, 609–613.

Horak, F., Hussarke, M., Jäger, S., & Skoda-Türk, R. (1979). Die Bestimmung der Aggressivität allergisierender Pollenarten. *Wiener klinische Wochenschrift, 92*, 161–164.

Horak, F., Stubner, P., Zieglmeyer, R., & Harris, A. G. (2003). Comparison of the effects of desloratadine 5-mg daily and placebo on nasal airflow and seasonal allergic rhinitis symptoms induced by grass pollen exposure. *Allergy, 58*(6), 481–485.

Italian Association of Aerobiology. (2002). An epidemiological survey of Cupressaceae pollinosis in Italy. *Journal of Investigational Allergology and Clinical Immunology, 12*, 287–292.

Jäger, S. (1998). Global aspects of ragweed in Europe. In F. Th. M. Spieksma (Ed.), *Ragweed in Europe* (6th International Congress on Aerobiology, Satellite Symposium Proceedings, Perugia 1998, pp. 11–14). Horsholm: ALK-Abello A/S.

Jensen, J., Poulsen, L. K., Mygind, K., Weeke, E. R., & Weeke, B. (1989). Immunochemical estimations of allergenic activities from outdoor aero-allergens, collected by a high-volume air sampler. *Allergy, 44*, 52–59.

Jones, A. M., & Harrison, R. M. (2004). The effects of meteorological factors on atmospheric bioaerosol concentrations: A review. *Science of the Total Environment, 326*, 151–189.

Jörres, R., Nowakm, D., & Magnussen, H. (1996). The effect of ozone exposure on allergen responsiveness in subjects with asthma or rhinitis. *American Journal of Respiratory and Critical Care Medicine, 153*(1), 56–64.

Juhász, M. (1998). History of ragweed in Europe. In F. Th. M. Spieksma (Ed.), *Ragweed in Europe* (6th International Congress on Aerobiology, Satellite Symposium Proceedings, Perugia 1998, pp. 11–14). Horsholm: ALK-Abello A/S.

Karrenberg, S., Kollmann, J., & Edwards, P. J. (2002). Pollen vectors and inflorescence morphology in four species of Salix. *Plant Systematics and Evolution, 235*(1), 181–188.

Kasprzyk, I. (2008). Aeromycology – main research fields of interest during the last 25 years. *Annals of Agricultural and Environmental Medicine, 15*(1), 1–7.

Kehrl, H. R., Peden, D. B., Ball, B., Folinsbee, L. J., & Horstman, D. (1999). Increased specific airway reactivity of persons with mild allergic asthma after 7.6 hours of exposure to 0.16 ppm ozone. *Journal of Allergy and Clinical Immunology, 104*(6), 1198–1204.

Kilpeläinen, M., Terho, E. O., Helenius, H., & Koskenvuo, M. (2002). Childhood farm environment and asthma and sensitization in young adulthood. *Allergy, 57,* 1130–1135.

King, T. P., Hoffman, F., Løwenstein, H., Marsh, D. G., Platts-Mills, T., & Thomas, W. (1994). Allergen nomenclature. IUIS/WHO Allergen Nomenclature Subcommittee. *Bulletin of the World Health Organization, 72,* 797–806.

Korsgren, M., Andersson, M., Borga, O., Larsson, L., den-Raboisson, M., Malmqvist, U., & Greiff, L. (2007). Clinical efficacy and pharmacokinetic profiles of intranasal and oral cetirizine in a repeated allergen challenge model of allergic rhinitis. *Annals of Allergy, Asthma & Immunology, 98*(4), 316–321.

Krug, N., Hohlfeld, J. M., Larbig, M., Buckendahl, A., Badorrek, P., Geldmacher, H., Behnke, W., Dunkhorst, W., Luettig, B., & Koch, W. (2003). Validation of an environmental exposure unit for controlled human inhalation studies with grass pollen in patients with seasonal allergic rhinitis. *Clinical and Experimental Allergy, 33*(12), 1667–1674.

Laurent, J., Lafay, M., Lattanzi, B., Legall, C., & Sauvaget, J. (1993). Evidence for chestnut pollinosis in Paris. *Clinical and Experimental Allergy, 23,* 29–43.

Lauzurica, P., Gurbindo, C., Maruri, N., Galocha, B., Diaz, R., Gonzalez, J., Garcia, R., & Lahoz, C. (1988a). Olive (*Olea europea*) pollen allergens-I. Immunochemical characterization by immunoblotting, CRIE and immunodetection by a monoclonal antibody. *Molecular Immunology, 25,* 329–335.

Lauzurica, P., Maruri, N., Galocha, B., Gonzalez, J., Diaz, R., Palomino, P., Hernandez, D., Garcia, R., & Lahoz, C. (1988b). Olive (*Olea europea*) pollen allergens-II. Isolation and characterization of two major antigens. *Molecular Immunology, 25,* 337–344.

Ledesma, A., Villalba, M., Batanero, E., & Rodríguez, R. (1998). Molecular cloning and expression of active Ole e 3, a major allergen from olive-tree pollen and member of a novel family of Ca2+-binding proteins (polcalcins) involved in allergy. *European Journal of Biochemistry, 258,* 454–459.

Lee, S. K., Yoon, S. H., Kim, S. H., Choi, J. H., & Park, H. S. (2005). Chestnut as a food allergen: Identification of major allergens. *Journal of Korean Medical Science, 20*(4), 573–578.

Levetin, E. (2001). Effects of climate change on airborne pollen. *Journal of Allergy and Clinical Immunology, 107,* 172.

Lierl, M. B., & Hornung, R. W. (2003). Relationship of outdoor air quality to pediatric asthma exacerbations. *Annals of Allergy, Asthma & Immunology, 90,* 28–33.

Linderholm, H. W. (2006). Growing season changes in the last century. *Agricultural and Forest Meteorology, 137,* 1–14.

Lipiec, A., Rapiejko, P., Samolinski, B., & Krzych, E. (2005). Correlation between conjunctival provocation test results and conjunctival symptoms in pollinosis – preliminary report. *Annals of Agricultural and Environmental Medicine, 12*(1), 17–20.

Lombardi, C., Penagos, M., Senna, G., Canonica, G. W., & Passalacqua, G. (2008). The clinical characteristics of respiratory allergy in immigrants in Northern Italy. *International Archives of Allergy and Immunology, 147,* 231–234.

Low, R. B., Bielory, L., Qureshi, A. I., Dunn, V., Stuhlmiller, D. F., & Dickey, D. A. (2006). The relation of stroke admissions to recent weather, airborne allergens, air pollution, seasons, upper respiratory infections, and asthma incidence, September 11, 2001, and day of the week. *Stroke, 37*(4), 951–957.

Lubitz, S., Schober, W., Pusch, G., Effner, R., Klopp, N., Behrendt, H., & Buters, J. T. (2010). Polycyclic aromatic hydrocarbons from diesel emissions exert proallergic effects in birch pollen allergic individuals through enhanced mediator release from basophils. *Environmental Toxicology, 25*, 188–197.

Manalai, P., Woo, J. M., & Postolache, T. T. (2009). Suicidality and montelukast. *Expert Opinion on Drug Safety, 8*(3), 273–282.

Mari, A., Wallner, M., & Ferreira, F. (2003). Fagales pollen sensitization in a birch-free area: A respiratory cohort survey using Fagales pollen extracts and birch recombinant allergens (rBet v 1, rBet v 2, rBet v 4). *Clinical and Experimental Allergy, 33*, 1419–1428.

Marsh, D., Déchamp, C., Cour, P., Bousquet, J., & Deviller, P. (1987). Study of the correlation between the atmospheric levels of the antigen Amb a 1 (Ag E) and the number of *Ambrosia artemisiifolia* in the Lyons area and the nearby areas. *Allergie et Immunologie, 19*(6), 238–243.

Matheson, E. M., Player, M. S., Mainous, A. G., King, D. E., & Everett, C. J. (2008). The association between hay fever and stroke in a cohort of middle aged and elderly adults. *Journal of the American Board of Family Medicine, 21*(3), 179–183.

Matthiessen, F., Ipsen, H., & Lowenstein, H. (1991). Pollen allergens. In G. D'Amato, F. Th. M. Spieksma, & S. Bonini (Eds.), *Allergenic pollen and pollinosis in Europe* (pp. 36–45). Oxford: Blackwell Scientific Publications.

Mitakakis, T. Z., Tovey, E. R., Xuan, W., & Marks, G. B. (2000). Personal exposure to allergenic pollen and mould spores in inland New South Wales, Australia. *Clinical and Experimental Allergy, 30*(12), 1733–1739.

Molfino, N. A., Wright, S. C., Katz, I., Tarlo, S., Silverman, F., McClean, P. A., Szalai, J. P., Raizenne, M., Slutsky, A. S., & Zamel, N. (1991). Effect of low concentrations of ozone on inhaled allergen responses in asthmatic subjects. *The Lancet, 338*(8761), 199–203.

Morgenstern, V., Zutavern, A., Cyrys, J., Brockow, I., Koletzko, S., Krämer, U., Behrendt, H., Herbarth, O., von Berg, A., Bauer, C. P., Wichmann, H.-E., Heinrich, J., Group, G. S., & Group, L. S. (2008). Atopic diseases, allergic sensitization, and exposure to traffic-related air pollution in children. *American Journal of Respiratory and Critical Care Medicine, 177*, 1331–1337.

Motta, A. C., Marliere, M., Peltre, G., Sterenberg, P. A., & Lacroix, G. (2006). Traffic-related air pollutants induce the release of allergen-containing cytoplasmic granules from grass pollen. *International Archives of Allergy and Immunology, 139*, 294–298.

Moverare, R., Westritschnig, K., Svensson, M., Hayek, B., Bende, M., Pauli, G., Sorva, R., Haahtela, T., Valenta, R., & Elfman, L. (2002). Different IgE reactivity profiles in birch pollen-sensitive patients from six European populations revealed by recombinant allergens: An imprint of local sensitization. *International Archives of Allergy and Immunology, 128*, 325–335.

Moverare, R., Kosunen, T. U., & Haahtela, T. (2006). Change in the pattern of IgE reactivity to timothy grass and birch pollen allergens over a 20-year period. *Journal of Investigational Allergology and Clinical Immunology, 16*, 274–278.

Namork, E., Johansen, B. V., & Lovik, M. (2006). Detection of allergens adsorbed to ambient air particles collected in four European cities. *Toxicology Letters, 165*, 71–78.

Negrini, A. C., Voltini, S., Troise, C., & Arobba, D. (1992). Comparison between Urticariaceae (Parietaria) pollen count and hay fever symptoms: Assessment of a "threshold-value". *Aerobiologia, 8*, 325–329.

Ng, D. K., Kwok, K. I., & Chan, C. H. (2008). The association between hay fever and stroke in a cohort of middle aged and elderly adults. *Journal of the American Board of Family Medicine, 21*(5), 477–478.

Niederberger, V., Pauli, G., Gronlund, H., Froschl, R., Rumpold, H., Kraft, D., Valenta, R., & Spitzauer, S. (1998). Recombinant birch pollen allergens (rBet v 1 and rBet v 2) contain most of the IgE epitopes present in birch, alder, hornbeam, hazel, and oak pollen: A quantitative IgE inhibition study with sera from different populations. *Journal of Allergy and Clinical Immunology, 102*, 579–591.

Okuda, M., Ohkubo, K., Gotoh, M., Hiroshima, K., Ishida, Y., & Hori, K. (2005). Dynamics of airborne pollen particles from inhalation to allergic reaction in the nose. *Rhinology, 43*(1), 29–33.

Ostroumov, A. I. (1989). The common ragweed as a source of mass allergic diseases. *Proceedings of Zoological Institute (Leningrad), 189*, 230–232 (in Russian).
Papa, G., Romano, A., Quaratino, D., Di, F. M., Viola, M., Artesani, M. C., Sernia, S., Di, G. M., & Venuti, A. (2001). Prevalence of sensitization to *Cupressus sempervirens*: A 4-year retrospective study. *Science of the Total Environment, 270*, 83–87.
Pascal, L. (2009). Effets à court terme de la pollution atmosphérique sur la mortalité. *Revue des Maladies Respiratoires, 26*(2), 207–219.
Peat, J. K., Toelle, B. G., Dermand, J., van den Berg, R., Britton, W. J., & Woolcock, A. J. (1996). Serum IgE levels, atopy, and asthma in young adults: Results from a longitudinal cohort study. *Allergy, 51*, 804–810.
Pehkonen, E., & Rantio-Lehtimaki, A. (1994). Variations in airborne pollen antigenic particles caused by meteorologic factors. *Allergy, 49*, 472–477.
Penel, V., De Clercq, G., & Déchamp, C. (2009). Is my hair a new model of pollen trap? *Ambroisie, The First International Ragweed Review, 26*, 60–68.
Pereira, C., Valero, A., Loureiro, C., Davila, I., Martinez-Cocera, C., Murio, C., Rico, P., & Palomino, R. (2006). Iberian study of aeroallergens sensitisation in allergic rhinitis. *European Annals of Allergy and Clinical Immunology, 38*, 186–194.
Perez-Badia, R., Rapp, A., Morales, C., Sardinero, S., Galán, C., & Garcia-Mozo, H. (2010). Pollen spectrum and risk of pollen allergy in central Spain. *Annals of Agricultural and Environmental Medicine, 17*, 139–151.
Postolache, T. T., Stiller, J. W., Herrell, R., Goldstein, M. A., Shreeram, S. S., Zebrak, R., Thrower, C. M., Volkov, J., No, M. J., Volkov, I., Rohan, K. J., Redditt, J., Parmar, M., Mohyuddin, F., Olsen, C., Moca, M., Tonelli, L. H., Merikangas, K., & Komarow, H. D. (2005). Tree pollen peaks are associated with increased nonviolent suicide in women. *Molecular Psychiatry, 10*(3), 232–235.
Postolache, T. T., Lapidus, M., Sander, E. R., Langenberg, P., Hamilton, R. G., Soriano, J. J., McDonald, J. S., Furst, N., Bai, J., Scrandis, D. A., Cabassa, J. A., Stiller, J. W., Balis, T., Guzman, A., Togias, A., & Tonelli, L. H. (2007). Changes in allergy symptoms and depression scores are positively correlated in patients with recurrent mood disorders exposed to seasonal peaks in aeroallergens. *The Scientific World Journal, 7*, 1968–1977.
Postolache, T. T., Komarow, H., & Tonelli, L. H. (2008). Allergy: A risk factor for suicide? *Current Treatment Options in Neurology, 10*(5), 363–376.
Pretorius, E. (2004). Asthma medication may influence the psychological functioning of children. *Medical Hypotheses, 63*(3), 409–413.
Prieto, L., López, M., Bertó, J. M., & Peris, A. (1994). Modification of concentration-response curves to inhaled methacholine after the pollen season in subjects with pollen induced rhinitis. *Thorax, 49*(7), 711–713.
Quiralte, J., Palacios, L., Rodríguez, R., Cárdaba, B., Arias de Saavedra, J. M., Villalba, M., Florido, J. F., & Lahoz, C. (2007). Modelling diseases: The allergens of *Olea europaea* pollen. *Journal of Investigational Allergology and Clinical Immunology, 17*(suppl. 1), 76–82.
Radauer, C., Bublin, M., Wagner, S., Mari, A., & Breiteneder, H. (2008). Allergens are distributed into few protein families and possess a restricted number of biochemical functions. *Journal of Allergy and Clinical Immunology, 121*(847–852), e7.
Rantio-Lehtimäki, A., Koivikko, A., Kupias, R., Makinen, Y., & Pohjola, A. (1991). Significance of sampling height of airborne particles for aerobiological information. *Allergy, 46*(1), 68–76.
Rantio-Lehtimäki, A., Viander, M., & Koivikko, A. (1994). Airborne birch pollen antigens in different particle sizes. *Clinical and Experimental Allergy, 24*, 23–28.
Rapiejko, P., Stanlaewicz, W., Szczygielski, K., & Jurkiewicz, D. (2007). Threshold pollen count necessary to evoke allergic symptoms. *Otolaryngologia Polska, 61*(4), 591–594.
Rasmussen, A. (2002). The effects of climate change on the birch pollen season in Denmark. *Aerobiologia, 18*, 253–265.
Raulf-Heimsoth, M., Kespohl, S., Crespo, J. F., Rodriguez, J., Feliu, A., Brüning, Th, & Ri¹ P. (2007). Natural rubber latex and chestnut allergy: Cross-reactivity or co-sensi¹ *Allergy, 62*(11), 1277–1281.

Reeves, G. M., Tonelli, L. H., Anthony, B. J., & Postolache, T. T. (2007). Precipitants of adolescent suicide: Possible interaction between allergic inflammation and alcohol intake. *International Journal of Adolescent Medicine and Health, 19*(1), 37–43.

Riccioni, G., Della Vecchia, R., Castronuovo, M., Di Pietro, V., Spoltore, R., De Benedictis, M., Di Iorio, A., Di Gioacchino, M., & Guagnano, M. T. (2002). Bronchial hyperresponsiveness in adults with seasonal and perennial rhinitis: Is there a link for asthma and rhinitis? *International Journal of Immunopathology and Pharmacology, 15*, 69–74.

Riedel, F., Krämer, M., Scheibenbogen, C., & Rieger, C. H. (1988). Effects of SO_2 exposure on allergic sensitization in the guinea pig. *Journal of Allergy and Clinical Immunology, 82*(4), 527–34.

Riediker, M., Koller, T., & Monn, C. (2000). Differences in size selective aerosol sampling for pollen allergen detection using high-volume cascade impactors. *Clinical and Experimental Allergy, 30*, 867–873.

Riedl, M., & Diaz-Sanchez, D. (2005). Biology of diesel exhaust effects on respiratory function. *Journal of Allergy and Clinical Immunology, 115*, 221–228.

Rodríguez, R., Villalba, M., Monsalve, R. I., & Batanero, E. (2001). The spectrum of olive pollen allergens. *International Archives of Allergy and Immunology, 125*, 185–195.

Roger, L. J., Horstman, D. H., McDonnell, W., Kehrl, H., Ives, P. J., Seal, E., Chapman, R., & Massaro, E. (1990). Pulmonary function, airway responsiveness, and respiratory symptoms in asthmatics following exercise in NO_2. *Toxicology and Industrial Health, 6*(1), 155–171.

Rogerieux, F., Godfrin, D., Senechal, H., Motta, A. C., Marliere, M., Peltre, G., & Lacroix, G. (2007). Modifications of *Phleum pratense* grass pollen allergens following artificial exposure to gaseous air pollutants (O_3), (NO_2), (SO_2). *International Archives of Allergy and Immunology, 143*, 127–134.

Rogers, C. A., Wayne, P. M., Macklin, E. A., Muilenberg, M. L., Wagner, C. J., Epstein, P. R., & Bazzaz, F. A. (2006). Interaction of the onset of spring and elevated atmospheric CO_2 on ragweed (*Ambrosia artemisiifolia* L.) pollen production. *Environmental Health Perspectives, 114*, 865–869.

Rottem, M., Szyper-Kravitz, M., & Shoenfeld, Y. (2005). Atopy and asthma in migrants. *International Archives of Allergy and Immunology, 136*, 198–204.

Rudell, B., Sandström, T., Hammarström, U., Ledin, M. L., Hörstedt, P., & Stjernberg, N. (1994). Evaluation of an exposure setup for studying effects of diesel exhaust in humans. *International Archives of Occupational and Environmental Health, 66*(2), 77–83.

Schäppi, G. F., Monn, C., Wuthrich, B., & Wanner, H. U. (1996). Direct determination of allergens in ambient aerosols: Methodological aspects. *International Archives of Allergy and Immunology, 110*, 364–370.

Schäppi, G. F., Suphioglu, C., Taylor, P. E., & Knox, R. B. (1997). Concentrations of the major birch tree allergen Bet v 1 in pollen and respirable fine particles in the atmosphere. *Journal of Allergy and Clinical Immunology, 100*, 656–661.

Schäppi, G. F., Taylor, P. E., Pain, M. C., Cameron, P. A., Dent, A. W., Staff, I. A., & Suphioglu, C. (1999a). Concentrations of major grass group 5 allergens in pollen grains and atmospheric particles: Implications for hay fever and allergic asthma sufferers sensitized to grass pollen allergens. *Clinical and Experimental Allergy, 29*, 633–641.

Schäppi, G. F., Taylor, P. E., Staff, I. A., Rolland, J. M., & Suphioglu, C. (1999b). Immunologic significance of respirable atmospheric starch granules containing major birch allergen Bet v 1. *Allergy, 54*, 478–483.

Schmid-Grendelmeier, P., Peeters, A. G., Wahl, R., & Wüthrich, B. (1994). Zur Bedeutung der Eschenpollenallergie. *Allergologie, 17*, 535–542.

Schumacher, M. J., Griffith, R. D., & O'Rourke, M. K. (1988). Recognition of pollen and other particulate aeroantigens by immunoblot microscopy. *Journal of Allergy and Clinical Immunology, 82*, 608–16.

Schwela, D. (2000). Air pollution and health in urban areas. *Reviews on Environmental Health, 15*(1–2), 13–42.

Singer, B. D., Ziska, L. H., Frenz, D. A., Gebhard, D. E., & Straka, J. G. (2005). Increasing Amb a 1 content in common ragweed (*Ambrosia artemisiifolia*) pollen as a function of rising atmospheric CO_2 concentration. *Functional Plant Biology, 32*, 667–670.

Smith, M., Emberlin, J., Stach, A., Rantio-Lehtimäki, A., Caulton, E., Thibaudon, M., Sindt, C., Jäger, S., Gehrig, R., Frenguelli, G., Jato, V., Rodríguez-Rajo, F. J., Alcazar, P., & Galán, C. (2009). Influence of the North Atlantic Oscillation on grass pollen counts in Europe. *Aerobiologia, 25*, 321–332.

Spieksma, F. T., Emberlin, J. C., Hjelmroos, M., Jäger, S., & Leuschner, R. M. (1995a). Atmospheric birch (*Betula*) pollen in Europe: Trends and fluctuations in annual quantities and the starting dates of the seasons. *Grana, 34*, 51–57.

Spieksma, F. T., Nikkels, B. H., & Dijkman, J. H. (1995b). Seasonal appearance of grass pollen allergen in natural, pauci-micronic aerosol of various size fractions. Relationship with airborne grass pollen concentration. *Clinical and Experimental Allergy, 25*, 234–239.

Stach, A., Garcia-Mozo, H., Prieto-Baena, J. C., Czarnecka-Operacz, M., Jenerowicz, D., Silny, W., & Galán, C. (2007). Prevalence of Artemisia species pollinosis in western Poland: Impact of climate change on aerobiological trends, 1995–2004. *Journal of Investigational Allergology and Clinical Immunology, 17*, 39–47.

Stahl, E., & Jork, H. (1964). Chemical Races in Medicinal Plants. I. Investigations on culture varieties originating from European parsley. *Archiv der Pharmazie, 297*, 273–281.

Sterk, P. J., Fabbri, L. M., Quanjer, P., Cockcroft, D. W., O'Byrne, P. M., Anderson, S. D., Juniper, E. F., & Malo, J. L. (1993). Airway responsiveness: Standardized challenge testing with pharmacological. Physical and sensitizing stimuli in adults. *European Respiratory Journal, 6*(Suppl. 16), 53–83.

Sunyer, J., Anto, J. M., Sabria, J., Roca, J., Morell, F., Rodriguez-Roisin, R., & Rodrigo, J. (1995). Relationship between serum IgE and airway responsiveness in adults with asthma. *Journal of Allergy and Clinical Immunology, 95*, 699–706.

Suphioglu, C. (1998). Thunderstorm asthma due to grass pollen. *International Archives of Allergy and Immunology, 116*, 253–260.

Swoboda, I., Jilek, A., Ferreiram, F., Engel, E., Hoffmann-Sommergruber, K., Scheiner, O., Kraft, D., Breiteneder, H., Pittenauer, E., Schmid, E., Vicente, O., Heberle-Bors, E., Ahorn, H., & Breitenbach, M. (1995). Isoforms of Bet v 1, the major birch pollen allergen, analyzed by liquid chromatography, mass spectrometry, and cDNA cloning. *Journal of Biological Chemistry, 270*, 2607–2613.

Taylor, P. E., Flagan, R. C., Miguel, A. G., Valenta, R., & Glovsky, M. M. (2004). Birch pollen rupture and the release of aerosols of respirable allergens. *Clinical and Experimental Allergy, 34*, 1591–1596.

Taylor, P. E., Jacobson, K. W., House, J. M., & Glovsky, M. (2007). Links between pollen, atopy and the asthma epidemic. *International Archives of Allergy and Immunology, 144*, 162–170.

Tejera, M. L., Villalba, M., Batanero, E., & Rodríguez, R. (1999). Identification, isolation, and characterization of Ole e 7, a new allergen of olive tree pollen. *Journal of Allergy and Clinical Immunology, 104*, 797–802.

Teranishi, H., Kenda, Y., Katoh, T., Kasuya, M., Oura, E., & Taira, H. (2000). Possible role of climate change in the pollen scatter of Japanese cedar *Cryptomeria japonica* in Japan. *Climate Research, 14*, 65–70.

Teuber, S. S., Comstock, S. S., Sathe, S. K., & Roux, K. H. (2003). Tree nut allergy. *Current Allergy and Asthma Reports, 3*, 54–61.

Thibaudon, M. (2003). Allergy risk associated with pollens in France. *European Annals of Allergy and Clinical Immunology, 35*, 170–172.

Thibaudon, M., & Lachasse, C. (2006). *Alternaria, Cladosporium*: Atmospheric dispersion, daily and seasonal rythmes. *Revue Française d'Allergologie et d'Immunologie Clinique, 46*, 188–196.

Thibaudon, M., & Sindt, C. (2008). Mesure des allergènes de pollens d'arbre dans l'air (bouleau, olivier). *Revue Française d'Allergologie et d'Immunologie Clinique, 48*(3), 179–186.

Thibaudon, M., Olliver, G., & Cheynel, A. (2008). L'index clinique: Outil d'évaluation de l'impact sanitaire du pollen. *Environnement, Risques and Santé, 7*, 411–416.

Traidl-Hoffmann, C., Kasche, A., Menzel, A., Jakob, T., Thiel, M., Ring, J., & Behrendt, H. (2003). Impact of pollen on human health: More than allergen carriers? *International Archives of Allergy and Immunology, 131*(1), 1–13.

Traidl-Hoffmann, C., Mariani, V., Hochrein, H., Karg, K., Wagner, H., Ring, J., Mueller, M. J., Jakob, T., & Behrendt, H. (2005). Pollen-associated phytoprostanes inhibit dendritic cell interleukin-12 production and augment T helper type 2 cell polarization. *The Journal of Experimental Medicine, 201*, 627–36.

Traidl-Hoffmann, C., Jakob, T., & Behrendt, H. (2009). Determinants of allergenicity. *Journal of Allergy and Clinical Immunology, 123*, 558–566.

Trigo, M. M., Recio, M., Toro, F. J., Caño, M., Dopazo, M. A., García, H., Sabariego, S., Ruiz, L., & Cabezudo, B. (1999). Annual variations of airborne *Casuarina* pollen in the Iberian Peninsula. *Pollen, 10*, 67–73.

Tunnicliffe, W. S., Burge, P. S., & Ayres, J. G. (1994). Effect of domestic concentrations of nitrogen dioxide on airway responses to inhaled allergen in asthmatic patients. *The Lancet, 344*(8939–8940), 1733–1736.

Vaidyanathan, V., Miguel, A. G., Taylor, P. E., Flagan, R. C., & Glovsky, M. M. (2006). Effects of electric fields on pollen rupture. *Journal of Allergy and Clinical Immunology, 117*, 157.

van der Heide, S., De Monchy, J., de Vries, K., Bruggink, T. M., & Kauffman, H. F. (1994). Seasonal variation in airway hyperresponsiveness and natural exposure to house dust mite allergens in patients with asthma. *Journal of Allergy and Clinical Immunology, 93*, 470–475.

Vandenplas, O., D'Alpaos, V., & Van, B. P. (2008). Rhinitis and its impact on work. *Current Opinion in Allergy and Clinical Immunology, 8*, 145–149.

Varela, S., Subiza, J., Subiza, J. L., Rodríguez, R., García, B., Jerez, M., Jiménez, J. A., & Panzani, R. (1997). *Platanus* pollen as an important cause of pollinosis. *Journal of Allergy and Clinical Immunology, 6*, 748–754.

Viander, M., & Koivikko, A. (1978). The seasonal symptoms of hyposensitized and untreated hay fever patients in relation to birch pollen counts: Correlations with nasal sensitivity, prick tests and RAST. *Clinical Allergy, 8*(4), 387–396.

Villalba, M., Batanero, E., Monsalve, R. I., González de la Peña, M. A., Lahoz, C., & Rodríguez, R. (1994). Cloning and expression of Ole e I, the major allergen from olive tree pollen. Polymorphism analysis and tissue specificity. *Journal of Biological Chemistry, 269*, 15217–15222.

Vinckier, S., Cadot, P., Grote, M., Ceuppens, J. L., & Smets, E. (2006). Orbicules do not significantly contribute to the allergenic micro-aerosol emitted from birch trees. *Allergy, 61*, 1243–1244.

Vitányi, B., Mamkra, L., Juhasz, M., Borsos, E., Béczi, R., & Szentpéteri, M. (2003). Ragweed pollen concentration in the function of meteorological elements in the south-eastern part of Hungary. *Acta Climatologica Chorologica, 36–37*, 121–130.

Wahl, R., Schmid-Grendelmeier, P., Cromwell, O., & Wüthrich, B. (1996). In vitro investigation of cross-reactivity between birch and ash pollen allergen extracts. *Journal of Allergy and Clinical Immunology, 98*, 99–106.

Waisel, Y., Mienis, Z., Kosman, E., & Geller-Bernstein, C. (2004). The partial contribution of specific airborne pollen to pollen induced allergy. *Aerobiologia, 20*(4), 197–208.

Waite, K. J. (1995). Blackley and the development of hay fever as a disease of civilization in the nineteenth century. *Medical History, 39*, 186–196.

Watson, H. K., & Constable, D. W. (1991). Allergenic significance of *Plantago* pollen. In G. D'Amato, F. Th. M. Spieksma, & S. Bonini (Eds.), *Allergenic pollen and pollinosos in Europe* (pp. 132–134). Oxford: Blackwell Scientific Publications.

Wiltshire, P. E. (2006). Hair as a source of forensic evidence in murder investigations. *Forensic Science International, 163*(3), 241–248.

Winkler, H., Ostrowski, R., & Wilhelm, M. (2001). *Pollenbestimmungsbuch der Stiftung Deutscher Polleninformationsdienst*. Paderborn: Takt Verlag.

Wopfner, N., Gadermaier, G., Egger, M., Asero, R., Ebner, C., Jahn-Schmid, B., & Ferreira, F. (2005). The spectrum of allergens in ragweed and mugwort pollen. *International Archives of Allergy and Applied Immunology, 138*, 337–346.

Zeghnoun, A., Ravault, C., Fabres, B., Lecadet, J., Quénel, P., Thibaudon, M., & Caillaud, D. (2005). Short-term effects of airborne pollen on the risk of allergenic rhinoconjunctivitis. *Archives of Environmental & Occupational Health, 60*(3), 170–176.

Ziska, L. H., & Caulfield, F. A. (2000). Rising CO_2 and pollen production of common ragweed (*Ambrosia artemisiifolia*), a known allergy-inducing species: Implications for public health. *Australian Journal of Plant Physiology, 27*, 893–898.

Ziska, L. H., Gebhard, D. E., Frenz, D. A., Faulkner, S., Singer, B. D., & Straka, J. G. (2003). Cities as harbingers of climate change: Common ragweed, urbanization, and public health. *Journal of Allergy and Clinical Immunology, 111*, 290–295.

Zureik, M., Neukirch, C., Leynaert, B., Liard, R., Bousquet, J., & Neukirch, F. (2002). Sensitisation to airborne moulds and severity of asthma: Cross sectional study from European Community respiratory health survey. *British Medical Journal, 325*, 411–414.

Chapter 7
Presentation and Dissemination of Pollen Information

Kostas D. Karatzas, Marina Riga, and Matt Smith

Abstract The aim of this chapter is to describe the ways that pollen information is being presented and disseminated to the general public, in various European countries and elsewhere, with the aid of information and communication systems and methods. For this purpose, the chapter firstly addresses the legal framework concerning the dissemination of environmental information and especially information concerning the quality of the atmospheric environment. In the next section, the production of pollen related information via monitoring systems and with the aid of appropriate models is addressed. Then, the chapter presents and analyses pollen information dissemination, including internet technologies as well as participatory sensing. Furthermore, on-line pollen information systems are investigated, and pollen information presentation as well as communication means are analysed. On the basis of the aforementioned investigations, the chapter then addresses the area of electronic information systems and services for quality of life. Such services are suggested as appropriate for disseminating environmental quality information related to pollen levels. Lastly, conclusions are drawn concerning information services that would include pollen related data and knowledge.

Keywords Participatory sensing • Pollen information • Quality of life

---

K.D. Karatzas (✉) • M. Riga
Informatics Systems and Applications Group, Department of Mechanical Engineering,
Aristotle University, GR-54124 Thessaloniki, Greece
e-mail: kkara@eng.auth.gr; mriga@isag.meng.auth.gr

M. Smith
Forschungsgruppe Aerobiologie und Polleninformation,
Medizinische Universitaet Wien, HNO Klinik, Waehringer Guertel 18-20, A-1090 Wien
e-mail: matthew.smith@meduniwien.ac.at

7.1 Introduction

Airborne pollen is a major problem for a considerable percentage of people, due to the allergic reactions that it triggers, and the health problems that it is related to. Therefore, it is evident that information concerning the existence of pollen grains in the air, as well as the appearance of allergic reactions and risks associated with specific pollen types, is of major importance for citizens (Gonzalo-Garijo et al. 2009). The need for pollen information provision has already been recognised, in the frame of various research initiatives concerning the dissemination of information concerning the quality of the atmospheric environment (Peinel and Rose 2004; ROADIDEA 2011). Such information should be collected, prioritised, disseminated and presented in a structured, easily understandable way, in order to increase the awareness of the sensitive parts of the population and allow them to take preventive or protective measures to minimise the allergic reactions and risks (Moriguchi et al. 2001; Neidell 2009).

Although airborne pollen, via its associated allergy, can considerably affect a sufferer's quality of life, it is not regulated in terms of: (a) pollen emission abatement, (b) obligation to monitor and map airborne pollen concentration levels, (c) agreed limit or target values and definition of associated exceedance episodes, (d) measures to prevent or alter allergic reactions and risks. Yet, it has been recognised that pollen is part of the environmental pressures that the atmospheric environment may pose to people (Kusch et al. 2004), and on this basis, the legal framework on environmental information dissemination as well as air quality management may be used as a starting point for pollen information dissemination and presentation.

7.2 The Legal Framework of Environmental Information Dissemination

The issue of access to environmental information and its dissemination towards the public has been on the agenda of various administrative and decision making bodies for many decades in Europe. The EU environmental legislation covers all major aspects of various fields of environmental quality, like air pollution, water quality, noise, waste management etc. Nevertheless, there is no specific regulation concerning pollen, although the latter may be considered as part of the 'air' domain, in line with the provisions of Dir. 90/313/EEC, which defines environmental information as any available information, in any form, concerning the state of the various elements of the environment. Thus, pollen does not appear explicitly in any part of the EU environmental monitoring and management legislation, most probably due to the fact that such legislation is used to address mainly man-made pollution or environmental pressures. This issue has been revised in the latest directive concerning the quality of the atmospheric environment (EC Dir. 2008/50/EC), where air pollutant contributions coming from natural sources and mechanists are explicitly mentioned (Article 2 (15) of the aforementioned directive). The inclusion of pollen in natural sources of air pollution (as part of the primary biological aerosol particles), has already been discussed in terms of providing guidelines for air quality assessment (Marelli 2007; EC 2011).

7.2.1 Information Access for the Atmospheric Environment

It has been recognized that citizens who are well informed on environmental issues can support the formulation and application of practices for the protection of the environment and the sustainable development, while being able to protect themselves from negative effects that the environment might have to their health (Karatzas et al. 2005). Yet, what is lacking is a model for effectively communicating environmental information to the public. Such a model – set of guidelines, may be formulated via the related legal framework. Those legal texts, in their turn, attempted to incorporate in the best possible way the scientific knowledge concerning the impacts of Air Quality (AQ) to human health and the environment, and introduced in our everyday vocabulary terms like assessment, limit values, target values, concentration and many others.

The first EU legislation concerning air quality information availability was Dir. 82/459 which was later replaced by Directive 97/101, and stated that Environmental Information (EI) should be made accessible to the public via an information system set up by the European Environment Agency (EEA), the European Air Quality Information System Airbase (http://air-climate.eionet.europa.eu/databases/airbase/).

The major change came with Directive on Ambient Air Quality Assessment and Management (96/62/EC) that required the development of action plans concerning zones within which concentrations of pollutants in ambient air exceed limit values. These limit values were established by so-called Daughter Directives, that replaced old ones. It is worth noting that within these Daughter Directives, the use of computer-network services is explicitly mentioned in order to provide the public with the appropriate air quality information, which should be up-to-date and routinely made available (Karatzas and Moussiopoulos 2000).

The latest Directive on public access to environmental information (2003/4/EC), declares that environmental information should be provided to the public on-line. This is in line with Article 7 (1) of this Directive which states that *'Member States shall take the necessary measures to ensure that public authorities organise the environmental information… with a view to its active and systematic dissemination to the public, in particular by means of computer telecommunication and/or electronic technology'*. Lastly, in the Directive 2008/50/EG on ambient air quality and cleaner air for Europe (so-called 'Air Quality Directive'), it is clearly stated in Article 26 on public information that *'The information shall be made available free of charge by means of any easy accessible media including the Internet or any other appropriate means of telecommunication'*. Concerning environmental information provision, this directive also includes forecasts for air quality information that should be disseminated to citizens, thus defining a timely and in advance information provision scheme. Moreover, the fact that such information should be made available on both an everyday basis and on the basis of an incident-event (i.e. exceedance of regulated levels), suggests a combined pull and push communication mode (Juvva and Rajkumar 1999).

In addition to the above, there is also the Directive on public participation in relation to environmental decisions (Dir. 2003/35/EC), as well as the Directive on the re-use of public sector information (Dir. 2003/98/EC), that dictate the distribution of environmental information to the citizens. Overall, there is a sufficient legal framework that supports the provision of environmental information to the citizens, and the dissemination of any information related to the quality of the atmospheric environment. Yet, it should be noted that pollen is not explicitly included in the list of environmental domains who's quality should be 'officially' monitored (like in the case of air or water quality), and for this reason it may be considered to be unclear whether it constitutes public sector information. On the other hand, any information that is related to the quality of the environment is, by definition of the relevant directives, environmental information, and thus related to the well being of the public. The way that pollen may currently be considered is via the related AQ directives and regulations. Even in this case, it is difficult to come up with a harmonised and homogeneous way (regardless of time, space and pollutants) to briefly and accurately describe AQ for the general public, and communicate related information, due to the complexity of the air pollution problem per se and the variations in the way that limit values and alert thresholds are defined and calculated. This difficulty is more pronounced in the case of pollen, due to the lack of any agreement or regulation concerning thresholds and limit values. In practice, the closest that pollen levels have come to being regulated, has been via pollen or allergy indexes that were proposed by experts and are being used to inform citizens concerning allergy risks and symptoms in various countries.

All the above demonstrate the differences in the way that aeroallergens and 'classical' air pollutants are treated in terms of the sensitivity of people and the impact they have to their health. It is evident that an aero-pollutant is considered to affect ambient air quality in a different way in comparison to 'classical' air pollutants, and these differences should be taken into account in the design of any AQ information and communication method. On this basis, the pollen related information to be made available to the public should take into account (i) the mandates of the legal framework, (ii) the prerequisites of the information content, (iii) current practices concerning air quality information dissemination (Karatzas 2009), and (iv) the recognised need for providing pollen forecast information (Mäkinen 1985). Following the aforementioned requirements, the environmental quality information to be made available to the public should consist of: (a) spatial and temporal pollen concentrations data, (b) pollen forecasts (c) measures and advices to decrease personal exposure, (d) guidelines for sensitive individuals and administrative details.

7.3 Pollen Information Sources

7.3.1 Pollen Monitoring

Aeroallergens (i.e. allergenic airborne pollen grain and fungal spore concentrations) are being observed with the aid of numerous technological methods, employing manual, semi-manual or fully automated procedures, all aiming at: (a) recording the various taxa of the pollen species that are present in a certain observation location, and (b) estimating the concentration of the pollen grains (usually in number of grains per cubic meter). In addition, there are observations coming from citizens, which report on the symptoms that they are experiencing and the possible cause of those symptoms, thus helping in mapping the pollen influence within the population of an area of interest, and estimating the potential impact of pollen concentrations (like in the case of the AllergieRadar system – www.allergieradar.nl). Extended versions of such information systems address various EU countries, and aim at collecting information on allergic symptoms and the pollen content of the air, fostering information dissemination on the basis of personalized records (www.pollendiary.com). Another type of system that is currently under investigation and prototype development, makes use of the technological capabilities of modern mobile phones, which are capable of taking good quality pictures of pollen producing plants, automatically geo-reference them, provide additional information and tagging, and posting them on a freely available geographic information space like Google Earth or Visual Earth. The aforementioned approaches are based on the concept of Participatory Environmental Sensing (PES), where the individual takes part in recording environmental status, reporting on environmental conditions and pressures, and exchanging information that is of importance for his/her health status and concerns (Burke et al. 2006; Goldman et al. 2009; Karatzas 2011).

7.3.2 Simulation and Modelling

Having multiple sources of pollen observations, it is evident the need to manage them properly and to efficiently extract knowledge from a vast amount of data. On the basis of a literature review conducted by the authors of this chapter, pollen data are being simulated and modelled with the aid of various computational and mathematical methods, such as regression and time series analysis, neural networks, fuzzy analysis, genetic algorithms, deterministic numerical models, etc. Yet it should be noted that despite the fact that prototype research and effective model development is accomplished, few of these approaches lead to operational use of models, as suggested by the CAFE (Clean Air For Europe) air quality Directive for the atmospheric environment.

The aforementioned literature review addressed 97 papers and revealed that there are certain pollen taxa which are ranked as the most 'popular' for model-based

Table 7.1 Results of a literature analysis on pollen forecasting models

Pollen Taxa	Daily values	Season start	Peak day	Season duration	Season severity	Phenology	Long term (1 month)	Numerical modelling	Total
Alnus	5	14	–	1	1	1	–	1	23
Ambrosia	1	1	–	1	–	–	–	–	3
Artemisia	–	1	–	–	–	–	–	–	1
Betula	9	20	1	3	5	–	–	2	40
Chenopodiaceae	–	–	–	–	–	–	–	–	0
Corylus	1	8	–	–	1	1	–	–	11
Cupressaceae	–	3	–	1	–	–	–	–	4
Olea	5	13	1	–	3	2	–	–	24
Platanus	1	6	–	1	–	1	–	–	9
Poaceae	10	17	1	4	8	1	1	–	42
Quercus	3	9	3	1	2	–	–	1	19
Urticaceae/ Parietaria	1	1	–	1	–	–	–	–	3
Total	**36**	**93**	**6**	**13**	**20**	**6**	**1**	**4**	**179**

The table reports on the forecasting goal for each one of 12 pollen taxa in Europe (note that more than one target may be mentioned per taxa)

forecasting, as reported in Table 7.1. According to the review results, *Poaceae*, *Betula*, *Alnus* and *Olea* are the most forecasted taxa, while the most popular forecasting goals are season start, followed by daily values, and when it comes to forecasting goals per pollen taxa, then *Poaceae* is the subject of forecasting for season start and daily values, accompanied by season severity.

This information should be complemented by the type of forecasting method being used for each one of the 12 pollen taxa, as reported in Table 7.2. It is evident that regression models are applied in the majority of cases (targeting mostly at *Poaceae* and *Betula*). The next most applied modelling method is the one based on the temperature sum, and then phenological – climatological modelling and chilling. These models were mostly applied for predicting season start of *Alnus*, *Betula*, *Corylus*, *Olea* and *Quercus*.

7.4 Pollen Information Dissemination

7.4.1 Pollen Information Type

It is very important to distinguish the difference between information about allergy potency and information about allergy risk; allergy potency is a parameter depending only of the pollen grain itself, and of its content in allergens, while allergy risk is another parameter that represents the expected health impact caused by airborne pollen. The allergy risk depends on heterogeneous parameters such as allergy potency of specific pollen, pollen counts, period of observation during pollen season,

7 Presentation and Dissemination of Pollen Information   223

Table 7.2 Literature results concerning forecasting methods applied for each one of 12 pollen taxa in Europe (note that more than one method may be used in one paper)

Methods used in forecast models (12 pollen types)	Alnus	Ambrosia	Artemisia	Betula	Chenopodiaceae	Corylus	Cupressaceae	Olea	Platanus	Poaceae	Quercus	Urticaceae/ Parietaria	Total
Agromet. coefficient	–	–	–	–	–	–	–	–	–	1	–	–	1
Chilling	9	–	–	6	–	3	1	3	2	–	5	–	29
Linear discriminant analysis	–	–	–	–	–	–	–	–	–	1	–	–	1
Dispersion model	–	–	–	1	–	–	–	–	–	–	–	–	1
Multiple component analysis	–	–	–	1	–	–	–	–	–	–	–	–	1
Artificial neural networks	–	–	–	1	–	–	–	2	–	3	–	–	6
Non-linear logistic regression model	1	–	–	–	–	–	–	–	1	1	–	1	4
Phenological/ Climatological methods	5	–	–	10	–	3	1	5	1	3	4	–	32
Poisson regression	1	–	–	–	–	–	–	–	–	1	–	–	2
Polynomic regression	–	–	–	–	–	–	–	–	–	1	–	–	1
Regression	7	2	–	16	–	4	1	9	3	19	7	1	69
Simulated annealing	1	–	–	–	–	–	–	1	1	–	–	–	3
Temperature SUM	9	1	–	9	–	4	2	9	3	1	7	–	45
Time series analysis	2	–	–	2	–	1	–	1	–	2	–	–	8
Atmospheric transport model	–	–	–	1	–	–	–	1	–	–	–	–	2
Total	**35**	**3**	**0**	**47**	**0**	**15**	**5**	**31**	**11**	**33**	**23**	**2**	**205**

geographical position of the trap, ethnic consideration and meteorological conditions. Although in most cases the allergy potency information (in terms of pollen grain concentrations) is the only information type provided to citizens, there are cases where allergy risk information is also calculated and provided, in terms of risk potential and possibilities to develop certain symptoms.

7.4.2 Pollen Information Sources and Dissemination Methods

The main issue about optimizing the dissemination process of pollen and quality of life information to users, is to precisely define: (a) what type of information will be destined for dissemination, (b) how it will be presented and visualized, (c) when it will become available, and (d) to which user group.

Usually, there are three levels of information coverage: national, regional and personal. National information consists of general reports written about today's observations or forecasting information of a country, commonly published on a website, on newspaper, radio etc. Regional information is related to certain geographic locations-areas, addressing the main pollen taxa being present in the air, and usually includes recommendations for quality of life reassurance. These information services and forecasts are usually published on different websites or other media outlets. Conversely, personal information delivers targeted information provision to every subscribed user of a pollen information/alerting system. Notifications of pollen data concerning user's geographical area, as well as alerts according his/her pollen sensitivity can be sent directly by email, SMS or become available on-line through particular ICT services.

7.4.3 Pollen Observation-Based Information System Examples

Polleninfo (http://www.polleninfo.org/) is a detailed, web-based pollen information portal, addressing national alerts and issues concerning pollen season, phenology, allergy risk, health recommendations and preventive measures. The platform also provides with pollen diary services, where users can record their daily allergy symptoms and compare this information with actual pollen load of the main allergenic plants. Polleninfo is the most comprehensive information system concerning pollen data at a European level, currently including 35 countries. Contributors are experts per country and members of the European Aeroallergen Network (EAN) that either run a pollen observation station/system, or are responsible for data handling. The portal provides with a European overview of aeroallergen levels, forecasts, distribution maps and timeseries charts per country and per pollen taxa, where available. Forecasting maps, giving information about today's and next day grass pollen counts, are derived in collaboration with the Finnish Meteorological Institute (FMI).

Coloured flow charts produced six times a day, give annual statements and analyze principal pollination period of different pollen types in certain geographical regions in Europe. Additionally, static colour-scaled maps of Europe show the intensity of pollen loads for every 10 days per month. These maps result from a detailed analysis of a 15-year-record of daily pollen counts coming from 300 monitoring stations. Furthermore, Polleninfo includes numerous supporting material and links, making it the most complete pollen information portal in Europe.

An example of a regional, comprehensive, web-based information system for information on allergenic pollen and spores in the air, is PIA (Point of Information on Aerobiology, http://lap.uab.cat/aerobiologia/en/), which is operated by the Autonomous University of Barcelona. PIA provides with pollen and spores data for various locations in Spain, based on observations. The portal also provides with weekly forecasts, that inform the public on the current levels as well as on the trend (increase, stable, decrease, attention) of air conditions, for 16 pollen and 2 spore types. The portal also includes pollen calendars for Catalonia and Tenerife.

Coming to observation-based information systems outside Europe, the official website of Albuquerque city in New Mexico may be used as a typical example (http://www.cabq.gov/airquality/pollen.html), which incorporates a regional information system that gives general information, alerts, as well as recommendations about air quality and correlated health issues. The system provides with a detailed record about today's and past concentration levels of allergic pollen types, but also gives useful information concerning the range of the highest daily pollen counts of primary pollen producers, the elaboration of pollen types, pollen seasons and the collection sample methodology. Operating as an alerting system, it provides pollen count observations on a daily basis. Information is visualized in a table format, comprising actual counts and rating of allergy risk per monitored pollen. Apart from the aforementioned information being available to every interested party through the website, email notification services are available for those who want to get informed directly through their mail-box. Internet access is prerequisite in order to accomplish communication between general public and the information authority.

Another example of a system providing with similar information and visualization methods, is the Forsyth county (Georgia, USA), which maintains a regional website (http://www.co.forsyth.nc.us/EnvAffairs/default.aspx) concerning environmental affairs. It provides, among other, online services, with daily air quality/pollen reports. These reports are updated each business day, on a settled time, while forecasts can be directly emailed to every subscribed user. In a detailed manner, it publishes pollen history diagrams, daily pollen reports and 3-days-ahead forecasts of three types of pollen emission sources: trees, grasses and weeds. Pollen measured data (density, as grains per cubic meter) are visualized in a table format, using a pollen rating coloured scale with five levels (absent, low, moderate, high, very high), where ranges are defined in correlation to pollen density and to scales suggested by the National Allergy Bureau (NAB) and the American Academy of Allergy, Asthma and Immunology (AAAAI, https://pollen.aaaai.org).

7.4.4 Pollen Model-Based Information System Examples

One of the most easily accessible model-based information systems for pollen in Europe is based on SILAM (System for Integrated modeLling of Atmospheric coMposition). The open-source Air Quality and Emergency Modelling System named SILAM, started back in 2005 and was developed under a research project of the Academy of Finland (http://silam.fmi.fi). Its functionality and services are built by adopting efficient Computational Intelligent and mathematical methods. The system combines various phenological models describing pollination season of different pollen types, with meteorological and aerobiological information. The system demonstrates emissions, concentration forecasts and spatial dissemination of pollen across Europe, giving daily updated predictions for specific pollen types (birch and grass), in a 5-day ahead forecasting horizon. Pollen concentrations are estimated as counts (grains per cubic meter), taking into account detailed grass land and birch forest maps. Results are represented on the basis of a coloured scale, visualizing the information as animated movies on geographical maps, where users can navigate through different prediction horizons. No estimated values of pollen counts are given to the public. All information is freely accessed through the corresponding website (http://pollen.fmi.fi/pics/index.html), while no personalized notification /alert services are available to the users.

Concerning model-based pollen information systems, the one launched in Denmark in 2005 may also be mentioned (http://www.pollenprognoser.dk; Rasmussen et al. 2008). The system gives detailed information on the expected pollen concentration of *birch*, *grass* and *mugwort* in Denmark every 6 h up to 2 days ahead. The presentations are put on maps using a colour scale for 6 levels from 'very low' to 'very high'. From 2010 the forecast is also made available as time series of hourly data for all Danish cities. The system's forecasts were originally based on a statistical model of pollen emissions and pollen concentrations utilising data from the meteorological numerical weather prediction model DMI-HIRLAM (High Resolution Limited Area Model, Korsholm et al. 2006) with a horizontal resolution of 5 km. In 2008 the existing modelling system was improved by the implementation of a new semi-operational dynamical modelling system for birch pollen covering the pollen emissions, atmospheric transport, dispersion and deposition of pollen particles. The system based on Enviro-HIRLAM is developed at DMI in cooperation with the FMI and other European partners. The dynamical model system is planned to run also for *grasses* and *mugwort* within 2011.

Moreover, and in addition to operational models, a vast number of modelling methods and tools have been employed for the production of pollen information. Such data are integrated into pollen information bulletins that are issued by aeroallergen societies and pollen research institutes. These bulletins, however, do not primarily target the citizen, as the information recipient, and thus are exempted from the current analysis.

7.4.5 Pollen Related Observations via PES

PES establishes a citizen's observations and information exchange. EDDMapS (Early Detection and Distribution Mapping System, http://www.eddmaps.org/) is a web-based system for documenting invasive species distribution that combines data from databases and organizations as well as users' observations to create a national network of invasive species distribution data that is shared with every interested party (citizens, scientists, researchers, educators, ecologists, etc.). All these data serve the foundation of a better understanding of invasive species distribution around the world.

The system requires registration to the website in order to become member and be able to report an invasive species occurrence. Simply by filling a form about location and infestation data, participants can submit their observations, upload captured images of species and then track results through interactive queries into the EDDMapS database. Users can also maintain their profile or personal records, and visualize data with interactive maps.

In the same line of action lies the so called 'What's Invasive!' system (http://whatsinvasive.com/), which supports citizens' participation for locating invasive weeds. Nevertheless, the additional potential given to participants is that, by the use of their mobile phones, they can capture and upload information about invasive species that they have witnessed. Since these data are immediately available to anyone, users can get automatically updated information about invasive species associated with their nearest location.

The most integrated solution on pollen observation and information systems based on the concept of PES is the Envitori project (http://knowledge.vtt.fi/eo/envitori.jsp), which aims to become an environmental information market place that collects, manages and shares environmental data among citizens, as well as private and public parties. Envitori combines strong expertise of ICT with environmental technology, by making use of Web 2.0 capabilities, advanced Computational Intelligence (CI) methods, wireless networks, low-cost sensors and mobile internet services in order to create new customized and personalized services, dependent on user, place and time.

The EnviObserver (https://knowledge.vtt.fi/eo/) is the operational system for mobile environmental monitoring in EnviTori. The system utilizes multiple media channels so that air quality and pollen information can be monitored on Google Maps, Google Earth or corresponding applications being available on mobile phones. A user can choose among various observations, such as air quality, weather conditions and pollen symptoms, all data coming either from stations or from user observations (e.g. users can report on various types of flowering plants that they have spotted and/or report on symptoms). The system's information dissemination utility about air quality and health risks can be targeted according to a particular place at a given moment (real-time), where the user is located. Generally, observations are being updated on a daily basis, with measurements provided to the public coming from the last 24 h recorded data.

7.5 Pollen Information Systems and Services

Pollen Information Systems may be considered as a type of an Air Quality Information System (AQIS). Although there are not as many PIS as air quality ones, those existing make use of internet technologies and in some cases of SMS messaging and smart phone services or communicating information and provide warnings to the recipients. Some systems are designed for and integrated in mobile phone applications, reinforcing the existence of ubiquitous pollen information and its direct provision to citizens, regarding the region of interest or even the exact location where the individual is living.

On a continental level, the European PIS is www.polleninfo.org, (already presented in detail in Sect. 7.4.3). From 2007, a scientific collaboration project (COST action) on the assessment of production, release, distribution and health impact of allergenic pollen in Europe has initiated its activities (www.eupollen.eu) and provides support concerning pollen information provision. In the frame of the current chapter, the authors, in collaboration with the participants of the aforementioned EUPOLLEN Cost action, undertook the initiative to compile an extensive list of on-line PIS and services that are currently in operation in Europe and also some of the systems that operate outside the European continent. The results of this inventory are presented in Table 7.3, in country-based alphabetic order. For each PIS or service, the following information is provided: country of origin, the information type provided, the information presentation method, the means of communication employed, the frequency of information provision, and the nature (commercial or free) of the PIS/service. Table 7.3 is accompanied by a detailed list of the internet addresses of the PIS/services, which is provided in the form of an Annex at the end of this chapter.

In some of the systems included in Table 7.3, meteorological models are combined with air dispersion models, that receive input from pollen emission models (in some cases in combination with pollen emission observations), for the calculation of grains of pollen per cubic metre on a daily basis. Thus, these computational information systems incorporate operational forecasting models, in order to provide with predictions of pollen count levels for different forecasting horizons. In most of the PIS of Table 7.3, data come from public authorities via networks of observation stations, while some others combine pollen observations with information such as reports on allergy symptoms, or flowering phenology, made by citizens utilising ICT device services. In the majority of these systems, web-sites enriched with computational results are freely open to the general public, while some of the systems require further registration to provide with more detailed and personalized information, or even some sort of monthly fee in order to obtain personalized services.

It should be noted however that some dedicated sites for the forecasting and information provision of pollen levels, and some related projects, have already appeared some years ago. A couple of such systems are briefly presented hereafter, for reasons of completeness concerning PIS/services:

- The *A.S.T.M.A.* project (ENV4-CT98-0755), developed between 1998 and 2001, provided forecast information on allergy risk regarding 4 aeroallergens

Table 7.3 Pollen Information Systems and Services

Name	Country	Information type	Presentation method	Means of communication	Frequency	Commercial availability
Weather Zone and pollen tracker	Australia	Local data/pollen index and weather conditions	Textual information/charts and coloured graphs	Website/desktop widget/iPhone application	Real time and 4-day forecast	Free and registered services (Weather Zone)
PollenWarn-Dienst.at	Austria	National pollen load/annual reviews/general info about pollen/estimated pollen season	Textual information/charts/maps	Website	Near real time	Free
ZAMG	Austria	Pollen load (no, low, medium, high pollen load)	Maps (linked to PollenWarnDienst.at for textual information)	Website	Real time	Free
AirAllergy	Belgium	National data/pollen concentration levels/pollen calendars	Textual information/graphs	Newsletters (bulletins)/mass media (TV, radio)	Daily/weekly	Free (Belgian Aerobiological Surveillance Network)
Polenes.cl	Chile	Local data (4 main cities)/pollen concentration levels	Coloured graphs/table format	Website/email notification	Daily/weekly data	Free
CAN (Croatian Aerobiology Network)	Croatia	National and Regional data/pollen concentration levels/pollen calendar	Textual information/table format/coloured schemes, graphs	Website, Newsletters (bulletins)/mass media (TV, radio)	Daily/3–4 days forecast	Free (City Government)
Czech pollen information service	Czech Republic	National data/pollen counts (Prague station)/pollen calendar/pollen bulletin	Textual information/graphs	Website/bulletin/mobile and iPhone application/newsletter	Daily/weekly (forecasts 1 week ahead)	Free weekly bulletin/SMS and iPhone application (paid – daily information)

(continued)

Table 7.3 (continued)

Name	Country	Information type	Presentation method	Means of communication	Frequency	Commercial availability
AAD (Astma-Allergi Danmark)	Denmark	National data/pollen and spore counts/pollen alerts/pollen calendar and statistics	Textual information/coloured graphs/maps	Website/e-mail notifications/magazine/telephone advice/mobile and iPhone application	Daily/2-days forecasts (updated 4 times a day)	Mobile and iPhone services to subscribed users with fixed cost (Asthma and Allergy Denmark and the Danish Meteorological Institute)
DMI	Denmark	Local data (2 main cities) and National data/pollen counts and warnings/pollen calendar/pollen history/weather data	Textual information/graphs	Website/RSS feed	Today's data/next day forecast (updated once a day)/2-days forecasts for cities (updated 4 times a day)	Free/RSS-feed services provided for a minor cost
PollenPrognoser	Denmark	National data/pollen counts/pollen concentration levels	Maps/coloured schemes	Website	2-days forecasts (updated 4 times a day)	Free (from the ADD – Asthma and Allergy Denmark – and DMI – Danish Meteorological Institute)
Polleninfo.org	Europe	Regional data (across Europe)/pollen counts/annual statements	Graphs/coloured schemes	Website/e-mail notification	Today's data	Free
Pollen Report in Finland	Finland	National data/pollen counts levels	Coloured schemes	Website	Today's levels	Free (Aerobiology Unit, University of Turku)

AFEDA (French Foundation of Ragweed Study)	France	Regional data	Textual information/coloured graphs/	Website/e-mail subscription/paper annual pollen calendar (to subscribed members)	1 week available/2 week evolution forecast	Free (Association Française d'Etude Des Ambroisies)
RNSA	France	National data/pollen concentration levels/pollen alerts/allergy risks	Textual information/graphs/coloured schemes	Website/bulletins/desktop widget/iPhone/Android/iGoogle/netvibes	Today's levels (on demand)	Free (Réseau National de Surveillance Aérobiologique)
Deutscher Wetterdienst	Germany	National data/pollen concentration levels/pollen annual history	Graphs/coloured schemes/table format/charts	Website/newsletter	3-day overview (today and 2 days ahead)	Free (supported by the German Weather Service, in cooperation with Foundation German Pollen Information Service)
Pollenflug	Germany	Regional data/pollen concentration levels	Textual information/coloured schemes	iPhone application	3-day overview (today and 2 days ahead)/updated 3 times daily	Free
PollenIndex	Hungary	Regional data/pollen index/blooming period/allergenic impact	Textual information/coloured schemes	Website	1 week available forecasts	Free (National Institute of Environmental Health and Hungarian Meteorological Service)

(continued)

Table 7.3 (continued)

Name	Country	Information type	Presentation method	Means of communication	Frequency	Commercial availability
Pollinfo.ini.hu	Hungary	National data (Hungary, Croatia, Romania, Serbia, Slovenia)/pollen counts/pollen calendar	Textual information/table format/coloured graphs	Website	Weekly data	Free
Irishhealth Pollen Alert	Ireland	National data/risk index/pollen concentration levels	Textual information/coloured schemes	Website/iPhone application	Real-time data and next-day forecasts/updated daily	Free (Asthma Society of Ireland)
AIA (Associazione Italiana Di Aerobiologia)	Italy	National data/pollen calendars	Textual information/coloured graphs	Bulletins	Pollen monitoring in a weekly basis	Free
Pollen Radar	Japan	National data/pollen counts	Textual information/charts/coloured schemes	iPhone application	Daily data	commercial (company Up-Frontier working with Japanese Weather Association)
Pollen Map	Korea	Regional data/pollen counts/allergy index/history data (complete database from years 1997 to 2010)	Textual information/table format/coloured schemes	Website	1 week available forecasts	Free (Korean Academy of Pediatric Allergy and Respiratory Disease)
Pollen Luxembourg	Luxembourg	Local data/pollen concentration levels/pollen thresholds/annual charts	Textual information/table format/coloured charts	Website	Daily report	Free (Aerobiology Station of the Ministry of Health)

7 Presentation and Dissemination of Pollen Information

Name	Country	Data	Format	Medium	Frequency	Cost
Allergie Radar	Netherlands	National data/pollen season/pollen counts/symptom levels	Textual information/interactive coloured schemes	Website/e-mail notification	Pollen history and daily data (based on personal complaints diaries)	Free
LUMC (Leids Universitair Medisch Centrum)	Netherlands	National data/daily pollen counts	Textual information/table format/coloured schemes	Website	Weekly report	Free
MetService	New Zealand	National data (specified by location)/pollen concentration levels	Textual information	SMS text message	Daily data and 2-day forecasts	Daily fee or specific cost per message (on demand)
Pollenvarslingen.no	Norway	National data/pollen concentration levels/pollen calendar	Textual information/table format/coloured graphs	Website/e-mail notifications/bulletins	Daily data and 2-day forecasts	Free/also supports registered users
Mobilink Pollen Services	Pakistan	Local (2 main cities) and National data/pollen counts	Textual information	SMS text message	Today's pollen data	Unknown cost (by Mobilink)
Ufone Pollen Count	Pakistan	Local data/pollen counts	Textual information	SMS text message	Daily data/alert data	Monthly fee (Ufone)
RPA (Rede Portuguesa de Aerobiologia)	Portugal	National data/pollen concentration levels/total pollen counts/pollen calendar/pollen annual history/advice	Textual information/colored graphs/maps/table format	Website/bulletin/newsletter (personalized e-mail alerts)/mass media (TV, radio)	Weekly forecasts/daily, weekly, yearly data	Free (by SPAIC – Sociedade Portuguesa de Alergologia e Imunologia Clínica)
Allergology	Russia	Local data (12 main regions)/pollen concentration levels/pollen season	Textual information/table format	Website	Current data/history data	Free (RAAKI – Russian Association of Allergology and Clinical Immunology)

(continued)

Table 7.3 (continued)

Name	Country	Information type	Presentation method	Means of communication	Frequency	Commercial availability
NSPolen	Serbia	National data/pollen concentration levels	Textual information/ coloured graphs	Website	Weekly forecasts	Free (Laboratory of palynology)
IVZ (Institut za Varovanje Zdravja)	Slovenia	Local data (Ljubljana, Maribor)/pollen concentration levels	Textual information/ charts	Website/e-mail notification/ newsletters (bulletins)/	Weekly forecasts	Free (Institute of Public Health of the Republic of Slovenia)
AeroUEx	Spain	National data/pollen concentration levels/ pollen history data	Textual information/ table format/graphs	Website/SMS text message	Daily data	Registered members
Informació del Pollen – Ajuntament de València	Spain	Local data (Valencia)/ pollen concentration/ pollen calendars	Textual information/ table format/	Website	Weekly observations	Free (AVAIC – Asociación Valenciana de Alergología e Inmunología Clínica)
PIA (Point of Information on Aerobiology)	Spain	Regional and National data/pollen and spores levels/history data/pollen calendars	Textual information/ graphs/indexed information	Website	Weekly observations and forecasts	Free (Laboratori d'Anàlisis Palinològiques)
Red Palinológica de la Comunidad de Madrid	Spain	Regional data/pollen concentration levels/ pollen alerts/allergy risks	Textual information/ maps/graphs/ coloured schemes	Website/	Daily forecasts and observations	Free
RAA (Andalusia Aerobiology Network)	Spain	National data (Andalusia region)/pollen concentration levels/ pollen alerts/allergy risks	Textual information/ maps/graphs/ coloured schemes	Website	Weekly forecasts and observations	Free

REA (Spanish Aerobiology Network)	Spain	National data/pollen concentration levels/pollen alerts/allergy risks	Textual information/maps/graphs/coloured schemes	Website	Weekly forecasts and observations	Free/Under subscription
Red AEROCAM (Red de Aerobiologia de Castilla – La Mancha)	Spain	Regional data/pollen concentration levels/pollen alerts/allergy risks	Textual information/maps/graphs/coloured schemes/table format	Website	Weekly forecasts and observations	Free
Red de Aerobiologia de Castilla – Leon	Spain	Regional data/pollen concentration levels/pollen alerts/allergy risks	Textual information/maps/graphs/coloured schemes	Website	Weekly forecasts and observations	Free
RGA (Rede Gallega de Aerobiologia)	Spain	Regional data/pollen concentration levels/pollen alerts/allergy risks	Textual information/table format/graphs/coloured schemes	Website	Weekly forecasts and observations	Free
SEAIC	Spain	National data/pollen concentration	Graphs/numeric data	Website	Daily/weekly/yearly data	Free (Comité de Aerobiología)
Sun Coast University of Málaga	Spain	Local data/pollen concentration levels/pollen alerts/allergy risks	Textual information/maps/graphs/coloured schemes	Website	Weekly forecasts and observations	Free/Under Subscription
Swedish Pollen Laboratories	Sweden	Regional data/pollen concentration levels	Textual information/table format/graphs	Website/newspapers, radio, TV/SMS text message	Daily data/2–3 days ahead forecasts	Free to the public (funded by regional authorities and media customers)/specific cost per SMS message

(continued)

Table 7.3 (continued)

Name	Country	Information type	Presentation method	Means of communication	Frequency	Commercial availability
MeteoSwiss	Switzerland	National data/pollen counts/pollen concentration levels	Textual information/coloured schemes	Website	3-day overview (today and 2 days ahead)/	Free/commercial for specific consumer applications
NPARU (National Pollen and Aerobiology Research Unit)[1]	UK	National data/pollen forecasts	Textual information/bulletins	Website/media mass, SMS text message	Various time scales and daily forecasts for the next 5 days	Commercial use for specific customers (measurements coming from the National Pollen Monitoring Network in UK)
NPARU (National Pollen and Aerobiology Research Unit)[2]	UK	National data/pollen concentration levels	Textual information/coloured schemes	Website	Today's levels and 2-days ahead	Free (National Pollen and Aerobiology Research Unit, sponsored by a pharmaceutical company)
Weather Channel	UK, Ireland	Local data/pollen concentration levels	Coloured schemes	Website	2-day overview (today and tomorrow)	Free (The Weather Channel)
Clarityn	UK	Local and national data/pollen concentration levels	Textual information/3D animated pollen representation/coloured schemes	iPhone application	Today's and tomorrow's forecasts	Free (pollen forecasts supplied by the National Pollen and Aerobiology Research Unit)

7 Presentation and Dissemination of Pollen Information

AccuPollen	USA	Regional data (New York, New Jersey)/pollen counts levels/predominant pollen	Coloured graphs/3D representation	Website	Daily data	Free (UMDNJ – Asthma and Allergy Research Center, Departments of Medicine and Pediatrics)
Allergy Alert[1]	USA	Local data (continental USA)/pollen concentration levels	Graphs/coloured schemes	Joomla module	Real time/4-day forecast	Required registration (Pollen.com)
Allergy Alert[2]	USA	Local data/allergy index levels	Graphs/coloured schemes	iPhone application	Real time/4-day forecast	Free (Pollen.com)
Allergy Alert[3]	USA	Local data/allergy index levels	Textual information	e-mail notification	Every morning (daily) and 2-day forecast	Free (Pollen.com)
Allergy Forecast	USA	Local data (based on zip code)/pollen concentration levels/pollen index history (last 30 days)	Graphs/coloured schemes	Website/Pollen Widgets (Google's Gadget, Yahoo's Widget)	Real time/4-day forecast	Free (Pollen.com)
BaltimoreSun	USA	National data/pollen concentration levels and weather conditions	Textual information	Website	Today's forecasts	Free
NAAC	USA	Regional data (Seattle)/pollen counts/pollen season	Textual information/coloured graphs	Website	Daily data	Free (Northwest Asthma and Allergy Center)
Nasal Allergies Resource Center	USA	National data/pollen and weather alerts/seasonal pollen data	Textual information/coloured graphs	Website/e-mail alerts	Daily data	Free (leaded by a pharmaceutical company)

(continued)

Table 7.3 (continued)

Name	Country	Information type	Presentation method	Means of communication	Frequency	Commercial availability
Online Pollen and Mold Center	USA	Local data (Saint Louis County)/pollen concentration levels	Textual information/table format	Website	Weekly summary report (for previous 7 days)/60 day summaries/all data updated every weekday	Free (Saint Louis County/Environmental Health Laboratory)
Pollen Journal	USA	Local data (based on zip codes)/pollen concentration levels/personal record of allergy symptoms and allergy history/pollen alerts	Graphs/coloured schemes	iPhone application	Real time/3-day forecast	Free (Ringful LLC)
Pollen Mobile web	USA	Local data/allergy index levels/pollen index history (last 30 days)	Graphs/coloured schemes	Mobile web application	Real time/4-day forecast	Free (Pollen.com)
WeatherBug	USA	Local and national data/pollen concentration levels	Coloured graphs	Website	4 day allergy forecast	Free (Pollen.com)
Weather Channel	USA	Local and regional data/weather and pollen alerts/pollen hotspots/pollen almanac/season trends	Textual information/graphs/coloured schemes	Website/SMS text message	Daily and 3-days forecast	Free (The Weather Channel)

WebMD	USA	Regional data (specified by location or zip code)	Coloured graphs/table format/maps	Website	Today's data	Free (AccuWeather, Inc.)
Wunderground	USA	Regional data (specified by location or zip code)/pollen index along with air quality index and weather data	Coloured graphs	Website	Daily data and 1-day before/1-day ahead	Free (Weather Underground)
National Allergy Bureau	USA, Canada, Argentina	Local data/pollen concentration levels	Graphs/coloured schemes	Website	Today's levels	Free and registered users (American Academy of Allergy, Asthma and Immunology)
Pollen Count Albuquerque	USA, New Mexico, Albuq-uerque	Regional data/pollen counts/concentration ratings/pollen production period	Textual information/ table format/ coloured schemes	Website/e-mail notification	Today's data/updated daily (pollen sampling every week)	Free (city of Albuquerque)

The internet addresses of the systems are provided in Annex 1

(*Cupressaceae, Olea, Poaceae* and *Parietaria*) in 4 regions: Nice (France), Emilia Romagna (Italy) and Andalucía and Catalonia (Spain). (www.enviport. com).
- Dynamic on-line forecasting system for cedar pollen in Kanto, Japan (Delaunay et al. 2002). In this system a meteorological forecast model was fed with pollen emission data and observations, for the provision of pollen concentration levels. Information was made available via the web, while WAP access was also being tested and used during the first years of operation.

7.6 Results and Discussion Concerning Pollen Information Dissemination

On the basis of the detailed review and analysis of the PIS included in Table 7.3, it was made evident that all the systems studied adopt, to some degree, Human-Computer Interaction prototypes, enabling people to be part of an 'observing-analyzing-disseminating' information chain. The presentation of the available data is made, in most of the cases, in an encoded and user-friendly manner, by using colour-scaled graphs, short messages, static or dynamic maps, with the resulting information displayed as a graph, or as an interactive, animated movie. In this way, pollen information provision may easier overcome any misunderstandings in scientific terminology while the said information is more easily perceived. In order to have a better overview of the characteristics concerning pollen information dissemination as well as PIS, the authors took also into account a survey that was conducted, with the aid of a questionnaire, at the end of 2008, in the frame of the EUPOLLEN COST Action (www.eupollen.eu). The survey included information from 19 EU countries (Fig. 7.1) on pollen information dissemination, and the various dissemination media that were identified are reported in Table 7.4.

By combining the information of Table 7.3 with that of Table 7.4, it is evident that electronic media (mainly the internet but also mobile phones) play an important role in the provision of pollen related information to the citizens. In addition, 'traditional' electronic media (TV and radio) and newspapers are also popular in providing pollen related information to the public. In most of the cases information is being available to everybody visiting the relevant web site, ('pull' communication mode), while in some others a kind of subscription (usually without any fee) or another type of user-oriented action was required in order to have access to the information ('push' communication mode). This demonstrates the fact that a wide range of information channels, acting complementary to each other, are required in order to be able to communicate with as many citizens as possible. Moreover, it was made evident that citizens prefer information that addresses everyday utility and habits, while they would like to receive pollen forecasts early in advance and not after an incident of pollen 'episode'.

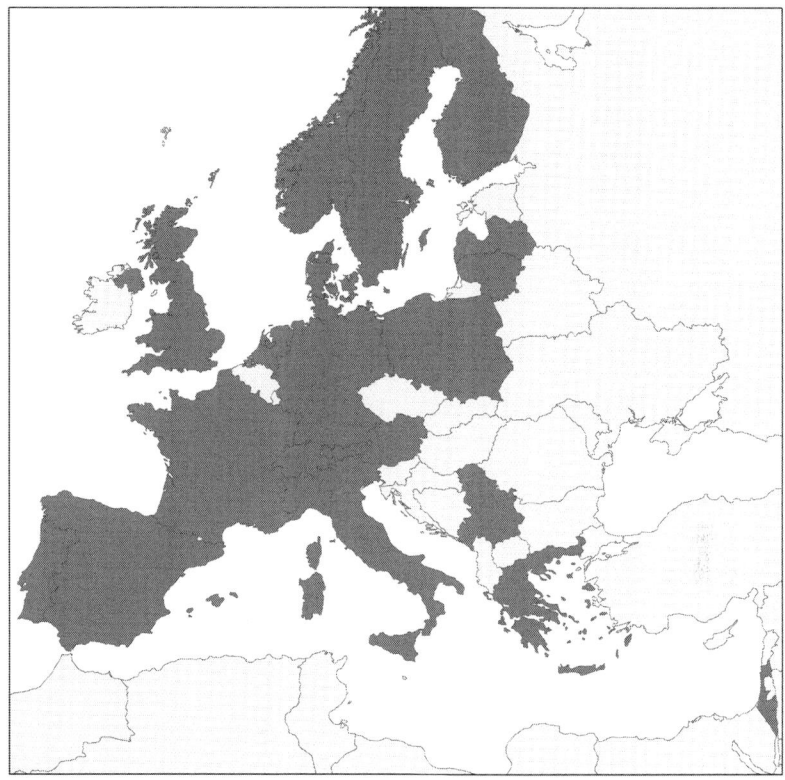

Fig. 7.1 The 19 EU countries that provided data on current pollen forecasting services and outlets in member (as reported in Table 7.4)

Table 7.4 Number of countries supplying pollen information to various information-media types

Services	Number of countries/19	Number of countries paid for service
Television	12	5
Radio	14	4
Newspapers: national	12	4
Newspapers: regional	13	6
Newspapers: local	12	6
Websites	17	7
Mobile phone/SMS	11	4

7.7 Conclusions

Information about aeroallergens plays an important role in the timing of prophylactic medication and in maintaining compliance in treatments among the sensitised population (Bousquet et al. 2001; Sabbah et al. 1999; Stern et al 1997). Such information is usually aimed at allergy sufferers directly so that they can plan their medication and activities in advance, or to medical professionals who plan treatment and schedule clinical trials. Pollen (aeroallergen) information, particularly forecasts, is also of value to those who produce and stock health care products (Emberlin et al. 1999). Information of this type comes mainly from observations, yet it is complemented in some cases by modelling, while the usage of participatory sensing methods has started to appear in recent years.

Pollen information systems are usually geared towards the internet, while they also make use of mobile phones and smart devices for the personalised provision of information and related advice. The latter is tailored with the aid of colour scales or other information representation techniques (graphics, etc), and in some cases it addresses possible symptoms and precaution measures. The analysis of the on-line information systems suggests that there is a strong interest for personalized provision of information, which addresses everyday life. Moreover, for those that suffer from allergies, it is of major importance to be able to receive information tailored to their own needs and everyday habits, and to be able to compare their health status with the health status of those that suffer from same or similar allergies, while also have access to historical as well as real time data and forecasts concerning pollen levels at geographic areas of interest. There is also a trend towards the development of information services based on social media, as the latter allow for the creation of interest groups that not only receive information from predefined sources, but also develop and exchange information and knowledge between them. On this basis, it is expected that the participatory sensing concept, in combination with social media, will play an important role in future pollen information dissemination systems and services, addressing quality of life.

Acknowledgements The authors would like to thank Prof. Jean Emberlin, formerly Director of the National Pollen and Aerobiology Research Unit at the University of Worcester, for conducting a questionnaire study for COST ES0603 that was used as a basis for some of the information presented in this chapter, i.e. Table 7.4. The authors also greatly acknowledge all members of the ES 0603 EUPOLLEN Cost Action for their collaboration and information provision, concerning the PIS reported in Table 7.3, as well as for providing suggestions for the improvement of the current chapter.

Annex 1

The internet addresses (URLs) of the online PIS included in Table 7.3

Websites in Europe

AAD (Astma-Allergi Danmark)
    http://hoefeber.astma-allergi.dk/hoefeber/pollen

7 Presentation and Dissemination of Pollen Information

AeroUEx
   http://www.unex.es/
AFEDA (French Foundation of Ragweed Study)
   http://afeda.assoc.pagespro-orange.fr/
AIA (Associazione Italiana Di Aerobiologia)
   http://www.ilpolline.it/
AirAllergy
   http://airallergy.wiv-isp.be/sites/airallergy/default.aspx
Allergie Radar
   http://www.allergieradar.nl/
Allergology
   http://www.allergology.ru/
Clarityn
   http://itunes.apple.com/WebObjects/MZStore.woa/wa/viewSoftware?mt=8&ign-lr=Lockup_r2c1&id=370539091
CAN (Croatian Aerobiology Network)
   http://www.stampar.hr/PeludIPeludna
Czech pollen information service
   http://www.pylovasluzba.cz/home
Deutscher Wetterdienst
   http://www.dwd.de/pollenflug
DMI
   http://www.dmi.dk/dmi/index/danmark/dagens_pollental.htm
Informació del Pollen – Ajuntament de València
   http://www.valencia.es/polen
Irishhealth Pollen Alert
   http://www.irishhealth.com/pollen_cnt.html
IVZ (Institut za Varovanje Zdravja)
   http://www.ivz.si/Mp.aspx?ni=130
LUMC (Leids Universitair Medisch Centrum)
   http://www.lumc.nl/con/1070/85683/105795/
MeteoSwiss (page not currently available - July 2012)
   http://www.meteosuisse.admin.ch/web/en/weather/health/pollen.html
NPARU (National Pollen and Aerobiology Research Unit)[1]
   http://www.worcester.ac.uk/discover/aerobiology.html
NPARU (National Pollen and Aerobiology Research Unit)[2]
   http://www.zirtek.co.uk/
NSPolen
   http://www.nspolen.com/en/home
PIA (Point of Information on Aerobiology)
   http://lap.uab.cat/aerobiologia/en/
PollenIndex
   http://www.pollenindex.hu/
Polleninfo.org
   http://www.polleninfo.org/

Pollinfo.ini.hu
   http://www.pollinfo.ini.hu/
Pollenflug
   http://www.apfelnews.eu/2010/03/19/pollenflug-info-furs-iphone/
Pollen Luxembourg
   http://www.pollen.lu/
PollenPrognoser
   http://www.pollenprognoser.dk/
Pollen Report in Finland
   http://aerobiologia.utu.fi/tiedotus/siitepolytiedote/polleninformation.html
Pollenvarslingen.no
   http://www.pollenvarslingen.no/forsiden/varsel.aspx
PollenWarnDienst.at
   http://www.pollenwarndienst.at/
REA (Spanish Aerobiology Network)
   http://www.uco.es/rea
Red AEROCAM (Red de Aerobiologia de Castilla – La Mancha)
   http://aerocam.uclm.es/niveles_de_polen
Red de Aerobiologia de Castilla – Leon
   http://www.sanidad.jcyl.es/polen
Red Palinológica de la Comunidad de Madrid
   http://www.madrid.org/polen/
RGA (Rede Gallega de Aerobiologia)
   http://www.usc.es/aerobio/index.htm
RNSA
   http://www.pollens.fr/
RPA (Rede Portuguesa de Aerobiologia)
   http://www.rpaerobiologia.com/?iml=EN
SEAIC
   http://www.polenes.com/en/index.html
Sun Coast University of Málaga
   http://webdeptos.uma.es/biolveg/02Aer/00HAer/01Aer.html
Swedish Pollen Laboratories
   http://www.nrm.se/sv/meny/faktaomnaturen.7036.html
Weather Channel
   http://uk.weather.com/health/index
ZAMG
   http://zacost.zamg.ac.at/phaeno_portal/

Other International Websites

AccuPollen
 http://njms2.umdnj.edu/pollen/
Allergy Alert[1]
 http://extensions.joomla.org/extensions/maps-a-weather/weather/climate/11581
Allergy Alert[2]
 http://www.pollen.com/iphone.asp
Allergy Alert[3]
 http://www.pollen.com/alert.asp
Allergy Forecast
 http://www.pollen.com/allergy-forecast.asp
BaltimoreSun
 http://www.baltimoresun.com/news/weather/
MetService
 http://vaac.metservice.com/default/index.php?alias=smsservices
Mobilink Pollen Services
 http://propakistani.pk/2009/03/17/mobilink-offers-pollen-updates-via-sms/
NAAC
 http://www.nwasthma.com/resources/pollen.asp
Nasal Allergies Resource Center
 http://www.nasonex.com/nasx/application?namespace=main&event=content_display&event_input=enterzip
National Allergy Bureau
 http://www.aaaai.org/nab/index.cfm?p=pollen
Online Pollen and Mold Center
 http://www.stlouisco.com/doh/pollen/Index.cfm
Pollen Count Albuquerque
 http://www.cabq.gov/airquality/pollen.html
Polenes.cl
 http://www.polenes.cl/
Pollen Journal
 http://itunes.apple.com/app/pollen-journal/id301787659?mt=8
Pollen Map
 http://www.pollen.or.kr/english/map/map_main.asp
Pollen Mobile web
 http://www.pollen.com/mobileweb.asp
Pollen Radar
 http://www.gizmag.com/iphone-app-pollen-radar-japan/14111/
Ufone Pollen Count
 http://www.ufone.com/mobPro_PollenCount.aspx
WeatherBug
 http://weather.weatherbug.com/pollen-forecast.html

Weather Channel
http://www.weather.com/activities/health/allergies/index.html
Weather Zone & Pollen Tracker
http://www.weatherzone.com.au/pollentracker/
WebMD
http://www.webmd.com/allergies/healthtool-pollen-counter-calculator
Wunderground
http://www.wunderground.com/DisplayPollen.asp

References

Bousquet, J., Van Cauwenberge, P., Khaltaev, N., Ait-Khaled, N., Annesi-Maesano, I., Baena-Cagnani, C., & Weiss, K. B. (2001). Allergic rhinitis and its impact on asthma. *The Journal of Allergy and Clinical Immunology, 108*, 147–334.

Burke, J., Estrin, D., Hansen, M., Parker, A., Ramanathan, N., Reddy, S., & Srivastava, B. (2006). *Participatory Sensing.* 4th ACM Conference on Embedded Networked Sensor Systems, Boulder, CO, USA. Retrieved from http://www.sensorplanet.org/wsw2006/6_Burke_wsw06_ucla_final.pdf

Delaunay, J.-J., Fedra, K., & Kubat, M. (2002). Cedar pollen forecasting in the Kanto region. *Archives of Complex Environmental Studies, 14*(34), 59–64.

EC. (2011). Commission Staff Working Paper establishing guidelines for demonstration and subtraction of exceedances attributable to natural sources under the Directive 2008/50/EC on ambient air quality and cleaner air for Europe, European Commission, SEC (2011). Retrieved April 15, 2011, from http://register.consilium.europa.eu/pdf/en/11/st06/st06771.en11.pdf

Emberlin, J., Mullins, J., Cordon, J., Jones, S., Millington, W., Brooke, M., & Savage, M. (1999). Regional variations in grass pollen seasons in the UK, long term trends and forecast models. *Clinical and Experimental Allergy, 29*, 347–356.

Goldman, J., Shilton, K., Burke, J., Estrin, D., Hansen, M., Ramanathan, N., Reddy, S., Samanta, V., Srivastava, M., & West, R. (2009). Participatory sensing: A citizen-powered approach to illuminating the patterns that shape our world. Wilson International Center for Scholars. Retrieved from http://wilsoncenter.org/topics/docs/participatory_sensing.pdf

Gonzalo-Garijo, M. A., Tormo-Molina, R., Silva Palacios, I., Pérez-Calderon, R., & Fermàndez-Rodrìguez, S. (2009). Use of a short messaging service system to provide information about airborne pollen concentrations and forecasts. *Journal of Investigational Allergology & Clinical Immunology, 19*, 414–442.

Juvva, K., & Rajkumar, R. (1999). A real-time push-pull communications model for distributed real-time and multimedia systems. School of Computer Science, Carnegie Mellon University, *Report Collection* (CMU-CS-99-107). Retrieved January 28, 2010, from http://reports-archive.adm.cs.cmu.edu/anon/1999/CMU-CS-99-107.pdf

Karatzas, K. (2009). Informing the public about atmospheric quality: Air pollution and pollen. *Allergo Journal, 18*(3), 212–217.

Karatzas, K. (2011). Participatory environmental sensing for quality of life information services. *Information Technologies in Environmental Engineering.* In P. Golinska, M. Fertsch, J. Marx-Gómez (Eds.), *Proceedings of the 5th International Symposium on Information Technologies in Environmental Engineering, Springer Series: Environmental Science and Engineering* (pp. 123–133), Poznan, 6–8 July 2011, ISBN: 978-3-642-19535-8, doi: 10.1007/978-3-642-19536-5_10.

Karatzas, K., & Moussiopoulos, N. (2000). Urban air quality management and information systems in Europe: Legal framework and information access. *Journal of Environmental Assessment Policy and Management, 2*(2), 263–272.

Karatzas, K., Endregard, G., & Fløisand, I. (2005). Citizen-oriented environmental information services: Usage and impact modeling. *Proceedings of the informatics for environmental protection- networking environmental information.* (Paper presented at 19th International EnviroInfo Conference (J. Hrebicek, J. Racek (Eds.)), Brno, Czech Rebublic, pp. 872–878).

Korsholm, U. S., Baklanov, A., & Mahura, A. (2006). DMI-ENVIRO-HIRLAM an on-line coupled multi-purpose environmental model. *Abstracts of the 19th ACCENT/GLOREAM workshop on tropospheric chemical transport modelling,* 11–13 October 2006, Paris, France.

Kusch, W., Fong, H. Y., Jendritzky, G., & Jacobsen, I. (2004). *Guidelines on biometeorology and air quality forecasts.* WMO/TD 1184. Retrieved November 26, 2010, from http://www.wmo.int/pages/prog/amp/pwsp/pdf/TD-1184.pdf

Mäkinen, Y. (1985). Forecasting the presence of atmospheric pollens. *Allergy, 40,* 40–42. doi:10.1111/j.1398-9995.1985.tb04297.x.

Marelli, L. (2007). Contribution of natural sources to air pollution levels in the EU – a technical basis for the development of guidance for the Member States. Institute for Environment and Sustainability, EUR 22779 EN, ISSN 1018-5593. Retrieved January 28, 2011, from http://www.lu.lv/materiali/biblioteka/es/pilnieteksti/vide/Contribution%20of%20natural%20sources%20to%20air%20pollution%20levels%20in%20the%20EU.pdf

Moriguchi, H., Matsumoto, M., Nishimoto, Y., & Kuwada, K. (2001). The development of a pollen information system for the improvement of QOL. *Journal of Investigate Medicine, 48,* 198–209. Retrieved November 26, 2010, from http://medical.med.tokushima-u.ac.jp/jmi/vol48/text/v48_n3-4_p198.html

Neidell, M. (2009). Information, avoidance behaviour, and health: The effect of ozone on asthma hospitalizations. *Journal of Human Resources, 44*(2), 450–478.

Peinel, G., & Rose, T. (2004). Dissemination of air quality information: Lessons learnt in European field trials. *Proceedings of the EnviroInfo* 2004 Conference, Editions du Tricorne Geneva, pp. 118–128. ISBN: 282930275-3.

Rasmussen, A., Mahura, A., Korsholm, U. S., Baklanov, A., & Sommer, J. (2008). The Danish operational pollen forecasting system. 4th ESA 2008, Turku, Finland, http://www.sci.utu.fi/projects/biologia/aerobiologia/4ESA2008/

ROADIDEA. (2011). Summary of all ideas created during the ROADIDEA and ROADIDEA-INCO Innovation Seminars in 2009-2010, Deliverable of the ROADIDEA project. Retrieved April 15, 2011, from http://www.roadidea.eu/

Sabbah, A., Daele, J., Wade, A. G., Ben Soussen, P., & Attali, P. (1999). Comparison of the efficacy, safety, and onset of action of mizolastine, cetirizine, and placebo in the management of seasonal allergic rhinoconjunctivitis. *Annals of Allergy, Asthma & Immunology, 83,* 319–325.

Stern, M. A., Darnell, R., & Tudor, D. (1997). Can an antihistamine delay appearance of hay fever symptoms when given prior to pollen season? *Allergy, 52,* 440–444.

Printed by Publishers' Graphics LLC
BT20121211.19.20.165